Tracy Frost

music of the spheres

Books by Guy Murchie

Men on the Horizon
Song of the Sky
Music of the Spheres

MUSIC OF THE SPHERES

The Material Universe—From Atom to Quasar, Simply Explained

By

GUY MURCHIE

With illustrations by the author

IN TWO VOLUMES

VOL. II. THE MICROCOSM:

Matter, Atoms, Waves, Radiation, Relativity

DOVER PUBLICATIONS, INC., NEW YORK

Published in Canada by General Publishing Company, Ltd., 30 Lesmill Road, Don Mills, Toronto, Ontario.
Published in the United Kingdom by Constable and Company, Ltd., 10 Orange Street, London WC 2.

This Dover edition, first published in 1967, is a revised version of the work originally published in a one-volume clothbound edition in 1961. This Dover edition is published by special arrangement with Houghton Mifflin Company, publisher of the original edition.

International Standard Book Number: 0-486-21810-4
Library of Congress Catalog Card Number: 67-22255

Manufactured in the United States of America
Dover Publications, Inc.
180 Varick Street
New York, N.Y. 10014

contents to part two

fields in space, deeps of time

fields in space, deeps of time

9. stuff of the worlds

IF MY STATION HERE in the black heights is good for stargazing, I think it may also offer a fresh perspective upon the fine texture of the smaller things of the universe. For we have a fairly complete little world of our own out here, populated with eighteen men, two women, four hamsters and a canary. To say nothing of probable bird lice and God knows how many other species of insects, microbes and viruses.

True, it is not as independent a world as the earth itself, but still, it is provisioned with ample atmosphere for its needs, with circulating liquids, and structural solids of many kinds, including all the common elements, most metals and virtually every important alloy and planetary compound — not to mention the intermittent meteoric intruders that suggestively rap and spark upon our outer shell like the symbolic knuckles of nature reminding us that she is at least as eager as we for a wider opening of the door of knowledge.

When I was a child I used to think that little things were simpler than big things, but one day, wandering in the woods, I suddenly understood that the smallness of an acorn may not really make it any simpler than the oak, for it as surely contains oaks as the oak contains acorns. And ever since then, whenever space outside our world of sense seems more important or more impressive than space within the atom, I can remind myself that the differences are only relative and almost certainly illusory. Are not the crystal world of the snowflake and the symmetrical lattice of metal as real as a comet or the Milky Way? And what of the wild microscopic jungles of yeasts and bacteria that have been making bread and cheese and wine since long before man could understand fermentation? Who are we to tell our genes what they may grow or our flesh its rate of metabolism? Can an emperor banish a case of sniffles? Is the elephant master of the mouse?

To grasp the meaning of size, one must consider the fact that outer space after all is made of nothing but inner space even as great Babylon was built of little bricks or a whale is outnumbered by its billions of invisible cells. Nor is inner space closer to our reach than outer space, paradoxical though this may appear, for its true dimensions and dynamic laws are even less understood than the more classic forces of the universe outside. In actuality, both kinds of space pervade our entire world, and as truly as the great suns of the remote sky radiate with the vibrations of their atomic parts do the orbits of our inmost structure add up to the amazing complexity and bulk of this material universe.

If we arrange a scale of sizes to give precise form to our thinking on this basic aspect of space, it turns out that a simple logarithmic spectrum or sequence of a hundred and fifty intervals just about covers our whole knowable world if we let each unit interval on the scale represent ten times the volume of the unit just below it and one tenth that of the unit next above. In linear dimensions thus each unit would come out $\sqrt[3]{10}$ or 2.15 times as long or wide or deep as the one below. And if we set the size of a grapeseed at magnitude 50, a third of the way up the scale, and the size of a solar system at magnitude 100, two thirds of the way, the entire deduced universe appears approximately at the 150 mark and the smallest size we could reasonably expect to measure (perhaps of some still unknown field entity) is somewhere around the zero point. Some physicists, indeed, would go so far as to say that such are the actual approximate limits of material reality: that there is in fact no distance as short as a quintillionth of an inch and literally no room for anything outside the finite but unbounded universe. In any case, the spectrum nicely encompasses and correlates our whole consciousness of space, with some to spare, balancing the macrocosm against the microcosm so neatly upon the pivot of Earth that we can mentally dance up its entire staircase ten steps at a time for a skip check of the key size stones of creation.

Starting down near step 10, for example, we are at about the size of an electron or proton, and perhaps other basic components out of which all atoms are made. At step 20 (10^{10} or ten billion times larger of volume) we can begin to see the inner shell of a small atom, at step 30 a molecule, at step 40 the virus or borderline of life, at step 50 a grapeseed and at step 60 a man. From the individual man we jump to step 70 for the sum bulk of all mankind (several billion individuals) conceived as concentrated into a cubic kilometer — and, passing the scale's asteroid-sized mid-point, to step 80, which is about the size of an average moon. Then on to step 90 for the sun, a medium star, to step 100 for a smallish planetary system around a star, to step 110 for a binary or multiple-star system (about one light-year in diameter), step 120 for a globular cluster of stars, step

SPACE SCALE OF THE UNIVERSE

Volumes in powers of a minimal unit volume	Object examples	Linear distances	Diameters in meters
150	possible universe (?)		10^{27}
	spherical "horizon" of knowledge		
140	a group of supergalaxies	one billion light-years . .	10^{24}
	supergalaxy		
	minor group of galaxies	one megaparsec	
130	large galaxy	one million light-years . .	10^{21}
	small galaxy		
	galactic satellite cluster	one kiloparsec	
120	globular cluster of stars	one thousand light-years .	10^{18}
	distance to Regulus		
	distance to nearest star	one parsec	
110	multiple star system	one light-year	10^{15}
	inner reservoir of comets		
	orbit of Pluto		
100	orbit of Jupiter	one billion kilometers. . .	10^{12}
	orbit of the earth		
	outer corona of the sun		
90	the sun (an average star)	one million kilometers . .	10^{9}
	Jupiter (a large planet)		
	the earth		
80	average moon	one thousand kilometers .	10^{6}
	large asteroid		
	medium asteroid or mountain		
70	all mankind (a cubic kilometer)	one kilometer . . .	10^{3}
	great pyramid		
	whale		
60	man (a cubic meter)	one meter	1
	grapefruit		
	cherry	one centimeter	
50	grapeseed (a cubic millimeter)	one millimeter . . .	10^{-3}
	flea or grain of sand		
	ovum or dust particle		
40	bacterium	one micron	10^{-6}
	virus		
	protein molecule		
30	sugar molecule	one millimicron	10^{-9}
	atom	one angstrom (10^{-10}m.)	
20	inner atom	one thousand fermis . .	10^{-12}
	atomic nucleus		
10	elementary particle	one fermi	10^{-15}
0	possible field entity (?)		10^{-18}

130 for a large galaxy, step 140 for a group of supergalaxies and step 150 for the deduced possible finite universe.

This scale is a fair approximation of reality according to our present state of scientific knowledge. It shows that a thimbleful of water is only midway between an H_2O molecule and the oceans, so that if you should dump a thimble of water into Liverpool Harbor today and wait a few years for thorough diffusion, you probably could not dip a thimbleful out of the Strait of Magellan or Tokyo Bay without its including at least a few molecules of the same water. It reveals that as the sun is to the moon, so does the mountain loom above the elephant and the whale appear to the rat, the flea to the amoeba, the bacillus to the virus, or the protein molecule to the atom.

But of course, these comparisons are only a starting point for the study of form and function, which relate in many curious ways to size. Perhaps you think a small model can be made to behave just like a full-scale machine. But have you ever wondered why pumpkins grow upon the ground instead of dangling on vines like grapes? Why bones are 18 percent of a man but only 8 percent of a mouse? The great Scottish biomathematician D'Arcy Thompson went so far as to say "the form of an object is defined when we know its magnitude," and he developed Galileo's famous "Principle of Similitude" to show how the geometrizing of God applies without exception to everything in nature. Galileo was probably the first man to observe that trees on Earth cannot grow more than about three hundred feet high nor animals more than about a hundred feet long, while terrestrial buildings and conveyances are similarly limited by the fact that their supporting surfaces, having only two dimensions, cannot increase as fast as their weights, which, having volume, must expand in three dimensions.

For this basic reason, all engineers know that under conditions of earthly gravity a plank that nicely spans a brook cannot be made to bridge a great river simply by increasing each of its dimensions a thousand times, not even if its wood be replaced with the strongest steel. For the solid rhombohedral weight would grow a thousand times more than the surface area and the span would inevitably bend and break under the burden of itself long before it attained the desired size. This is why great bridges must be so strictly limited in weight, so artfully designed to place their maximum strength precisely where strength is most needed, so delicately attuned to the graceful dictates of material magnitude.

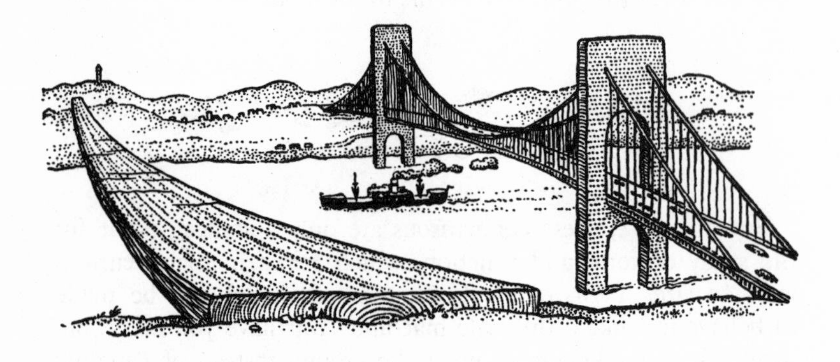

For the same reason, a paper model of an airplane that flies beautifully may turn into a flop when similarly made five times longer and out of cardboard. And it could well become a vehicle of tragic death in full-scale metal if it were not carefully adapted to its increased relative weight. There is no end to applications of the basic principle of similitude, which limits falling raindrops to a quarter-inch in diameter, keeps stars between 10^{32} and 10^{35} grams in mass, and prevents the elephant (if he ever should jump) from jumping higher than the flea. Great worlds, we have found, collide on the planetary scale not with a thud but a ponderous splash, and on the stellar scale with a long-drawn flash. Yet because gravity and other natural forces have very different values on different scales, a raindrop is not made round by the same influence that bulges the earthly oceans.

Indeed, things as small as raindrops begin to follow noticeably the principles of the lesser worlds where electromagnetism ultimately replaces gravity and where friction and surface tension and molecular vibrations grow rapidly from minor annoyances to overwhelming forces as we approach the utterly fantastic realm of the atom. That is why a moon can burst and crumble so easily through its weak surface, while most cells of your body, similarly round, are virtually indestructible within their tough integument, knit together by practically nothing but surface energy.

The surface of a great tree, by contrast, is quite relaxed in its widespread leaves, which are made necessary by its having far too much mass to be able to absorb enough sun energy through a simple spherical surface. Thus its form and beauty are geometric aspects of its magnitude, as is the human lung an effective means of greatly increasing the oxygen-absorbing surface of your body.

Such complicated solutions to the energy problem become less and less necessary as bulk diminishes. Insects do not need lungs, nor do algae require leaves. When an animal is smaller than a flea, its body tends quite noticeably toward the spherical, the shape with the simplest and smallest surface — a shape that is in large measure due to the increasing importance of surface tension as size diminishes. It is here in the upper microscopic size range that surfaces get to be the major structural members of all creatures and objects, right down to the world of molecules. Moreover, surfaces include internal ones such as the inner surface of a cell as well as its outer, both of which are intense zones of energy. A particularly surprising manifestation of surface tension is the fact that waves and ripples on the ocean move more and more slowly as they diminish in size down to a wave length of about three fourths of an inch, but below that critical point the relative power of their surface forces begins to move them faster and faster the tinier they become!

Surface tension is perhaps best described as the kind of energy

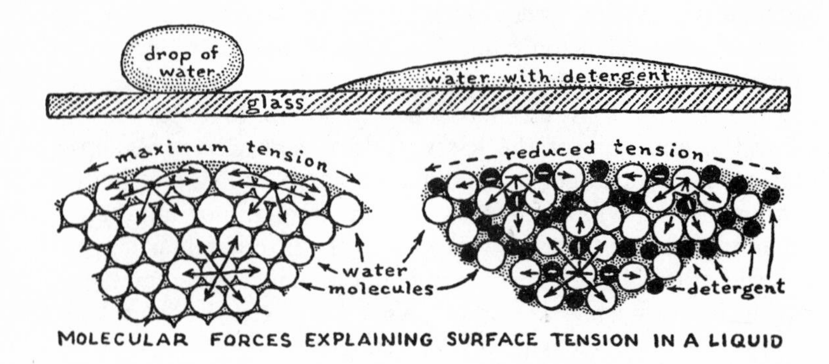

MOLECULAR FORCES EXPLAINING SURFACE TENSION IN A LIQUID

that gives exaggerated strength to any interface between a liquid, a gas or a solid and enables some bugs to walk on water, imprisons others in a drop and makes possible the capillary action of wicks and blotters. It was first described by Laplace as "the cumulative effect, the statistical average, of countless molecular attractions." More specifically, it is now known to result from the fact that the molecules at a surface are pulled harder toward the sides of their fellow molecules than toward the remoter alien molecules opposite them, thus giving them a kind of biased or polarized energy that might be compared to the fervor of patriotism at a national frontier in time of war. This surface stress naturally reaches its maximum influence in the size range of the colloids, between the magnitudes of bacteria and viruses, where the clannishness of these small-village-size gatherings of molecules attains a peak of intensity—where matter behaves in such a different way that a special field of study called colloid chemistry has arisen to deal with it. We must encounter these peculiar little gelatinous particles, in fact, whenever we look at the basic nature of life, for not only are colloids among the most complex of phenomena in the known inorganic world but the human body itself has been described as consisting essentially of "bundles of colloids soaked with water and strung around a bony framework to give form and support."

Although surface tension is a vital force below the range of 1/600,000 of an inch, and films of oil knit together by it have been measured to be as thin as 1/25,000,000 of an inch, the surface energy of a spherical body close to molecular size varies almost in proportion to its radius and tends to vanish when the radius of a drop or particle is less than 1/2,500,000 of an inch — just as the pressure of group loyalty may diminish in the case of human gatherings of subfamily size. Smaller than this, any material particle increasingly takes on the character of an isolated individual molecule and eventually of a single atom, where surface may no longer have much more meaning than does the surface of a solar system.

Long before this range of fineness is reached the macromolecular vibration known as Brownian movement must be reckoned with. Studied a little over a century ago by Robert Brown, the English botanist, the mysterious quivering motion he descried among crushed pollen grains in water and which others have seen in milk, ink and in floating flakes of smoke, has since proved to be due to the natural bombardment of molecules upon bodies so small that these impacts do not average out. It is a basic restlessness of matter in the size range that is still too large to reveal the unimaginably frequent molecular collisions yet small enough to feel their varying residual effects. Beginning just below naked-eye visibility, the motion is more and more noticeable right down to its maximum violence in the bacteria and colloid magnitudes, where particles are slightly smaller than one μ (mu) or micron (a millionth of a meter or about 1/25,000 of an inch) in diameter.

In his famous book *On Growth and Form* D'Arcy Thompson wrote that, in addition to their random quivering, Brownian particles exhibit a straight back-and-forth motion as well as a rotation that increases in rapidity with smallness, particles 13

μ in diameter turning on an average through 14° a minute while the 1μ particles rotate as fast as 100° a second. "The very curious result appears," he summed up, "that in a layer of fluid the particles are not evenly distributed, nor do they ever fall under the influence of gravity to the bottom. For here gravity and the Brownian movement are rival powers striving for equilibrium."

Along with this revelation of gravity's dependence on magnitude has come our new understanding of such concepts as the lower limit of flight. For just as there obviously is an upper limit to the size of birds and airplanes, which must in general double their airspeed to stay aloft after each quadrupling of their length, so does flight become more and more uncontrollable as any winged creature approaches the lowly size ranges of the hectic Brownian activity. The semimicroscopic fairy-fly, for example, probably the smallest living earthly flyer, has wings (or are they propellers?) only a millimeter long, made of nothing but a few hairs that whip the sticky molecules of air like eggbeaters in syrup. Although his weight is so slight in proportion to his surface that he probably notices no more pull of gravity than a man would notice while living on an asteroid, the fairy-fly must in effect fly through a kind of dust storm of swirling nitrogen and oxygen granules that are a ponderable obstacle to progress. And the invisible blizzard, of course, would mount relatively with any diminution in his magnitude, soon swallowing up the last traces of gravity and turning flying into swimming and swimming ultimately into digging at some stage far below the magnitudinal level where less-belabored progress could by any stretch of meaning be classified as flight.

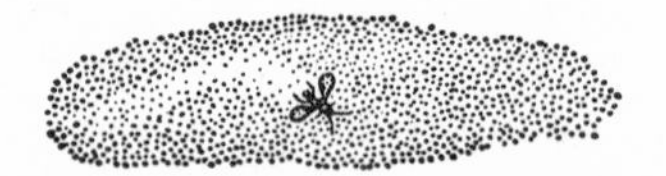

With an effort of imagination we can thus picture a corpuscle or a molecule as living in the midst of a vast, bustling crowd, where temperature is expressed only in the average speed or excitement of the individuals around, and light shows itself in giant rainbows sweeping continuously over the surging populace in dazzling floods of pure color. From here, continuing on downward past this vibrant molecular range into atomic dimensions, we find great fields of electromagnetic pressure dominating the "planetary systems" of electrons and other atomic particles so completely that, according to modern calculation, the electrical force between a proton and an electron is 10^{39} (a thousand million million million million million million) times the gravitational force, the latter being far too feeble at this level to be measurable by any means yet known.

Thus the very laws of nature (in effect) turn out to change with magnitude, introducing almost incredible regions of the very big and the very small that are far stranger than anyone realized a century ago and that have scarcely begun to be understood even today. Although there are striking similarities between a plasma of whirling particles viewed through a microscope and a galaxy of whirling stars viewed through a telescope, this is probably just a rare parallel of form between one thing and another some 10,000,000,000,000,000,000,000,000,000,000,000,000,000,000,000,000,000,000,000,000,000,000,000,000,000,000,000,000,000,000,000,000,000 times bigger. Certainly it does not similarly happen that the wild beauty of a Himalayan glacier is noticeably suggested by the stylized charm of a magnified snow crystal, even though the glacier be materially composed of 1,000,000,000,000,000,000,000,000 such symmetric gems of frozen H_2O. For not only is each individual crystal unique unto itself, but its very time-space relationship is so different from that of something a septillion times bigger that a glacier and a snow crystal must be to each other in many unobvious ways quite literally alien worlds.

In the magnitudes where life is apparent, time-space disparities, of course, express themselves quite eloquently in various rates of living. Not only does the velocity of a fish or a bird or a running mammal tend "to vary directly as the square root of its linear dimensions" by Froude's law of the correspondence of speeds, but the rhythm and tempo of its limbs and organs must change inversely with size in accord with Galileo's earlier principle of similitudes. The shortened pendulum of mouse muscle thus quickens to its smaller task, the whale's 1000-pound heart pumps at a sedate eight beats a minute, the mayfly's lifetime may be bounded by a day and the marriage of the hummingbird is consummated in the wink of a human eye. One can measure such differentials also in oxygen consumed per body ounce per second or in relative production rates of carbon dioxide, both inverse to magnitude, while obviously they all abide by the general physical law of conservation of angular momentum which enlivens Saturn's inner ring, "fixes" the stars and in effect allows the quantitatively slower speed of a spinning propeller to be much too fast to see.

As speed is the quotient of distance and time, here might be an appropriate place to introduce a time scale to match our scale of sizes: another simple logarithmic spectrum to show the basic temporal range of our world. This scale, as you can see, holds some fifty units of duration, each unit ten times as long as the unit just below it and extending about equally upward and downward from its mid-point: one second. Thus the scale's bottom is below a single proton revolution and its top somewhere above practically every estimated age of the universe. Since human consciousness has little natural awareness of time's shorter intervals, you may possibly be startled to notice that there is fully as much time difference between the durations of a light vibration and of a sound vibration as between a minute and a million years.

Before going deeper into the mysterious inner nature of our world, it may help to step back into the well-worn sandals of Leucippos, who lived in Abdera in Thrace in the fifth century before Christ, and consider his meditations while strolling upon a gray Aegean strand. He is said to have wondered aloud to his young pupil Democritos whether the water of the sea, which appears continuous in structure, could really be composed of

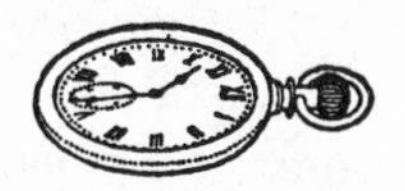

TIME SCALE OF THE UNIVERSE

Durations in seconds	Ages or periods of time	
10^{25}	unknown outer limits of time	
10^{20}	possible age of the universe (?)	
	age of the earth (5 billion years)	
	one revolution of the sun around the galaxy	
10^{15}	age of younger mountain systems	
	duration of human race (a million years)	
	written history	
10^{10}	age of a nation	
	a year	one revolution of earth around sun
	a month	one revolution of moon around earth
10^{5}	a day	one rotation of the earth
	an hour	eating of a meal
	a minute	taking of a breath
1	a second	a heartbeat
	blink of an eye	
	vibration period of audible sound	
10^{-5}	a flash of lightning	
	duration of a muon particle	
	time for light to cross a room	
10^{-10}	vibration period of radar	
	time for an air molecule to spin once	
	vibration period of infra-red radiation	
10^{-15}	vibration period of visible light	
	vibration period of x-rays	
10^{-20}	vibration period of gamma rays	
	time for a proton to revolve once in the nucleus of an atom	
10^{-25}	unknown inner limits of time	

separate, extremely tiny grains like the beach, which, at first glance, likewise appears continuous.

"I can divide it into drops," he observed, "and then I can divide each drop into smaller drops. Is there any reason why this process of subdivision cannot continue forever?" If it continues without end, of course, there will be no end of drops or smaller things — but at some degree of smallness the things must pass from the known into the unknown, then from the knowable into the unknowable or, depending on definition, from the tangible into the intangible, mayhap even from the concrete into the abstract, yet not — heaven preserve our reason — not quite from the something into the nothing.

In any case, it was the kind of question Greek philosophers liked to discuss, for they were strangely drawn to the fundamental mysteries. But no one knew of any experiment by which to test it. Instead, by deep intuition alone, Leucippos, and later Democritos, concluded that there must be a limit to the subdivision of any material — that somewhere there must be "parts which are partless" and that the world is therefore made of ultimately indivisible or "a-tomic" grains that are nothing but themselves in a state of constant motion and which, by being at various times and places packed densely together or spread thinly apart, compose and decompose the four classic elements of fire, air, water and earth, besides all the compound materials of the world.

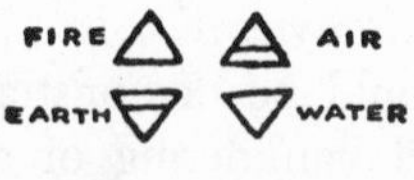

FIRE AIR
EARTH WATER

Leucippos assumed that the forms of such atoms were infinite in number since there was "no reason why they should be of one kind rather than another" and because he observed an "unceasing becoming and change in things." Democritos more specifically defined them as identical in substance though different in shape, order and position — three differences which he attributed to rhythm, interconnection and spin respectively. This amazingly modern hypothesis included also the principle of conservation of matter, since the atoms themselves were considered hard to the point of being absolutely indestructible, and all chemical changes in all minerals, vegetables and animals were therefore due to the unceasing aggregation and disaggregation of these constituent grains in their limitless combinations.

A growing tree, to Democritos, must thus somehow join together the earth atoms and water atoms of the soil with fire atoms from sunrays to produce the more highly organized material of wood — a line of reasoning quite close to the modern, as it turns out, even though he underrated the air, which, as we have recently discovered, gives all plants the great bulk of their growing material in the form of oxygen and carbon. Certainly the Greek atomic school with its abstract imagination was more advanced than the Chinese science of the period, which did not consider air as part of the material world at all, having adopted instead the five rather arbitrary elements of fire, water, wood, earth and metal.

Of course, comprehension of the inner nature of anything as subtle as wind or fire could not be expected before a long period of accumulating odd facts and testing wild surmises. Even Galileo two millenniums later, with his new microscope that made fleas look "as big as rabbits," was not able to contribute anything significant to atomic knowledge. It was not until the work of Christian Huygens, Wilhelm Leibniz, Robert Boyle and particularly Daniel Bernoulli, who published his kinetic theory of gases in the famous book *Hydrodynamica* in 1738, that the reality of atoms and molecules could be demonstrated mathematically as *pressure* — the combined hammering of a "practically infinite" number of invisibly small corpuscles composing a gas as they are "driven hither and thither with a very rapid motion," their myriad collisions literally adding up to a steady outward push in all directions. And heat, Bernoulli showed, was just another aspect of the same motion, the temperature always rising in direct proportion to the "increasing internal movement of the particles."

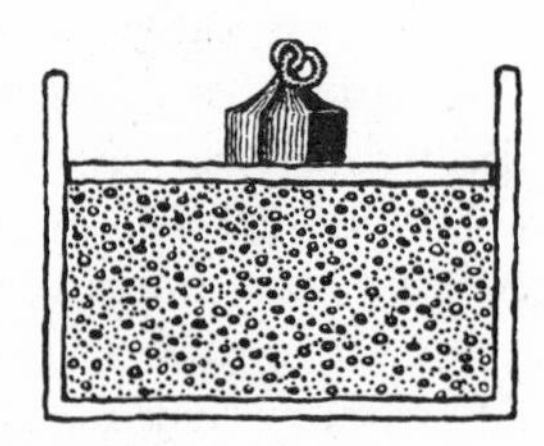

GAS MODEL FROM BERNOULLI'S HYDRODYNAMICA 1738

This was a significant break in the old stagnation of alchemy, that shadowy pre-science with which soothsayers and witch doctors, borrowing freely from astrology, religious ritual and hieroglyphic symbols, had long influenced kings and generals, searching deviously for the elixir of life and the inner relationships between sunshine and gold, for the seeds of silver in moonlight. One enthusiastic alchemist is reported to have experimented two full years with a thousand eggs, trying to synthesize gold, methodically writing down the data of every tested combination in mystic symbols only he could understand.

By the 1770's, however, the growing list of known chemically irreducible elements numbered nearly twenty, and the fundamental stuff of the world was increasingly being sorted out. To the classic materials of gold, silver, copper, iron, mercury, lead,

As Cu Au Fe Pb Hg Ag S Sn

tin, sulfur and carbon, a German physician had added flaky, metallic antimony in 1492 and another German described the reddish-white metal, bismuth, in 1530. These were followed by zinc, phosphorus (seen glowing in urine), arsenic and cobalt (named for the German mine goblin known as a *kobold*). In 1735 a strange heavy nugget from Colombia was designated as platinum; later, a lustrous metal was isolated as nickel.

It was not long after this that an astonishingly buoyant and invisible gas was discovered by Henry Cavendish in England while he was dissolving metal in acid. Described first as "inflammable air," then later named hydrogen, it paved the way for the age of ballooning, by which time another gas, much heavier and evidently endowed with the mysterious essence of life, had been separated from air by Carl Wilhelm Scheele of Sweden in 1771 and Joseph Priestley of England in 1774. Discovery of the latter vital element, soon to be named oxygen, inevitably revealed also the large inactive residue in air that we now know as nitrogen.

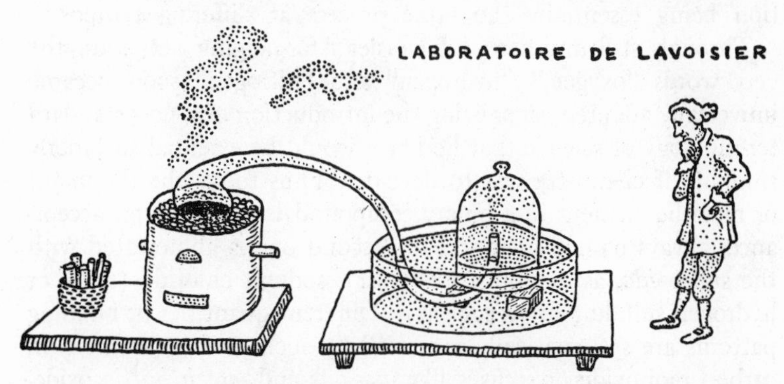

More important still, in the laboratory of a brilliant young French chemist, Antoine Laurent Lavoisier, who was studying combustion and the "calcination" (oxidation) of metals, all three of these strange invisible gases revealed their true places in the elemental scheme of nature. Learning from Cavendish that the burning of hydrogen produced pure water and knowing from his own experiments that combustion is the chemical joining of a burning substance with oxygen, Lavoisier correctly reasoned that water must be a compound of the two invisible elements, hydrogen and oxygen.

Unbelievable as it seemed at first, this conclusion threw so clear a light into so many old dark corners that it was speedily accepted by leading scientists everywhere. By the 1780's it had revolutionized related fields so completely that without any doubt it formed the keystone to a whole new edifice of science. Ultimately, it was to earn Lavoisier world fame as the father of modern chemistry, although by 1794 he had suffered the ironical fate of death under the guillotine at the hands of the fanatical Revolutionary Tribunal, whose members were incapable of appreciating the greatest scientist their country had ever produced.

Not least of his discoveries about oxygen, I must mention, is the fact that animal heat is generated by breathing, which produces a kind of living combustion whose rate is intermediate between the much faster burning of ignited fuel and the vastly slower rusting of iron or rotting of wood, all four forms of oxidation being essentially the same process at differing tempos.

Through still another of Lavoisier's far-ranging activities, the very words "oxygen," "hydrogen" and "nitrogen" soon became universally adopted, signalizing the introduction of a new standard terminology of science that he knew would be essential to orderly thinking if chemistry was to develop. Thus today the dominant or metallic element in a binary compound is, by common acceptance, always named first, and the second one is abbreviated with the suffix *-ide,* as in iron oxide (rust), sodium chloride (salt) or hydrogen sulfide (rotten-egg gas). Different quantities or bonding patterns are systematically indicated by such prefixes as *mon-* in carbon monoxide or suffixes like *-ic, -ous* and *-ate* in nitric oxide, ferrous ammonium sulfate, and so forth, each term part of the integrated system of nomenclature that renders its relationships obvious to anyone with a modicum of chemical acquaintance.

By the turn of the nineteenth century, the list of elements had been further enriched by the discovery of chlorine, manganese, tungsten, chromium, molybdenum, titanium, tellurium, zirconium and uranium. And by 1869, a total of 63 elements had been isolated and described, including one that Joseph Norman Lockyer had found the year before by spectroscope not on Earth but, surprisingly, on the sun. Called helium after the Greek "sun," this inert gas, which is the second lightest of all elements, was not to be found in our world until William Hillebrand of the United States Geological Survey came across it in the rare mineral cleveite in 1890.

⊙ ⦶ ● ○ ⦵ ⊕ ⊛ ◎

Meantime, by a feat even more astonishing than learning about earthly material from our parent star 93 million miles away, three completely unknown elements were predicted in detail by a wild-haired Russian in 1869, simply through abstract deduction from the generalized relationships of the 63 already accepted elements. His name was Dimitri Ivanovich Mendeleyev, and he announced from Siberia, "There is an element as yet undiscovered. I have named it eka-aluminum. By properties similar to those of the metal aluminum you shall identify it. Seek it, and it will be found."

This sounded like an arrogant guess to most scientists at the time, even though a London consulting chemist named John

Dimitri Ivanovich Mendeleyev
1834-1907

Newlands had recently formulated an intuitive "law of octaves" to explain why the elements, when numbered in the order of their atomic weights, tended to repeat fairly similar properties at every seventh element like notes in the musical scale. But Mendeleyev had not reverted to the mystic methods of Pythagoras nor was he swayed by any suggestive analogy to the "seven planets" or the days of the week.

His music of the elements was tuned instead to quite modern concepts of observation, experiment and deduction. Ever since the famous John Dalton had established in 1807 that every chemical combination takes place only in its particular and precise weight proportions, the reality of atoms of different weights (as implied by Leucippos) had been accepted by science. The atoms of the known elements had been carefully weighed and measured by such means as hammering gold leaf to a thinness of one atomic diameter or spreading films of oil to their own definite limits, on the logic that no layer of material can become less than one atom or molecule (or other basic unit) thick and remain the same material. How could a layer of alarm clocks or chicks, for example, be reduced to half a clock or a half-chick in depth without a very drastic change in its nature? Thus a drop of oil one millimeter (about 1/25 of an inch) in diameter was found able to spread out over an area of nearly one square meter (a million square millimeters) before its film began to break, showing its molecules must be around one millionth of a millimeter thick.

This and more advanced types of reasoning had inevitably led Mendeleyev to realize that the most abundant elements (oxygen, silicon, aluminum, iron, calcium, sodium, potassium, magnesium, hydrogen, etc.) have small atomic weights and that those chemically similar to each other have weights either close in value (like platinum, iridium, osmium) or increasing in regular octaves (such as potassium, rubidium and cesium). From here it was just a matter of time for him to work out a Periodic Table of the elements that so clearly revealed harmonic relationships that he could confidently predict what several of the still undiscovered ones must be like. Even though Julius Lothar Meyer, a German chemist, had also (as often happens) independently conceived the same Periodic Law in almost identical form at about the same time, it was Mendeleyev's dramatic forecast of three unknown elements that won him immortal fame as the prophet of periodic chemistry who had enabled this strange new microscience to challenge respectable astronomy in oracular potency. For, sure enough, in the 1870's all of the Russian's prophecies came true as three new elements were discovered in France, Germany and Scandinavia on Mendelevian clues and named for their respective birthplaces gallium, germanium and scandium.

From then on the knowledge of our world stuff has rapidly accumulated. Chemists everywhere were awed to find that all the very active alkaline metallic elements forming group one of the Periodic Table (lithium, sodium, potassium, etc.) always united with oxygen in the strict proportion of two atoms to one, all the second group of elements oxidized atom for atom, the third group joined oxygen at the rate of two atoms to three—and each group in general became more acid and less metallic, going toward the seventh group of so-called halogens (fluorine, chlorine. . .). All of which expressed a mystic inner order of nature more beautiful than anyone could have anticipated—and which became increasingly obvious even though no scientist could yet say what gave elements their valence (combining power) or what an atom was really like.

Then as the number of known elements approached 80, the mystery was suddenly heightened in 1894 when two Englishmen, following a hint from Henry Cavendish, turned up a whole new group of invisible, impalpable elements that baffled even Mendeleyev. Later named the Zero Group, these queer inert, tasteless gases turned out to be the most unsociable of all elements and were pronounced uncombinable with either the "ideal mixer," potassium, or with fluorine, most violent of the nonmetals. Although helium, the first of them, had been known since 1868 as an unaccountable orange-yellow spectral line recorded in the sun's chromosphere during an eclipse and had been discovered on Earth by the aforementioned Hillebrand in 1890, it took William Ramsay to positively identify it as the "sun gas." Then he and his co-worker, John Rayleigh, liquified 120 tons of air and, using a microbalance that "could detect a difference in weight of one fourteen-trillionth of an ounce," painstakingly isolated the rest of what have come to be known as the "noble gases": argon (the lazy), neon (the new), krypton (the hidden), xenon (the stranger) and, working with an exclusive millionth of a gram of it, radon (the radiant).

With the discovery of this new family of aloof gases, whose very gaseousness seemed to be caused by their unsociability, the Periodic Table attained a kind of expectant maturity that might require only a few more key ideas to clear up the whole mystery of matter. At least that was how it looked to the British physicist Joseph John Thomson, who secured his place in history by making so bold as to divide the supposedly indivisible atom.

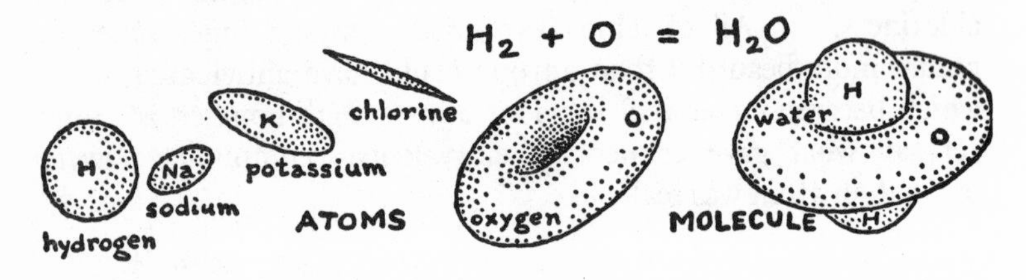

In the early 1890's, theorists had been still speculating about atomic shapes along the same macrocosmic lines used in ancient Greece. Hydrogen atoms were postulated as spherical; those of sodium and potassium were considered elongated ellipsoids. Oxygen atoms, on the other hand, were believed to be shaped like fat doughnuts so that two hydrogens could fit like cherries into the two ends of any doughnut hole to turn it into the neat three-part water molecule H_2O, Neptune's mystic trident out of which all the oceans are built. Even the stability of salt water could be plausibly explained with this model as a slipping of some of the bean-shaped sodium atoms or the pointed chlorines into the doughnut holes ahead of the cherries.

Ingenious as such mechanical postulations were, they could hardly begin to explain the real and increasing complexities of chemistry, and it wasn't until Thomson began to think of an atom as not a single particle but a conglomeration of much smaller entities, perhaps moving around in relation to each other, that a significant conceptual advance could be made. Thomson, of course, knew of Michael Faraday's classical hypothesis that the electrical charges of atoms are *always* a multiple of some minimum elementary quantity of electricity. But he went far beyond Faraday in considering these definite elementary charges as individual parcels of something that might somehow be extracted from atomic bodies and weighed and measured. He even accomplished the astonishing feat of shooting beams of negative charge units or "electrons" out of hot electric wires and through space between the positive and negative electrodes of a condenser so he could weigh these fantastic somethings by their deflection in a known electric field. And he found, to his amazement, that the mass of one electron is only 1/1,840 of the mass of a single hydrogen atom! This would make an electron about as much smaller than a pea as a pea is smaller than the earth.

Since Thomson regarded an atom's positive charge as distributed uniformly throughout its body, which in turn was swarming with negative electrons, his atomic model could well be compared to a handful of peas or beans swirling about in a pot of soup. And this beanpot atom dominated physics for about fifteen crucial years until another great British physicist, Ernest Rutherford from New Zealand, demonstrated in 1911 that an atom cannot really be anywhere near as solid or substantial as a beanpot, but must be literally like a miniature solar system with its entire positive charge as well as virtually all its mass concentrated in an extremely small nucleus located in the very center like a sun, and the rest almost complete emptiness!

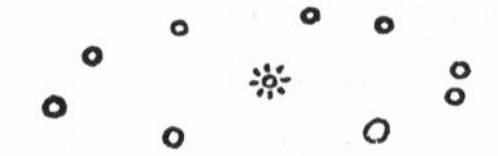

This new concept was even more revolutionary than that of the beanpot atom, and its demonstrable truth (particularly as explained by Niels Bohr, Rutherford's young Danish assistant) gave the scientific world a shock it has hardly yet recovered from. Using the heavy, positive, so-called alpha (α) particles emitted by such radioactive elements as the newly discovered radium or uranium, which are so concentrated and massive that they ram through electrons like bullets through snowflakes, Rutherford proved that the α particles can also pass through a sheet of "solid" aluminum foil with such strange deflections that the pattern of their scattering could only be explained by the fact that the positive parts of the aluminum were just about as concentrated as the α particles themselves: in other words, an atom's nucleus seemed to be millions of times smaller in volume than any whole atom. Describing later how amazed he was to find that a few of these enormously energetic particles should bounce back from the foil that the others penetrated like air, he said, "It was quite the most incredible event that ever happened to me in my life. It was almost as incredible as if you had fired a fifteen-inch shell at a piece of tissue paper and it came back and hit you."

Thus, by a series of classic experiments leavened with brilliant mathematical deductions, the atom was exposed and confirmed as a desolate waste of enormous emptiness — a microfirmament of space every bit as awesome in its vastness as the heavens outside us, and no less mysterious in its ultimate dimensions and significance.

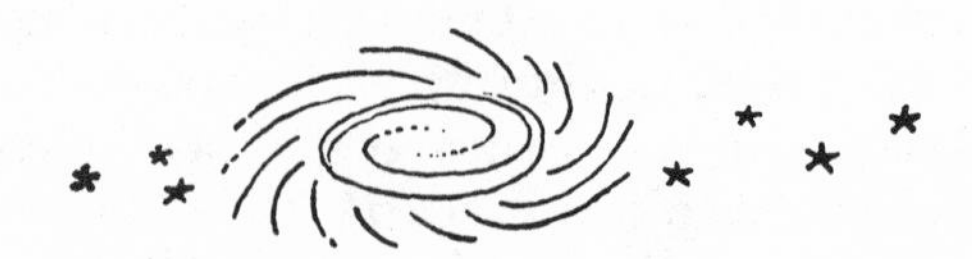

The easiest way to begin visualizing even the simplest of atoms (that of hydrogen) might be to think of the central positive particle or proton as a very dense grapeseed (made of material much heavier than lead) with a puff of smoke, the electron, whizzing around it at an unearthly speed and at varying distances up to five miles, moving so fast, in fact, that you could never actually see it but only know by statistical records what space its total and ever-changing orbit must occupy.

Although such a scale model is not very satisfying, it is about as realistic as is possible using familiar materials. And if you can accept the fact that the dense grapeseed is really 1,840 times as heavy as the whole puff of smoke, you will realize the rough basic resemblance to the solar system, in which our sun is 768 times as heavy as the sum of all his planets. You also will notice that the smoky electron is removed by about the same number of times its own uncertain diameter from the grapeseed proton as are Mercury and Venus from the sun. Most significant of all, the electromagnetic attraction between proton and electron obeys the same Newtonian inverse-square law that defines the force of gravity between sun and planets.

All in all, it is not a bad analogy and, as we look at bigger and more complicated atoms than the hydrogen, with increasing numbers of electrons upon larger and larger orbits, to say nothing of correspondingly bigger and more complex nuclei, we even find a kind of Bode's Law of harmonic intervals that gives a strict and beautiful order to all matter and explains the "octaves" of the Periodic Table in a way that would have delighted Pythagoras or Kepler or Mendeleyev.

The harmonic basis of chemistry, of course, is not as simple as the music of a harp or a piano, where each note is produced by the vibrations of a string of different length or thickness. Yet each element is actually made by a different kind of atom, and

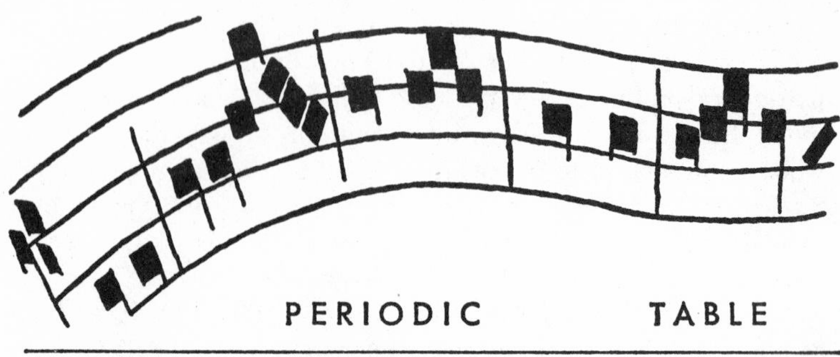

PERIODIC TABLE

	first octave	second octave	third octave	fourth octave

△ 🜄 🜁 ▽

☊ ♀ ☋ ♂ ♄ ☿ ☽ ♆ ♃

Group				
I	1. H (hydrogen) ⊙	3. Li (lithium)	11. Na (sodium)	19. K
II		4. Be (beryllium)	12. Mg (magnesium)	20. Ca
III		5. B (boron)	13. Al (aluminum)	21. Sc
IV	●	6. C (carbon)	14. Si (silicon)	22. Ti
V	⦶	7. N (nitrogen)	15. P (phosphorus)	23. V
VI	○	8. O (oxygen)	16. S (sulfur)	24. Cr
VII		9. F (fluorine)	17. Cl (chlorine)	25. Mn
0	2. He (helium)	10. Ne (neon)	18. A (argon)	26. Fe

Fourth octave (continued): 27. Co, 28. Ni, 29. Cu, 30. Zn, 31. Ga, 32. Ge, 33. As, 34. Se, 35. Br, 36. Kr

these are lanthanons (rare earths)

58. Ce (cerium)	63. Eu (europium)	68. Er (erbium)
59. Pr (praseodymium)	64. Gd (gadolinium)	69. Tm (thulium)
60. Nd (neodymium)	65. Tb (terbium)	70. Yb (ytterbium)
61. Pm (promethium)	66. Dy (dysprosium)	71. Lu (lutetium)
62. Sm (samarium)	67. Ho (holmium)	

OF THE ELEMENTS

fourth octave	fifth octave	sixth octave	seventh octave
(potassium)	37. Rb (rubidium)	55. Cs (cesium)	87. Fr (francium)
(calcium)	38. Sr (strontium)	56. Ba (barium)	88. Ra (radium)
(scandium)	39. Yt (yttrium)	57. La (lanthanum)	89. Ac (actinium)
		*58–71. Lanthanons: see opp. page	
(titanium)	40. Zr (zirconium)	72. Hf (hafnium)	90. Th (thorium)
(vanadium)	41. Nb (niobium)	73. Ta (tantalum)	91. Pa (protoactinium)
(chromium)	42. Mo (molybdenum)	74. W (tungsten)	92. U (uranium)
(manganese)	43. Tc (technetium)	75. Re (rhenium)	93. Np (neptunium)
(iron)	44. Ru (ruthenium)	76. Os (osmium)	94. Pu (plutonium)
(cobalt)	45. Rh (rhodium)	77. Ir (iridium)	95. Am (americium)
(nickel)	46. Pd (palladium)	78. Pt (platinum)	96. Cm (curium)
(copper)	47. Ag (silver)	79. Au (gold)	97. Bk (berkelium)
(zinc)	48. Cd (cadmium)	80. Hg (mercury)	98. Cf (californium)
(gallium)	49. In (indium)	81. Tl (thallium)	99. E (einsteinium)
(germanium)	50. Sn (tin)	82. Pb (lead)	100. Fm (fermium)
(arsenic)	51. Sb (antimony)	83. Bi (bismuth)	101. Md (mendelevium)
(selenium)	52. Te (tellurium)	84. Po (polonium)	102. No (nobelium)
(bromine)	53. I (iodine)	85. At (astatine)	103. Lw (lawrencium)
(krypton)	54. Xe (xenon)	86. Rn (radon)	*these are actinons* (2nd series of rare earths)

VARIOUS GRAPHIC ATTEMPTS, IN TWO AND THREE DIMENSIONS, TO CLARIFY THE PERIODIC TABLE OF ELEMENTS

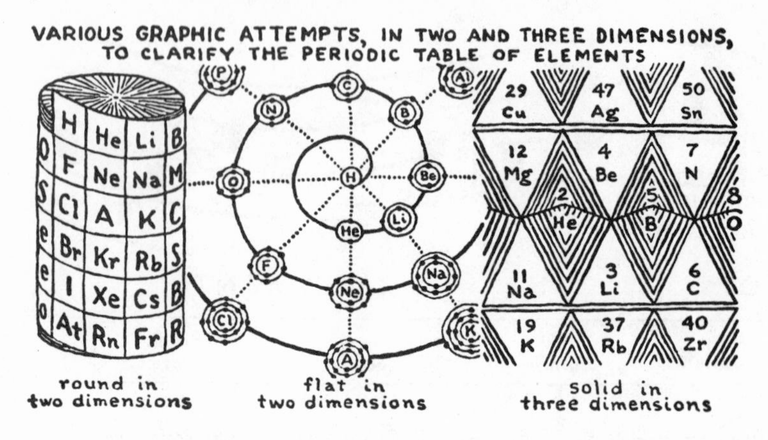

the atoms range progressively in size and weight up the scale, octave by octave, totaling only a few more than the eighty-eight keys of a piano. And the distinctions between the atoms of the various elements are ascribed primarily to the different numbers of negative electrons revolving around each positive nucleus, which in turn must be composed of an equal number of protons, since all whole atoms are found to be electrically neutral. Thus an atom of hydrogen has one electron vibrating about one proton, an atom of helium two electrons around two protons, a lithium atom three and three, beryllium four, boron five, carbon six, nitrogen seven, oxygen eight, and so on up to the heaviest natural element, uranium with its 92 electrons. And even beyond uranium, modern alchemy has artificially constructed still more complex elements: number 93, neptunium; 94, the appropriately named plutonium, used in atomic bombs; 95, americium; 96, curium; 97, berkelium; 98, californium, which has recently been detected as a major ingredient of supernovae; 99, einsteinium; 100, fermium; 101, mendelevium; 102, nobelium; and so forth.

This amazing array of elements may be compared to the letters of the alphabet, which can be put together into never-ending combinations to form words. For similarly do the atoms join in even greater variety to compose all the wordlike molecules and genes of this material universe, which, in its dynamic entirety, is to be presumed a much more basic medium for meaning than any or all the languages of man.

Not only are atomic combinations more basic but they are also so much more complex that even the simplest compounds like water, salt, carbon dioxide and ammonia could not be understood until modern times. And the mysteries in such common substances as wood, leather, rubber, oil and milk are even today still in process of being cleared up. What makes a cloud form in a clear sky? Why is oil slippery, glue sticky, rubber bouncy? Why does one kind of metal ring clearly in a bell, another make a good watch spring, another carry an electric current, another take on magnetism?

All of these questions are being studied intensively in laboratories. All relate to fundamental states of matter or the shape, behavior and interrelation of molecules. A molecule is defined as the smallest particle any compound material can be divided into without changing into a different material and, as such, it is both the next stage larger than an atom and so fantastically rich in its varieties that it easily serves as the common building unit of our world.

Consider a molecule of octane, the familiar constituent of gasoline. Its formula is C_8H_{18}, which is an abbreviated way of saying that it is a flexible, centipedelike chain of eight carbon atoms in zigzag formation, each with two tiny hydrogen atoms attached like legs, and an extra one at each end for head and tail, making eighteen in all — roughly representable as

But if you break up the chain, as may be done in a cracking plant, you do not get two half-molecules of octane. For, strictly speaking, there is no such thing as a half-molecule of any material. Anything less than a molecule of octane cannot be octane. Instead the fractions of octane may include any of many lesser molecules or even single atoms of carbon. It is simple to figure out the possibilities by adding up the atoms in the smaller hydrocarbon molecules. Thus octane (C_8H_{18}) might be a sum of methane (CH_4), ethane (C_2H_6) and pentane (C_5H_8), or of carbon (C), propane (C_3H_8) and butane (C_4H_{10}), of menthane (CH_4) joined with hexane (C_6H_{10}) in the proportion of two to one, of two-atom hydrogen molecules (H_2) united with four times as many of ethylene (C_2H_4), of methane, ethylene and butane, or any of various combinations of hydrogen and ethylene, propylene, butylene, and so on. Furthermore, the nature of a molecule depends on the *arrangement* of its atoms, as well as on their kinds and numbers, so that an oxygen atom and 2 carbons surrounded by 6 hydrogens (for example) may form a molecule of alcohol or a molecule of ether, depending merely on whether the oxygen is beside or between the carbons.

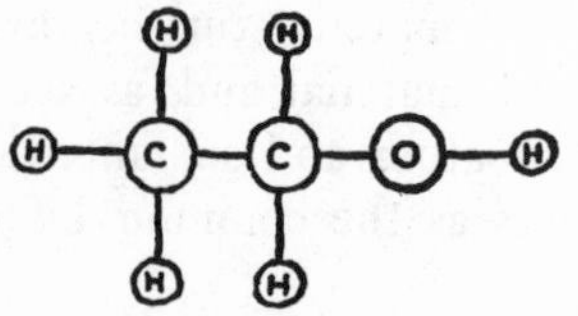

ALCOHOL 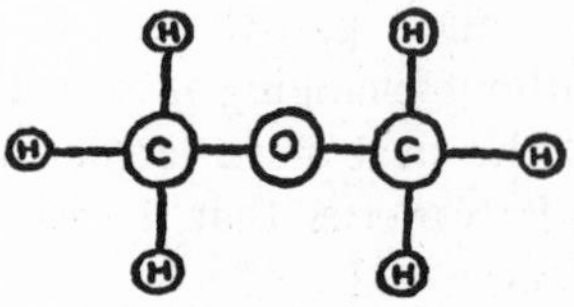 ETHER

If you are not a chemist, all this may appear fantastic, not to say confusing. But it is really simple, solid fact and not just theory or guesswork. Molecules in the quiet state can be seen clearly through electron microscopes and photographed. Even their constituent atoms can thus be definitely recognized and the precise ways they fit into one another. Besides "visual" observation, careful measurement, long experimentation and several lines of mathematical analysis have reliably predicted their behavior and proved their nature way beyond reasonable doubt.

In fact, one can literally subdivide a substance, say, water, into its ultimate grains with amazing ease. Put a drop of it in warm, dry air and, molecule by molecule, it will evaporate. If you had a sensitive enough balance-scale, and could read it fast enough, you could actually measure the water's weight from moment to moment and see that evaporation does not proceed continuously like an ebbing tide but jerkily in "jumps," whole molecule by whole molecule. Of course, the fact that each H_2O molecule weighs only .00000000000000000000000106 ounces shows how delicate this experiment would have to be. And the hopelessness of ever completing it for an entire drop of water is obvious from the fact that water molecules are only a hundred-millionth of an inch in diameter and therefore so numerous that if even those in one drop were placed end to end in a chain it would be twelve billion miles long or enough to reach to the moon fifty thousand times over!

One of the better ways of checking the size of molecules, incidentally, is by their collision rate as they bounce off one another in a gas, for obviously the smaller they are the less often will they hit each other and in a definite proportion. Naturally, such collisions cannot be counted except by some sort of statistical calculation, for not only are individual molecules almost inconceivably small and numerous but they move literally as fast as bullets. In the air of an ordinary room, for instance, the separate molecules of oxygen (O_2) and nitrogen (N_2) vibrate on their submicroscopic zigzagging courses at an average speed of around a thousand miles an hour — a third faster than sound. This relationship between molecules and sound is not accidental, for sound is nothing more than waves of disturbance among molecules propagated from each to the next by successive collisions between them. If all molecules traveled at the same speed and in the same direction, sound would have to move exactly at molecular speed, and it is only the fact that molecules move in all directions, including crosswise, that sound's net forward velocity works out at some 30 percent less. This 30 percent less, however, is a variable speed, for it is a fixed fraction of the collective rate of molecular and atomic motion — in other words, temperature — which, of course, is a varying quantity.

Perhaps a good way really to acquaint ourselves with molecules is to look at them where they are simplest and most scarce: out here in so-called "empty" space. And farther out, between the stars or, better still, between the galaxies, one might encounter scarcely a single star, planet or even a comet for hundreds of light-years in any direction. The rarity of molecules in such remote reaches may be hinted at by the remark of one astronomer that if so much as a faint whiff of cigarette smoke were diffused through each and every cubic mile of space, the stars would be invisible. In justification he had calculated that even the average of about a hundredth of an ounce of matter that is believed to be distributed as gas and dust through each hundred million cubic miles of space in the more crowded regions of the Milky Way is enough to cut the energy of starlight in half during every two thousand light-years of its passage. Which is the main reason why we cannot "see" many parts of our galaxy except through nonvisual (radio) telescopes.

Most intergalactic regions of space, of course, are clearer and presumably emptier than the Milky Way. And so an average molecule in all space would have to be a kind of tiny hermit world that could be compared to a peanut drifting somewhere deep in the Pacific Ocean with virtually no chance of encountering another nut (the nearest molecule) floating at random a few thousand miles away.

By transposing each thousand miles down to an inch, we would then have the proportionately reduced peanut representing the common hydrogen molecule (H_2) in space, something that in reality must have been created by an almost incredible fluke: the earlier collision of two smaller hydrogen atoms that met after months of flying about through millions of miles of real space (visualizable only as some such less improbable rendezvous as that of, say, two grapeseeds after a billion years of drifting in New York Harbor).

Yet this sort of thing must be repeatedly happening throughout the universe. And once two hydrogen atoms do meet at a permissive speed, they cling to each other like lovers in the natural magnetic embrace between each positive proton and each negative electron. While the nuclei (protons) of the two hydrogens keep their distance, being of similar (+) charge, there is room for the two electrons (−) to circulate around them both, perhaps in symmetrical figure-eight orbits. Thus the electrons are shared (probably alternately) by the protons — "in valence" as the chemists say — for it happens that two, and only two, electrons can revolve equidistantly and harmoniously around the immediate vicinity of an atomic nucleus. And this in turn suggests how the energy radiated by the hydrogen electron, vibrating back and forth between its two alternative positions in the hydrogen atom 500,000,000 times a second, broadcasts the common 21-centimeter waves that our radio telescopes receive from space.

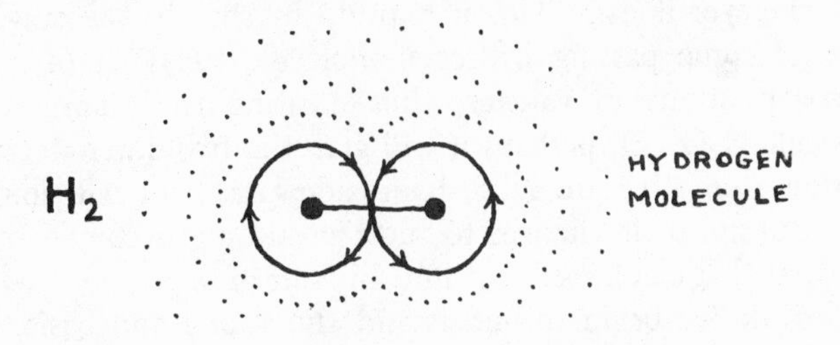

It is in accordance also with the strange law of chemical nature, first proposed by Niels Bohr of Denmark in 1913, that all electrons tend to occupy "shells" around the atomic nuclei, the first shell (mentioned above) holding only two electrons, the second shell (farther away and larger) eight electrons, the third eighteen, the fourth 32 and so on.

Nobody has completely explained why electrons like to follow such rules, but there is no doubt that they do follow them and, in doing so, they throw light on the harmonic mystery of the Periodic Table, though not without adding still another beautiful enigma (the mathematical series of doubled squares) to those already uncovered: 2×1^2 equaling the first shell of two electrons; 2×2^2 representing the second shell of eight (used by heavier atoms); 2×3^2 the third shell of eighteen (by still greater atoms); 2×4^2 the fourth shell of 32; and so on.

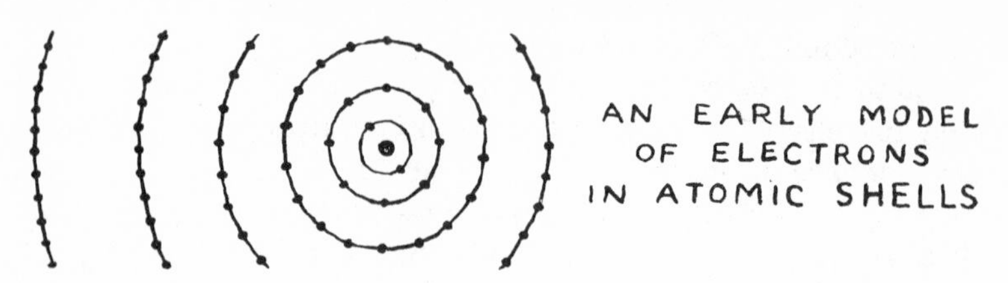

Thus an oxygen atom, for example, having eight electrons orbiting around its nucleus of eight protons, uses an inner first shell of two electrons plus six more in a second shell. But since the second shell needs eight electrons for harmonic completion, oxygen is inherently unstable by itself—and inevitably must be yearning for two more electrons which it will pick up whenever and wherever it can. This it may do by sharing the single electrons of some passing hydrogen molecule (H_2) or of any two hydrogen atoms in valence, thus creating the common water molecule H_2O. Or perhaps it will grab the two outer electrons of an iron atom to form ferric hydroxide (rust), or will join some other substance it chances to meet possessing a couple of likely electrons that can be seduced into any sort of oxide.

Thus do we begin to understand the strong and basic, almost sexual, craving of abundant oxygen to unite with other elements: to oxidize iron, copper, zinc, tin, aluminum, chromium, lead, magnesium, mercury, carbon, sulfur, sodium or other available materials; to burn, to ferment, to corrode, to ravish or devour or rot things away by combining with them chemically, after which they cannot be the same again.

And how many are the other elements that have it as bad or worse than oxygen! Only they are not as abundant, at least not on Earth. Carbon, the most versatile joiner of all, in fact specializes as the most essential chemical factor in the complexity needed for life. Sodium and potassium, each with one lonely outer electron, suffer from a built-in itch to find some body with an electron missing (like chlorine with its expectant seven electrons in the outer shell) and, until they join it, they are as restless as randy stallions.

This in essence is what all chemistry is about. Chemistry is the science of elemental combinations, of the behavior and interrelation of molecules. And, as the study of biology is making increasingly clear, the phenomenon of sex undoubtedly had a chemical origin.

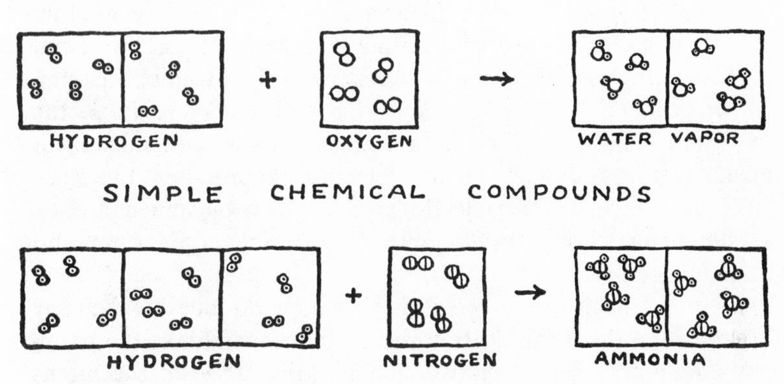

Now to get back to the simplicity of space, there is a constant struggle going on to determine what atoms will mate with what other atoms — a war between the vibrations that keep atoms apart and the magnetic forces that tend to draw them together. Although this conflict is at its minimum in intergalactic space, it has been calculated that atoms never anywhere get more than a few inches apart, and some physicists think that all the space between them, not to mention inside them, is also jam-packed with things that may ultimately be defined as matter. So the conflict of material forces is probably very real everywhere and needs to be understood if we are going to comprehend this impalpable, palpable world.

Besides the factor of atomic form already discussed (the electron shells), there is the factor of velocity. This can make all the difference in what joins what. Just as skaters in a roller-skating rink do not easily catch hold of other skaters moving at speeds much faster or slower than themselves, atoms in the void are more hospitable to the subtler approach. A violent collision of atoms will be as socially inconsequential as a corresponding clash of solar systems, for two speeding nuclei or suns have less than a ghost of a chance of touching each other, while the almost empty orbits of surrounding electrons or planets sweep through one another so swiftly that there is no time for any appreciable interplay. Even on those very rare occasions when some such intruding electron actually hits another, perhaps knocking both out of position, the effect is usually short-lived because ions (atoms with missing or extra electrons) are unstable and therefore so impulsive that they either sow their wild-oat electrons or are sown by someone else's at the slightest opportunity, quickly restoring equilibrium and normality to both sides.

More moderate atomic collisions, which do not involve any interpenetration of the electron shells, are comparable to the harmless bounding together of two billiard balls. Indeed, the atoms in real colliding billiard balls are believed to resist each other primarily through the negative-to-negative repulsion that arises when their outer electrons come into "contact."

It is only the very slow coming together of atoms that normally produces a deep and large-scale interaction between them. I mean, for example, the relatively slow bonding of two pieces of ice pressed together (which can unite them solidly even while immersed in warm water), the more leisurely setting of glue or cement or the still slower sticking of smooth, unoiled, metal surfaces as in long-idle machinery or old, unused hinges. Here something happens that must be a little like the action of one of those swinging gates that at first swings to and fro without locking but, when the motion has slowed down enough, suddenly drops its latch into place, with only a split second of rattling before all motion ceases. Of course, the outer shell of a slow-moving atom corresponds to the latch in this analogy, a shell of electrons that are repelled again and again by the similar charges of other electrons on both sides of their potential resting place in some nearby molecule. But this repulse is not quite as relentless as it seems at first and, if the shell of electrons has plenty of time to "swing" back and forth and to feel the varying pressures at different points on the molecule, it may suddenly discover some attractive gap and, still avoiding most of the worst repulsion, slip triumphantly into it.

The chances of attaining such a satisfying union, like a sound marriage, are normally enhanced by an unhurried courtship followed by some sensitive arranging and preparing. The congenial tempo of the whole affair, indeed, is of its essence and, from a larger view, is describable as not only a matter of speed but, even more, of temperature — the average excitement of hundreds or billions of molecules being what the word temperature actually means. It is by realizing this that we get an understanding of why temperature is such a critical factor in any chemical reaction — how it is the key to freezing, cooking, molding, cracking, welding and a prime clue to all the states of matter.

An apparently simple question here may help show what temperature and heat really measure. How cold is interstellar space? Strangely enough, this is a question that has no proper answer. Strictly speaking, the question itself does not even make sense. Temperature is only science's term for collective molecular motion such as exists everywhere on earth — underground, at sea and in the lower atmosphere. It is a statistical concept, the average of speeds of innumerable particles that continuously influence each other through perpetual collisions. But most molecules in interstellar space are no more collective than peanuts drifting a thousand miles apart. They react on each other probably less than our sun reacts on Alpha Centauri or Sirius.

For such independent particles, temperature (if the term can be used) means exactly the same thing as velocity, and it varies with every particle. The hydrogen atoms and molecules, impelled by starlight and obstructed by nothing, may be just as "hot" as the surface of the average star: some 20,000° F. On the other hand, grains of dust, being relatively large, rare assemblages of solid ice or maybe silicon or iron that cool themselves by radiating heat continuously, probably get fairly close to absolute zero, the limit of cold, which is 460° F. below zero. Thus we have in space a combination of a lot of independent particles of hot gas interspersed with a few bitter-cold chunks of dust with which they cannot really be said to mix, although it seems to be true that the rare collisions between dust and gas particles are what gradually builds up stars — the gas losing its velocity (heat) on sticking to the dust until a large enough body is amassed to generate heat by pressure.

The curious relativity of heat may be further illustrated by the apt phrase that an airplane (like a molecule) is coming in "hot" when it lands fast. Therefore, airplanes and molecules can be hot in relation to one object and cool in relation to another that happens to be moving along at nearly their own speed and direction, just as cars on a road are "hotter" to opposing traffic than to traffic going their way. This is the fundamental concept of temperature as *relative* velocity.

It is hard to imagine or appreciate the great difficulties men have had to go through to dig out such subtle bits of abstract truth. Almost up to the nineteenth century, heat was considered to be a mysterious, weightless fluid known as "phlogiston" or "caloric," which was created by burning and could flow invisibly from one body to another. But Benjamin Thompson of Woburn, Massachusetts, who became Count Rumford of the Holy Roman Empire and married Lavoisier's widow, had a profound flash of insight while directing the boring of cannon for the government of Bavaria in 1794. He suddenly saw a connection between the mysterious loss of energy expended in the drilling and the equally magical gain in "caloric" wherever the moving drill touched the metal. In short, he realized for the first time that "caloric" or heat is not an entirely separate thing-in-itself but is just energy that has changed its form. When Sir James Prescott Joule carefully measured and proved the equivalence of work and heat a century later, he established the first law of thermodynamics: the now well-known principle of conservation of energy. And this, logically, led to the famous second law (mentioned on page 214): the law of entropy or disorder which says, in effect, that random molecular motion (heat) can never be totally converted back into mechanical energy or work — that, in other words, there is bound to be some net flow of energy in any closed system from more orderly to more disorderly forms.

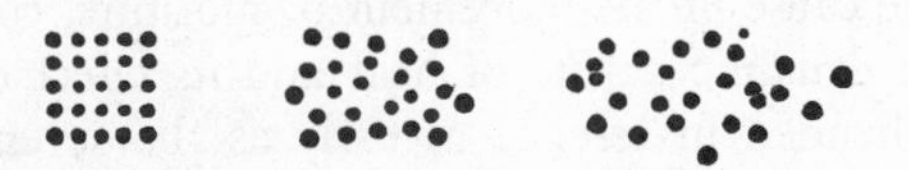

These thermodynamic laws come up frequently in any study of basic matter and help us to understand the strange behavior of substances in all their states from the maximum rigidity of absolute zero, where atoms and molecules are locked into the condition of least possible (not necessarily zero) motion, through solidity, plasticity, liquidity, gaseousness, atomic ionization or plasma, to the explosive fission of billions of degrees of heat where not only molecules and atoms have long since disintegrated but even the atomic nuclei (which we will look at in the next chapter) must burst violently asunder.

To bring all this down (or up) to Earth, molecules in their solid state can be compared to the leaves on a tree in summer. Green leaves obviously can move a little (depending on circumstances) but are securely anchored in their appointed places

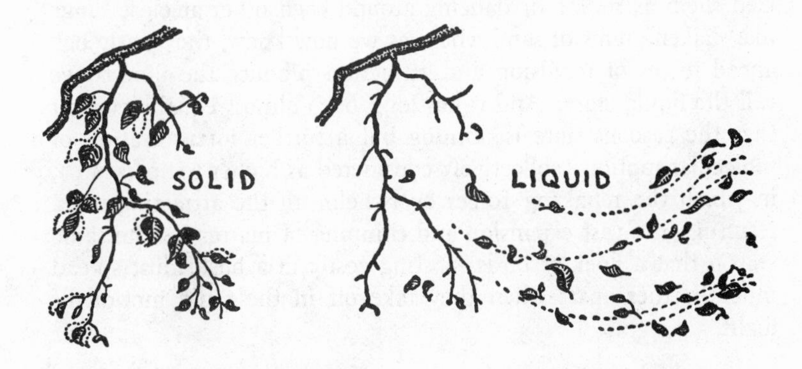

according to a fixed, crystallized pattern. When something solid melts and turns into liquid, it is as if an autumn breeze (heat) comes along to flap the leaves (molecules) harder and harder until eventually they snap loose and start to flow through the air completely free from their former fixed pattern.

The basic idea of this process may well have been felt intuitively by Leucippos or Democritos, and it was Leonardo who wrote, "If heat is the cause of the movement of moisture, cold stops it." Certainly the dynamic nature of heat and its effect on materials were comprehended in essence as early as the seventeenth century, when Robert Hooke, the English physicist and colleague of Newton, wrote,

> What is the cause of fluidness? Let us suppose a dish of sand set upon some body that is very much agitated and shaken with some quick and strong vibrating motion . . . By this means the sand in the dish, which before lay like a dull and unactive body, becomes a perfect fluid; and ye can no sooner make a hole in it with your finger but it is immediately filled up again and the upper surface of it levelled. Nor can ye bury . . . a piece of cork under it but it presently swims on the top; nor . . . a piece of lead on the top of it but it . . . immediately sinks to the bottom. Nor can ye make a hole in the side of the dish but the sand shall run out of it to a level . . . And all this merely caused by the vehement agitation . . . for by this means each sand [grain] becomes . . . a dancing motion.

Of course, Hooke could hardly have guessed the fantastic shapes of real molecules, but he seems at least to have correctly visualized them as rolling or dancing around each other at close range like shaken grains of sand when, as we now know, the gently balanced forces of repulsion and attraction produce the fluidity we call the liquid state. And it has long been almost equally evident that the gaseous state is nothing but a further intensification of molecular motion (collectively considered as higher temperature), in which the repulsing forces overwhelm all the attracting ones, resulting in a vast expansion and thinning of matter for the same reason that a flock of birds roosting cosily in a bush must spread much farther apart when they take off in the rapid motion of flight.

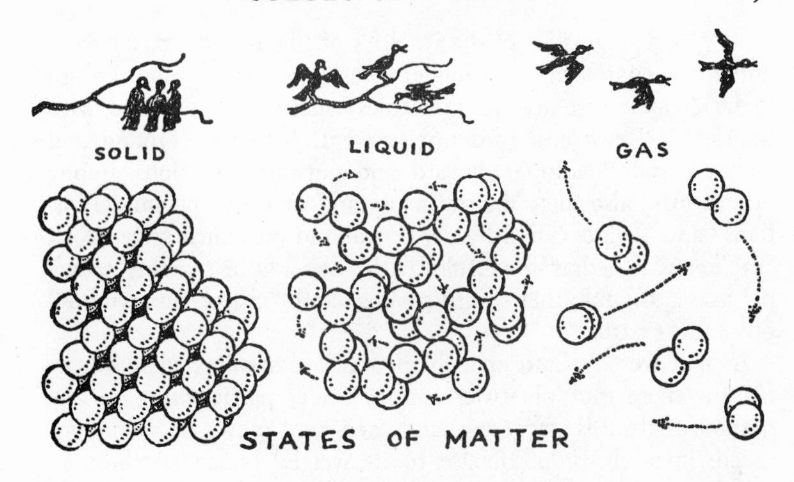

Another, and more precise, way of explaining gas is to say that the atmosphere is only 1/800 as heavy as water per unit volume. This, of course, is an expression of the amazing general vitality of molecules which, once free of the magnetic sociability of the solid or liquid state, suddenly become as independent as soaring eagles. The nature of their gassy emancipation, indeed, has been measured so precisely that it is now known that the first one thousandth of a degree Fahrenheit above absolute zero would give simple molecules on the average just enough vibration to make them bounce to a level of about seven inches above the ground. Such a temperature has actually been attained on a small scale in a modern laboratory and helps one to realize that the air in a room would literally freeze solid into a kind of powder and fall to the floor if the room could be made cold enough. Yet only a few thousandths of one degree of heat provide enough energy to keep the molecules bouncing off all the walls and ceiling, easily relegating gravity to inconsequence. And from there on up to normal room temperature, the speed and frequency of impact and recoil mount steadily, the molecules maintaining their unsociable gaseousness despite such a crescendo of excitement that their spritely dance is soon turned into a ferocious battle of billions of blows per molecule per second.

To give a slight idea of the varieties of this motion, particularly under the complexities of higher density, there are now known to be "planar systems" of molecules whose atoms vibrate with translatory (back and forth) movement, torsional (around and around) and breathing (inward and outward) motions among others. And also there are stationary and rotating "pyramidal systems" and "cubic, cylindrical, spherical and polygonal systems" of "cyclic and dihedral" molecules with "one-way or two-way principal axes" to guide their "turn groups" and "mirror turn groups" of spinning parts.

As if it weren't hard enough mentally to swallow such goings-on, the mere multiplicity of gas molecules participating in any natural mixture is staggering, and each kind moves at a different speed, inversely proportionate to its weight (since all those in contact must possess the same average energy), just as would interacting billiard balls if made out of hollow celluloid, cork, wood, aluminum, iron, silver, lead, gold . . . In such a melee of mixed and moving spheres, one could let something like pingpong balls represent the extremely light hydrogen molecules (H_2), which are actually batted about much faster than anything else: literally a mile per second in normal air. A cork ball, next in speed, might represent a stray methane molecule (CH_4) at ⅖ mile per second; a wooden ball, either nitrogen (N_2) or carbon monoxide (CO) at 3⁄10 m.p.s.; an aluminum ball, oxygen (O_2) at ¼ m.p.s.; perhaps an iron ball, carbon dioxide (CO_2) at ⅕ m.p.s.; a silver ball, chlorine (Cl_2) at ⅙ m.p.s.; a lead ball, bromine (Br_2) at ⅑ m.p.s.; a gold ball, mercury vapor (Hg) at 1⁄10 m.p.s. and so on.

Something along these lines, only immensely more complicated, must certainly be going on in the endless super-billiard game played by ordinary air molecules, and this is a good part of what makes the wind blow and what stokes your lungs with the oxygen pressure of life or enables a child to suck lemonade through a straw. For, in case you did not know, the drawing in of breath or the pulling of lemonade by suction up a straw is more realistically describable as the *pushing* than the *pulling* of molecules. Gas molecules just do not have any hooks or magnets capable of pulling each other, and suction will no more pull them

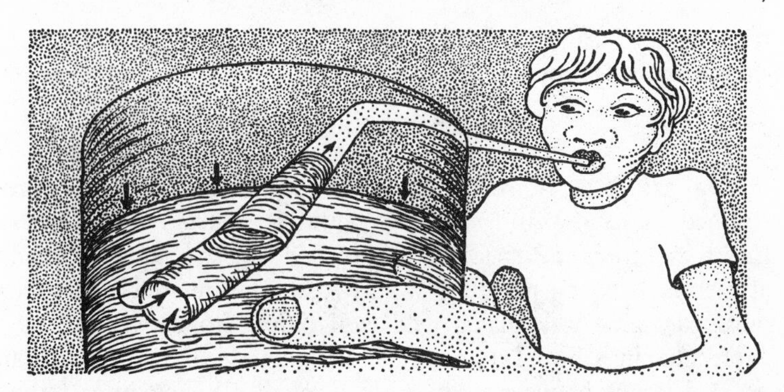

than opening a bird-cage door will pull the bird out. Like a bird, a molecule is powered from within. So, by expanding the hollow of his sealed-off mouth, the sucking child is really only pushing some atmosphere away from his body and letting the increased volume of his mouth harbor so many air molecules from among those flying in and out of the straw that the pressure inside the straw drops far enough below the pressure of the outside air to cause that air to push the lemonade upward from below, thus restoring the balance of nature.

Pressure, as we have said, is closely related to temperature, being essentially just a different aspect of the same molecular motion. Pressure is the collective outward force of moving molecules, while temperature is the motion itself. Thus the difference between hot gas at low pressure and cold gas at high pressure is about like the difference between a room containing a few frightened jay birds that fly to and fro at high speeds and a similar

HIGH TEMPERATURE
LOW PRESSURE

HIGH PRESSURE
LOW TEMPERATURE

room full of calmly roosting hens, pressed so tightly together they can hardly move.

If a gas is a frenzied free-for-all battle of relatively unrestrained molecules and atoms, a liquid is a social dance where the temptation to speed is controlled by an equally strong attraction between partners. The two states thus differ in somewhat the same way that a wild wolf differs from a trained dog. Molecules in a liquid, in fact, are disciplined enough to keep an exact distance between each other through all their unpatterned gyrations. This confines them at definite limits, such as the surface of the ocean or the surface of a dewdrop or at either surface of a bubble.

Yet "liquid discipline" is only a relative term, and a certain number of the more active liquid molecules, not held firmly enough by their attractions, are forever jumping the fence and escaping or, more exactly, evaporating into some gaseous outer wilderness. This naturally happens at any gas-liquid surface where, in the constant interchange of motion, outlying molecules of liquid are always getting knocked across the frontier — inevitably, some so hard they do not find their way back again. That is why a tub of water left in a closed dry room will inevitably raise the air's humidity in the form of more and more H_2O molecules flying about among the nitrogens and oxygens. But the humidity will not go on climbing forever for, as the number of free water molecules increases, ever-larger numbers of them are bound to strike the liquid surface, re-entering it as others leave it until, when

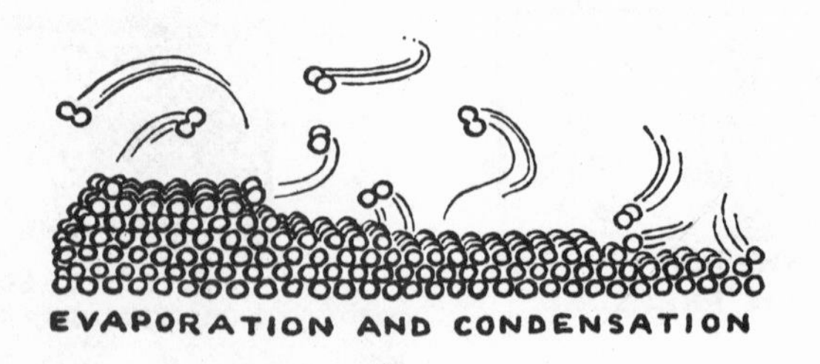

EVAPORATION AND CONDENSATION

the air is fully saturated, the exchange is equal and the level in the tub stands still.

If you have wondered why evaporation has a cooling effect, as is demonstrated in such familiar devices as the common porous water jug of the tropics, it is because the evaporating molecules are inevitably the faster (hotter) ones — their heat gives them take-off priority — and any continuous departure of heat from anything in any state must tend to leave it cooler.

Judging the effect on temperature of a combination of materials in different states is not always easy, however, as may be gleaned from the discovery that adding ice to water in certain cases can raise its temperature many degrees. This is because freezing is a crystallizing process which depends on more than temperature. Very pure, still water can be cooled to as low as 37° F. below zero without freezing, because ice is very reluctant to form unless microscopic impurities of some sort serve as nuclei to start the crystallization. But as soon as you trigger the freezing by touching a sliver of ice or so much as a single speck of dust to such "supercooled" water, the whole mass of the liquid will start crystallizing instantly by chain reaction. This is how an airplane flying through an innocent-looking supercooled cloud may ice up with fatal rapidity. And since freezing necessitates the release of energy (heat) which must somehow be conserved (first law of thermodynamics), the released energy just naturally radiates outward from the growing ice into the surrounding supercooled water, raising its temperature. Thus one finds that 32° F. is not necessarily the temperature at which water freezes. It is rather the temperature of a mixture of water and ice.

Of all liquids on Earth, water is far and away the most common but also, very likely, the most peculiar. Thales of Miletos considered water the basic stuff of which all else is made, including not only steam, snow, ice and clouds but "by absorption" the earth itself, even stones, trees, flesh and bones. Although such an idea is, to say the least, an exaggeration in the light of modern knowledge, water in some form is found nearly everywhere in our world and, as the mother of life and the queen of solvents, it is unique. It is the only substance we are thoroughly familiar with in all the three states of gas, liquid, solid — to say nothing of having a separate name for each. And, almost alone among materials, it is denser as liquid than solid — a fact of vast importance to life on this planet where ice's buoyancy keeps it floating on the surfaces of lakes and streams for the maximum insulation and preservation of aquatic creatures below during the severest periods of winter.

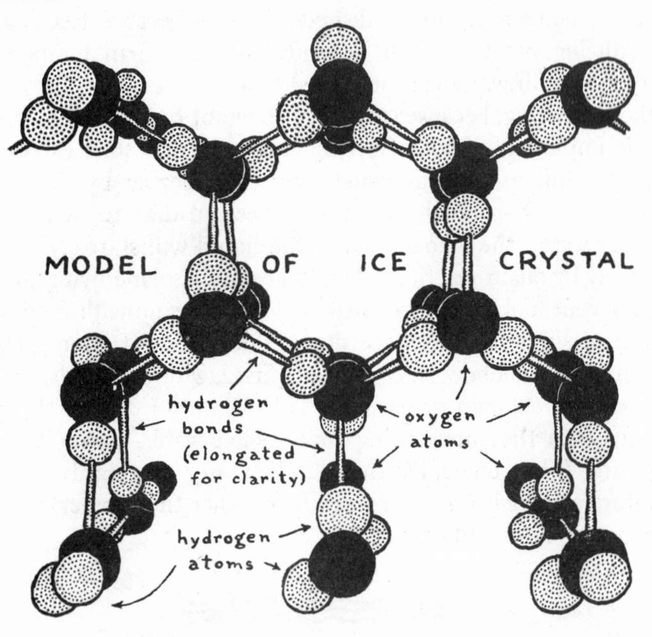

Probably the best approach to an understanding of the complex and dynamic structure of water, indeed, is to look first at its rigid crystal form in snow and ice. Through X-ray analysis and other means, the hexagonal tendencies that build the beautiful snowflake patterns are now fairly well understood. Each H_2O molecule of ice is held at the center of a kind of open tetrahedron (pyramid-shaped 3-D triangle) formed by its four nearest neighbors, one at each corner. The distance from molecule to molecule averages exactly 2.72 angstrom units — an angstrom being one hundred-millionth of a centimeter or (in simpler mathematician's notation) 10^{-10} meters. By a moderately strong attraction called the "hydrogen bond," these molecules are kept in firm relation to each other in the crystal lattice of ice. The hydrogen bond, incidentally, is the direct magnetic yoke between a positive hydrogen nucleus of one molecule and one of the negative outer electrons of another and, as no nucleus but the hydrogen nucleus is exposed enough for such an unvalanced bond, this tie is unique in chemistry and fundamentally different from the closer, stronger interlinking of the three atoms in each H_2O molecule itself. In fact, it is this special hydrogen bond that undoubtedly makes ice stick to ice. Even more significantly, it exexerts a steady inward pressure that is responsible for some of water's strange properties and, in ice, gives the lacelike hexagonal trestlework a stress like that of a bridge under a heavy load. When the ice's temperature rises, vibrating and shaking the bridge harder and harder, the load helps to bend, break and finally crumble the whole intermolecular structure at the melting point, which, as is well known, can be lowered many degrees in temperature by putting the ice under extra pressure. Conversely, the melting point can undoubtedly be raised if ice's internal pressure is reduced. For calculations indicate that, if this pressure should be eliminated entirely, as in icy meteorites in space, ice would not melt below 59° F.

As you may have surmised by now, melting quite literally means a stretching as well as a flexing and loosening of the solid, intermolecular bonds. And sure enough, as observed and analyzed with the help of the electron microscope, when ice bursts loose into water its hydrogen bonds stretch out an additional .18 angstroms and its molecules move just that much farther apart. One would expect water in consequence to be lighter than ice, just as lava is lighter than rock, but in water's case it so happens that the freedom of fluidity permits more H_2O molecules to occupy a given volume of space at a separation of 2.9 angstroms than the rigidity of ice permits at 2.72 angstroms. It is as if a battalion of a thousand soldiers completely filled a barrack square while standing stiffly at attention, shoulder to shoulder, elbow touching elbow, but, on "melting" and adjusting themselves into "at ease" positions, they suddenly discover they not only have a few inches between each other but, surprisingly, enough room for a hundred more men in the square. That such an analogy almost literally holds water is indicated by the strongly confirmed evidence that an H_2O molecule *always* has five close neighbors when it is water, while only four when it is ice.

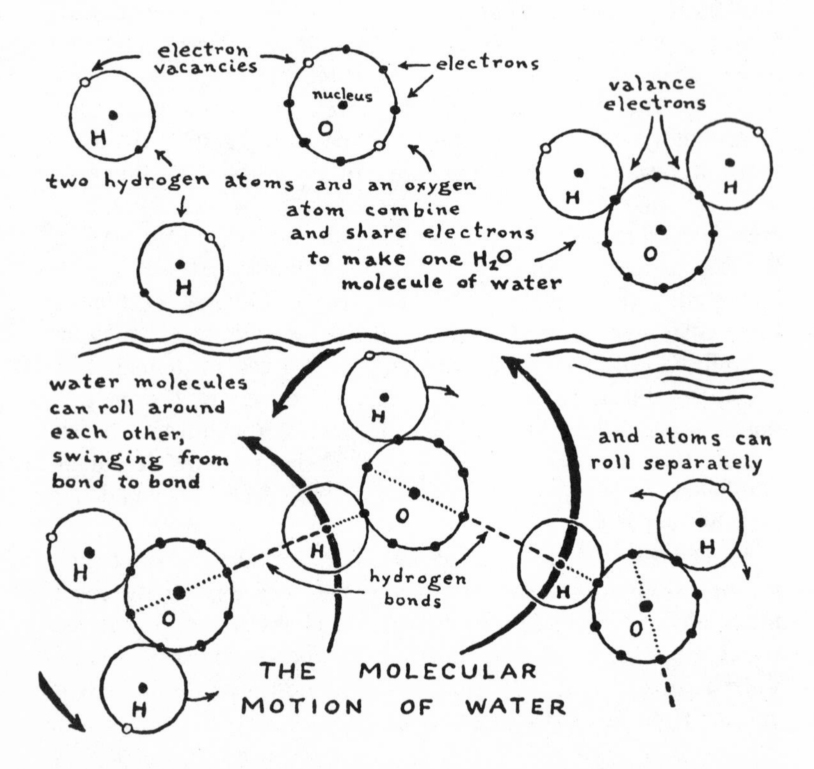

The hardest thing to visualize or describe about water is the chaotic rolling motion and interchange of its molecules in their liquid excitement. This is a highly complex turbulence not noticeably suggested by the gentle, macrocosmic rhythm of the barcarole. Although water's hydrogen bonds strictly enforce the law of 2.9-angstrom separation between molecules, the individual hydrogen atoms themselves are bandied about irregularly among the larger oxygens like coins in the marketplace but without ever forgetting their marital proportion of two hydrogens to one oxygen. To accomplish this, the single H electrons must keep popping in and out of the two vacancies in the outer oxygen shells like pistons in an engine even if, unlike pistons, few of them seem to hit the same engine twice. And all the time the hydrogens keep rolling from oxygen to oxygen in their peculiar chain-reaction, zipperlike motion. Of course, it hardly needs mentioning that the angle between the two H atoms in the H_2O molecule cannot remain a definite 105°, as in ice, but becomes variable, adding greatly to water's flexibility. And the excess of energy that gives water so much of its vitality is suggested not only by the explosive combustibility of hydrogen in the presence of oxygen — which is what destroyed all the great European airships — but also by water's eager circle of surplus hydrogens crowding around every oxygen atom while, paradoxically, only a few angstroms away three frustrated oxygens may fight over a single little hydrogen.

Naturally, this means that every atom in water sooner or later has valence intimacy with countless other hydrogens and oxygens, in effect making the whole body of the water, chemically speaking, a single large molecule. By this reasoning a microscopic water droplet floating in a cloud is not just a swarming quintillion of H_2O molecules but one single $H_{2,000,000,000,000,000,000}O_{1,000,000,000,000,000,000}$ molecule. And all the oceans of the earth can be scientifically approximated as simply $H_{2\times10^{46}}O_{10^{46}}$.

There are a few other reasons why H_2O alone does not fully express the nature of water. Water has turned out to be immensely more involved than the ancient mariner dreamed possible

— and is getting to appear more so every year. Probably its most revolutionary complexity turned up in 1934 when Harold Urey discovered "heavy water." He found that the purest water is made of something besides hydrogen and oxygen, something with an atomic weight of two, or double that of hydrogen. This substance, called deuterium, is a second form of hydrogen and is now known as an isotope. To explain isotopes, we must introduce a new elementary particle beside the proton and electron: the neutron. A neutron has essentially the same mass as a proton but it is electrically neutral. It is a basic part of the nuclei of all atoms except ordinary hydrogen and, although it has little chemical effect, the number of neutrons present in any atom determines what isotope or variety of the element it is. The deuterium nucleus contains one proton plus one neutron, around which circulates the usual lone electron as in ordinary hydrogen.

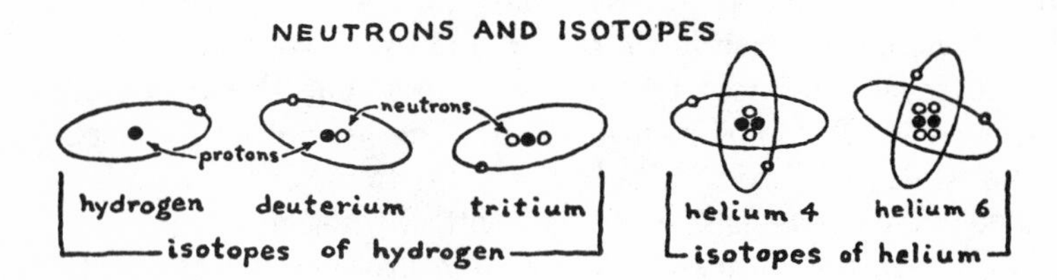

Deuterium or "heavy hydrogen" atoms are comparatively rare, accounting for only one out of about five thousand hydrogens, but these few also can combine with oxygen to form the compound D_2O. A drink of D_2O tastes almost exactly like H_2O but, mysteriously, seeds will not sprout in it and rats, watered with it, die of thirst. And now a third isotope of hydrogen, called tritium, is known, as well as the three isotopes of oxygen: O^{16}, O^{17} and O^{18}; these six variations of the two elements being combinable in eighteen different ways. When the several sorts of water ions (created by out-of-place electrons) are added, pure water is composed of at least 33 substances. And this is without even considering the foreign elements in ordinary water such as sodium, chlorine, carbon, calcium, sulfur, fluorine, potassium, nitrogen, phosphorus, magnesium, aluminum, lithium, boron, rubidium, zinc, iron, copper, lead and many others with their isotopes, ions and unnumbered compounds.

It is probably these alien complexities that are mainly responsible for the terrifying tricks that turbulent water full of bubbles can do. As every engineer knows, such water can, and often does, chew up a ship's propeller. It can punch holes in a water main until it looks like cheese or blast the valves of a giant dam as destructively as a high-explosive bomb.

You wouldn't think offhand that so much havoc could be wrought by nothing but water and bubbles. But the denseness of water makes it remarkably incompressible, and modern 20,000-frames-per-second photography has revealed that the walls of a collapsing bubble smack together with such a shattering impact that, delivered simultaneously by millions of bubbles and repeated trillions of times, it can make bubbles quite measurably eat away the hardest steel. Hidden behind this potent froth lies the mysterious phenomenon called cavitation, the curious hollowing out of water at its weak points. These microscopic flaws, we now know, are the real seeds of bubbles. Countless and ever-moving, they are the unassimilated oases of gas or solids that naturally flow throughout the dynamic homogeneity of the liquid.

How much actual difference they make is shown not only in the virtual elimination of bubbles when water is purified but also in the resulting increase of water's tensile strength. For, unlike gases which cannot pull liquid up a straw, pure water with its continuous bonds of hydrogen has the pulling power of a strong cable. Indeed, it is not only incompressible but virtually unstretchable. This pulling power explains how the tops of 300-foot sequoia trees get their moisture from the ground when transpiration from their leaves generates negative pressure in the sap arteries. The atmosphere does not push the water up, nor could water be pulled nearly so high if it had enough impurities in it to break the rolling continuity of linked hydrogen-oxygen-hydrogen-oxygen-hydrogen atoms. But the magical chains of extremely pure H_2O accomplish the lift easily and, according to laboratory tests by Dr. Lyman J. Briggs, could ideally pull their own weight vertically as high as 9,000 feet (nearly two miles) before snapping. In the words of Dr. Robert T. Knapp of the hydrodynamics laboratory at the California Institute of Technology, "You can hang heavy weights on an open tube of such water without breaking it. How? Just take a [large-sized] tube with both ends open, fill it with this water, and fit leakproof pistons or plungers into each end. Then . . . hang the top piston from the ceiling, and hang five hundred pounds on the bottom piston. The weight will be suspended there as if glued."

Most people feel they know the differences and changes in states of matter pretty well, especially in the familiar case of steam-water-ice, but even here things are not as simple as they seem. At least three distinct types of boiling have been observed and are now known as nuclear (the familiar burbling of separated bubbles in your teakettle), transition (the hotter, louder boiling where slugs of vapor form explosively over the whole heating surface at once), and film boiling (where a still hotter surface is blanketed with a thick, smooth, transparent film from which vapor rises in a steady roar). And there are at least six kinds of freezing, one of which produces an improbable ice of cubic molecular structure and another, under very high pressure, an ice so dense it sinks like glass in water.

There are also the partial freezings that produce so-called liquid crystals which look liquid but are revealed by polarized light to be organized as solid in one or two dimensions, their molecules forming either long brittle chains (as in some proteins) or filmy plates (as in graphite or silica gel) that can still slide or flow about in their unlocked second or third dimensions. And there are the uncrystallized, apparently frozen liquids like hard shoemaker's wax, which can be molded into a serviceable tuning fork but which, if a musician leaves it on his shelf for a few years, will be found to have flowed into a puddle of "honey." Glass is an extremely slow-flowing liquid of this type, tar a faster one, and the very hard, hot, dense, molten material inside the earth a particular case we have just begun to measure directly.

The basic difference between the liquid and solid states, of course, is always crystallization, the liquids having random, unfixed molecules that can slip or slide past each other even if (as in glass) the frictional forces between them reduce the sliding to "slower than molasses," while the true solids have molecules that are locked into a regular crystal lattice whose forces not only prevent sliding but tend to restore any dislocated molecules to the original positions in the pattern.

Water, it must be remembered, is only one particular kind of liquid. Most of the others do not have its peculiar rolling motion of hydrogens and oxygens, but rather a sliding motion like a mass of lively snakes. The oils are good examples of this, for they are composed of a wide variety of such centipedelike molecules as the octane (C_8H_{18}) described earlier, or nonane (C_9H_{20}), decane ($C_{10}H_{22}$), nonadecane ($C_{19}H_{40}$) and much longer ones. You might suppose that these slithering serpentine members would lie fairly flat, belly-to-ground fashion, at least when oil is stretched out into its thinnest films, but surprisingly, due to lateral magnetic polarity, the oil snakes habitually stand on their heads, side by side, something like blades of grass clinging to metal in the manner of weeds on the bottom of an old boat. The evidence is that the hydrogen bonds at the "head" end of each molecule bite into unstable outer electrons of nearly any solid surface like so many leech mouths, while their "tail" ends swing free, permitting encroaching material (such as the ends of other oil molecules either similarly attached to near-by solids or part of loose intervening films) to slip past them without friction, willy-nilly. Hence derives oil's familiar oiliness as the classic lubricant, the slipperiest of the slippery, that makes machinery workable and eases the wheels and axles of the world.

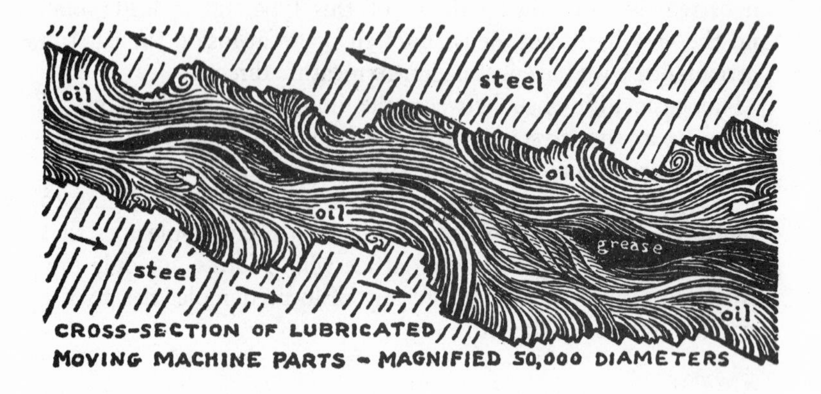

CROSS-SECTION OF LUBRICATED MOVING MACHINE PARTS – MAGNIFIED 50,000 DIAMETERS

Greases and fats and soaps are all made out of long chain molecules more or less the same as oil — and each kind works in its own surprising way. Soap, for instance ($C_{17}H_{35}COO\text{-}Na$), has a snaky molecule of some seventeen carbon atoms, each with two little hydrogens attached, an extra hydrogen on one end for a tail and a large head of oxygen, carbon and sodium (Na). It so happens that the sodium of the head has an affinity for water and for that reason it reaches out toward the nearest H_2O molecules whenever the soap is wetted as in washing. The hydrogen tail, on the other hand, is repelled by water and tends to avoid it by clinging to anything else it can find around such as molecules of grease or other bits of foreign matter which, in such a context, are considered as dirt. Thus each microscopic speck of dirt is

CROSS-SECTIONAL VIEW

SOAP AT WORK

progressively caught by the tails of all the soap molecules that reach it until it is surrounded by thousands or millions of them all facing outward, completely covering and insulating it from the water. That is soap's age-old secret, its technique of cleaning. It swallows the dirt as a mermaid swallows oysters and, mermaidlike, its long molecular tresses swirl outward into the water in cascades of microscopic loveliness, creating an exotic world of significance hitherto undreamed in your washtub.

In order to explain water's action as a solvent, it is helpful here to introduce a basic ingredient of the earth's oceans and a real common denominator of blood, sweat and tears: salt. Although salts are of many varieties, which can be defined in general as compounds of acid and basic radicals (made of excess positive and negative ions respectively) whose magnetic interaction gives them their characteristic salty tang, we will consider just simple sodium chloride (NaCl) or table salt. If you can abide a bawdy little illustrative analogy, let NaCl stand for an eager sailor named Nate (Na) who has just come down the gangplank to take in tow an attractive and well-stacked dame called Chloe (Cl). Such a briny combination of male and female elements, one of whom can receive exactly as much as the other can give, closely parallels the union of sodium (Na because the Romans called it natrium), its lone (eleventh) electron protruding outside its second shell, and chlorine (Cl), wanting a single (eighteenth) electron for fulfillment in its third shell. Almost as soon as the two atoms touch each other, inevitably Nate's prodigal electron finds a welcome haven while Chloe's need is precisely filled, and the resulting compound (salt) is strongly held together in a very stable cubic lattice of alternating sodium and chlorine atoms that are true atoms no longer.

Chemists sometimes describe the reaction in this pseudo-romantic valence as the "burning" of sodium in chlorine. Certainly it is violent enough to be considered a kind of combustion, and the powerful magnetic bonds thus created obviously derive from the fact that Nate is not quite all of himself afterward while Chloe has become more than she was and, as the new NaCl molecule, they are the salt of the earth both to each other and to the world — literally not independent atoms any more but just two linked ions: the male sodium ion (Na^+), who has become positive by losing his negative extra electron in chlorine, and the female chlorine ion (Cl^-), who has become negative by accepting the same negative electron from sodium. Thus the eternal masculine and feminine principles repeat themselves in the atomic realm as positive and negative ions, Nate and Chloe, ever distinct yet ever yearning to be one — even when dissolved in the sea.

This brings up the unique solvency of water, the ancient mystery of the deep, where salt is so thoroughly dissolved it never sinks to the bottom and the problem of economically separating it from the H_2O has become a major object of modern research. Water's extraordinary solvency, indeed, stems from its high "dielectric constant," exceeding that of any other known liquid, which can reduce the attraction between oppositely charged ions in solution to scarcely 1 percent of their original strength. Water molecules do this by ruthlessly swarming over and between the embracing ions in their midst like ants upon their prey, effectively "swallowing" the ions in the manner of soap molecules swallowing dirt. Since electrons shared between the oxygen and hydrogen atoms in water (H_2O) are perceptibly closer to the oxygen atoms, the oxygens have a slight net negative charge while the hydrogens are measurably positive. This is why the oxygen ($^-$) parts of water molecules cluster around each positive ion, such as sodium (Na^+), and why the hydrogen ($^+$) parts crowd in upon each negative ion, such as chlorine (Cl^-). It is water's own magnetic formula for dissolving ionic matter — by insulating and neutralizing its parts. It is why the oceans have become steadily saltier for billions of years and must go on getting saltier — even as the Great Salt Lake and the Dead Sea.

The simple, cubic, lattice structure of solid salt is a good introduction to the formation of crystals, which are the only completely rigid solids. Dry salt (NaCl), in which all lattice angles

are right angles (90°) and where each sodium ion is surrounded by six equidistant chlorine ions (north, south, east, west, up, down) and vice versa, is like the basic, cubic girder outline of a steel office building — the simplest possible repeating pattern of form. And it is not held together just by magnetic force, like a nail that can cling at any angle to a bar magnet or swing from a single point, but is in effect riveted and cross-linked at several points with each neighboring member. Its strength is shown by its high melting point, below which even the most violent thermal vibrations of the molecular girders are not enough to break the rivets.

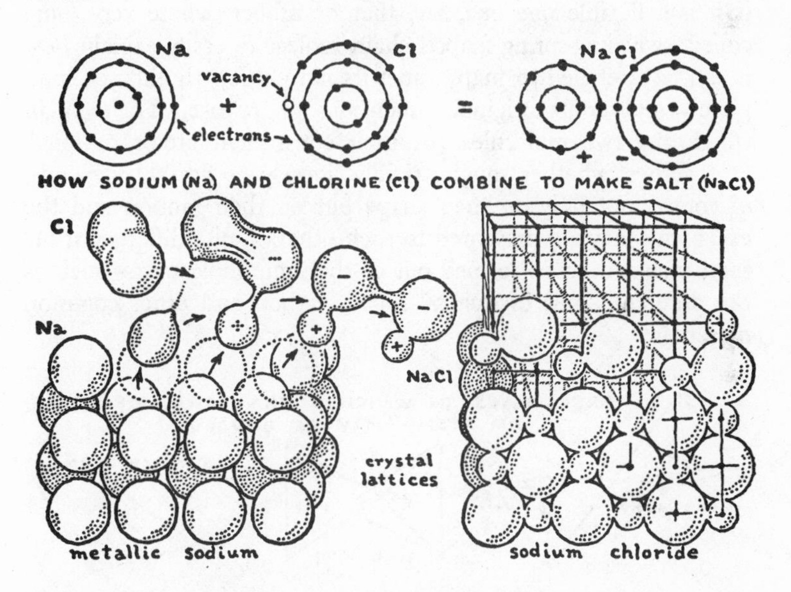

And yet the simple regularity of the salt crystal is formed automatically, by nature's way of least resistance, either when sodium "burns" in chlorine or when salt water evaporates, leaving solid salt behind. In either case, the sodium and chlorine ions just naturally group themselves alternately like boys and girls at a party where every boy wants a girl on either side and every girl wants to sit between two boys — only the "sex instinct" of salt ions is so much stronger that it is virtually impossible for two similar ions to touch each other at all, with the result that they form a completely regular three-dimensional pattern of plus-minus-plus-minus-plus ions, the salt lattice, simplest of crystals.

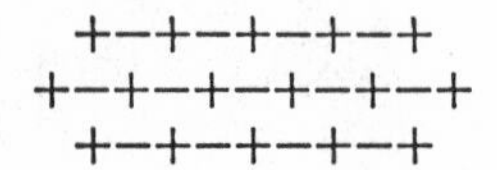

Just as other crystals are more complex than that of table salt, however, so are most molecules more irregular and complicated in form and nature. Yet all have some sort of definite shape, even if it is a flexible one like, say, that of rubber, whose very long, coiled, swiveling-spring-shaped chain molecules easily explain how it can be stretched to many times its natural length and yet completely recover its original dimensions on release. The way in which any two molecules (or their constituent atoms or ions) fit together, whether snugly, rigidly, loosely or flexibly, depends, of course, not only on their shape but on their motion and the exact way they are presented to each other. Quite different structures may indeed be formed out of the same molecules — such as the different kinds of quartz, sulfur, carbon and other common materials.

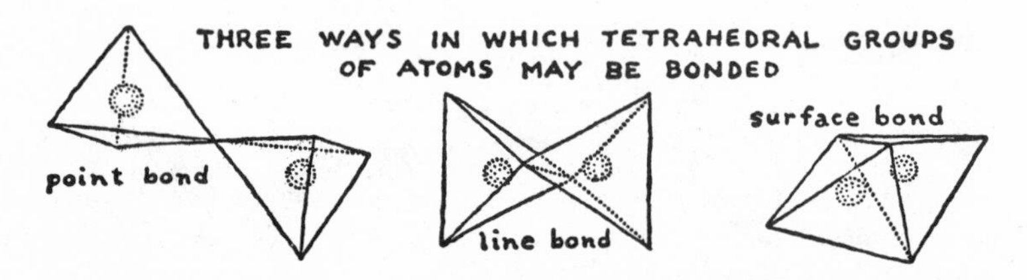

All this becomes increasingly understandable as you look deeper into molecular shapes, which, as they become more complex, tend to produce more lacelike or spongelike forms with many long narrow struts and stays like a cantilever bridge or like any sort of elaborate scaffolding. The emptiness of such trestlework accounts for the low density of the very complicated organic molecules, few of which are much heavier than water. The simplicity of the little two-atom molecules, on the other hand, produces denser stuff such as rubies or iron pyrites and the simplest-of-all single-atom units of pure elements make the even denser iron, silver, lead or gold, where the struts are no more than knobs, permitting the bridgework to pack down into something resembling a neat stack of bricks.

If the lattice patterns of such crystals are even more countless than wallpaper designs or musical themes, it is not because they are any less bound to repeat themselves in their three-dimensional way. For in doing just that, they make themselves the third stage or size order above the atoms and molecules of which all solid matter is constructed: an order now usually known as the crystal unit, which seldom has a definite place of beginning or ending but, like wallpaper, has a fixed magnitude to its repeating pattern no matter where you begin measuring it. By way of example, there are *atoms* called silicon and oxygen, and a *molecule* of silicon dioxide made of one silicon atom joined in a particular way to two oxygens. And now we discover there is also a *crystal unit* named quartz, which consists of three molecules of silicon dioxide arranged in a certain spiral or screw form. Like the flowers on a papered wall, you can mark a piece of quartz anywhere (at any molecule) and the pattern will repeat after every unit distance (the length of three silicon dioxide molecules) as far as the crystal extends. Each such unit has all the basic properties of quartz. In fact, it *is* quartz. It is the smallest possible piece of quartz. For nothing less has ever been demonstrated to be quartz. A single molecule of silicon dioxide quite definitely is not quartz. Even two such molecules, however aligned, cannot be quartz because quartz is actually a spiral lattice, and it takes three molecules to create its essential screw pattern.

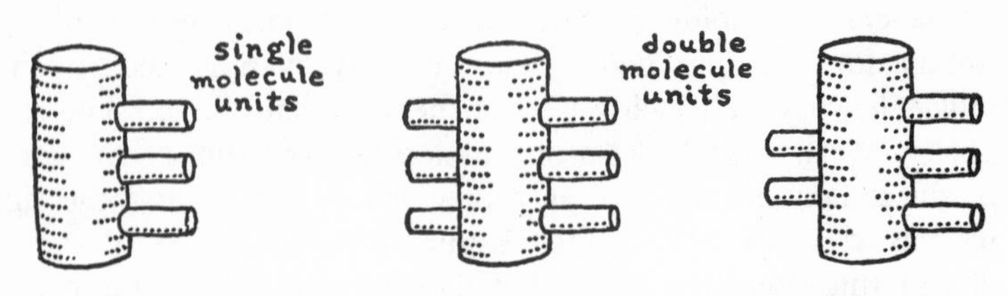

BASIC MOLECULAR PATTERNS IN CRYSTALS

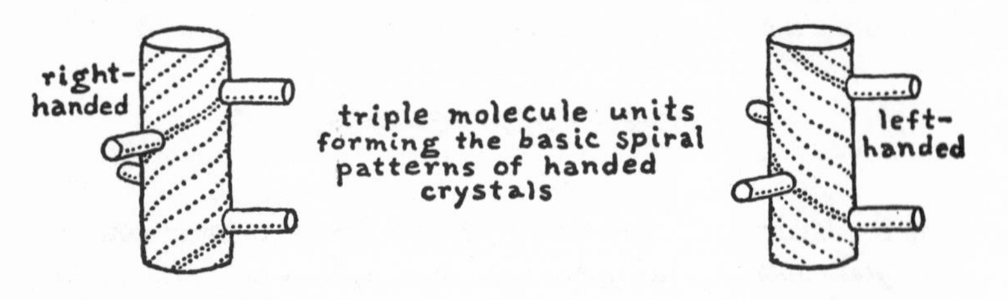

This truth is so important and characteristic of matter that it is worth the effort of visualization. Each molecule can be considered as a peg sticking out of a piece of broomstick. If the unit of pattern is one molecule, then every peg must exactly copy the peg next to it, all sticking out the same side of the broomstick in an even row. If the unit is two molecules, the pegs can stick out two sides of the stick, perhaps alternately. And it is only when the unit reaches three that they can occupy three sides of the stick, making possible the forming of a kind of rudimentary spiral staircase which has the option of turning either to right or left, clockwise or counterclockwise, as it goes down the broomstick. Something like this in essence is accepted as the explanation of the two kinds of quartz, often called right- and left-handed, whose crystal faces, observed in sequence, spiral off to right or left and correspond to quartz's well-known power of rotating the plane of polarization of light in either direction.

Such handedness (right or left), of course, is possible only to a perfect crystal whose lattice forms a single and continuous system. The growth of such a regular lattice, as crystallographers have long known, must start from a single nucleus, such as a microscopic dust particle or some other irregularity or happenstance that somehow brings the first molecules together into the key relationship. From here on, succeeding molecules will naturally follow the pattern already established as the line of least resistance, layer after layer of them continuously attaching themselves to the growing crystal in the correct order, settling out of formless gas or liquid into regimented solid, like the spool-shaped snow crystal that literally begins with an invisible dust mote in the sky. Obviously, only a very slow and peaceful accumulation will give the lattice flawless regularity — a circumstance nature in the raw can seldom provide — which accounts for the rarity of large naturally perfect crystals like the Hope diamond. So most crystalline solids turn out to be conglomerations of many small, if not microscopic, individual lattices that sprang from separate nuclear seeds, like large snowflakes which are the agglutinations of hundreds or thousands of tiny hexagonal crystals that collided by chance during their hours or weeks of development in the sky.

This is not to say that the joining of any kind of molecules into a crystal lattice cannot be studied from an ideal or abstract view, and the mathematicians have actually been ahead of the physicists and chemists in working out many of the laws of crystallography. Long before Euclid, some geometer asked himself, can I pave a floor with regular-shaped tiles of any number of sides? The answer, of course, quickly proved to be no. Only three kinds of regular tiles will fit to cover the whole area: the triangle, the square and the hexagon. This seemed quite obviously a divine restriction. Centuries later, Plato was thrilled by the revelation

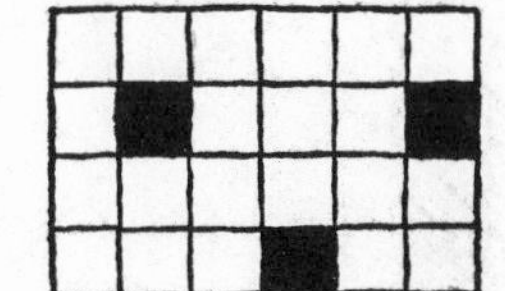

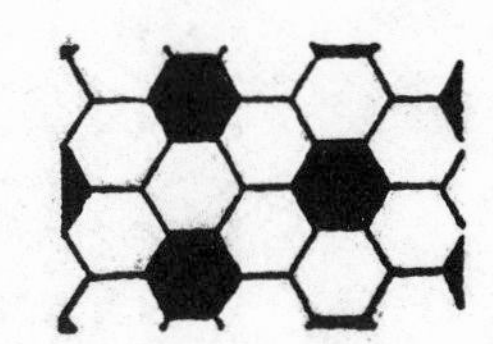

that a comparable decree limited the number of regular convex solids to five — the same five that Kepler would one day use to test his planetary spheres.

Although this did not have an obvious bearing on natural things of the earth as then understood, a time was to come after nearly two thousand years when a Danish bishop called Steno began to experiment scientifically with the "regular solids" of crystals, a research which culminated in 1782 when the Abbé René Just Haüy, a French crystallographer, discovered that the curious regular angles of any known crystal could be mathematically explained by the simple assumption that the crystal was composed of many very small identical regular "bricks." Haüy's Law of Rational Indices was not soon accepted, but it marked the first actual measurement of atoms or molecules and, perhaps more importantly, gave the first concrete understanding of space quantization as applied to the microscopic world — an answer to the ancient Greeks who had pondered the ultimate nature of matter and more than a hint that the same music of the spheres which separates the rings of Saturn also regulates the orbits of entities so tiny they have never been directly observed.

Haüy's Law was another way of saying that the lattice spacings of atoms permit certain particular crystal symmetries — and no others. A square made up of evenly spaced dots, for instance, cannot be turned into a regular octagon (figure of eight equal sides) no matter how you slice it, because that would mean dividing each side of the square into three parts in the proportion of $1:\sqrt{2}:1$ — which is impossible with integral dots because $\sqrt{2}$ is not an integer.

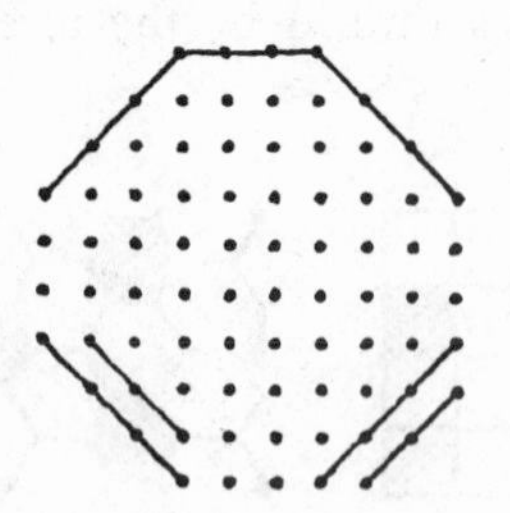

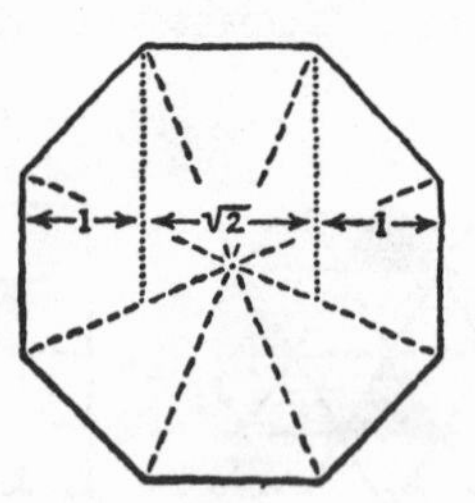

You probably did not learn this sort of thing in school, of course, because it is not included in Euclid's geometry of *lines*. Crystals instead follow the earlier geometry of Pythagoras, a geometry of *points*, from which evolved the modern Theory of Discontinuous Groups, which has already discovered much more than physics has been able to use. One of its simplest theorems, for example, proves that "the symmetry elements of crystals can be grouped in 32 different ways, and no others."

The crystallographers have long since then sorted every known crystal, according to its outer shape, into one of these 32 classes. But the mathematicians did not pause at the outer crystal and, imagining the atoms and molecules inside might also have characteristic shapes, they had proved by the end of the nineteenth century that "there are just 230 different ways of distributing identical objects of arbitrary shape regularly in space," a triumph of abstract theoretical logic that unexpectedly took on great practical importance in 1912 when x-ray analysis of the interior of crystals first made possible their actual sorting into the 230 space groups already divined.

Just as the Periodic Table of elements was found to follow a kind of insistent, sagacious harmony, the natural arrangement of atoms thus proved to be tuned to a deep but fathomable music of its own — if one more peremptory yet perhaps also one more beautiful than any contemplated by Pythagoras.

Such modern melodies of the spheres may, I think, be most simply introduced as the keys to something formal and familiar like those historic piles of spherical shot heaped up beside the century-old cannon of war memorials. For those neat pyramids of cannonballs represent exactly how the atoms are packed in the lattices of gold, silver, copper, aluminum and other elemental crystals. If you study such a pile carefully, you will find that any ball in the interior of it touches exactly twelve other balls: six around its equator at the same level and three around each temperate zone, above and below. There is obviously no closer way to pack spheres, but there are two forms of this twelve-point packing. After the second layer of cannonballs has been placed upon the first with each ball nesting in a hollow between three below it, the third layer can be so started that none of its balls will be directly above any of the lower balls (as in the piles of ancient shot) or so that every ball is exactly above one of the first-layer balls. These two basic alternatives are possible because there is room for a ball only in every second hollow of each layer and, in the choosing of which set of hollows to use, it is the third layer (like the third molecule in quartz) whose position in relation to the other two makes the crucial difference.

It so happens that the first of these packing alternatives tends to make a large lattice of spheres form a cube, and probably for this reason grains or flakes of gold, silver, copper or aluminum are normally cubic or squared off in 90° angles as seen under the

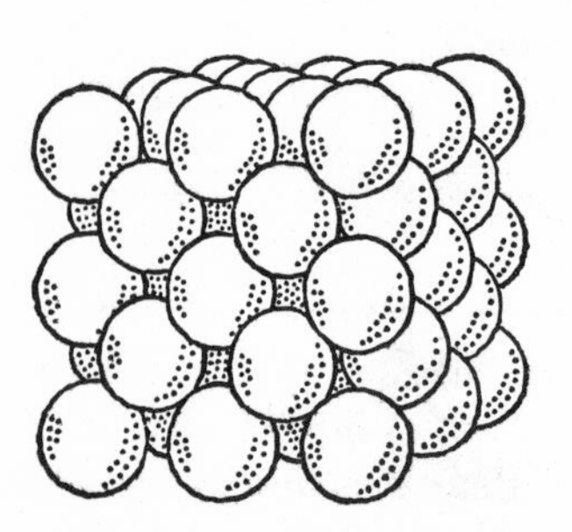

The atomic packing of gold tends toward a cubic form.

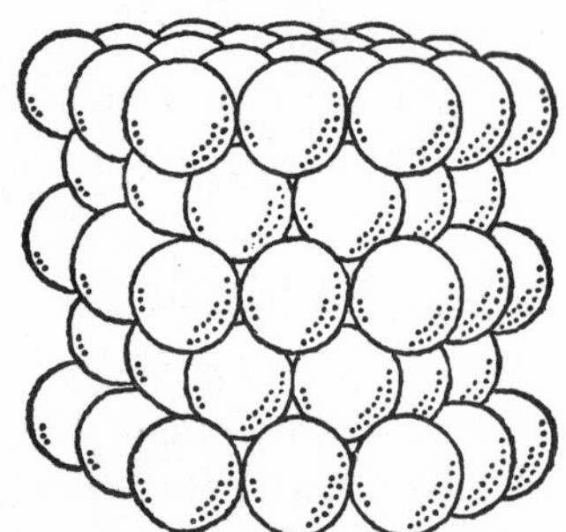

The different packing of magnesium tends toward a hexagonal spool.

THE PACKING OF SPHERES

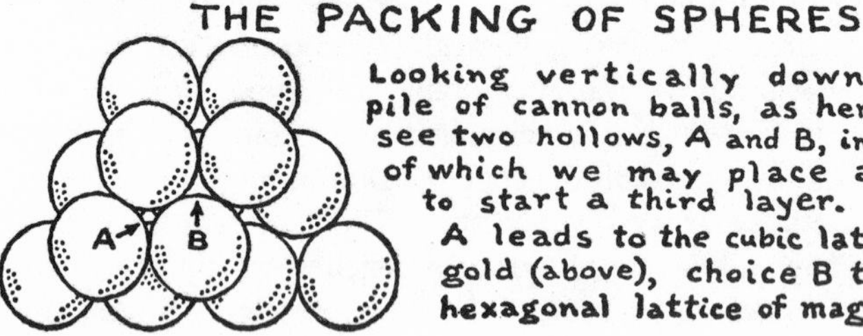

Looking vertically down on a pile of cannon balls, as here, we see two hollows, A and B, in either of which we may place a ball to start a third layer. Choice A leads to the cubic lattice of gold (above), choice B to the hexagonal lattice of magnesium.

microscope. The second packing, on the other hand, tends to produce a hexagonal spool of 60° angles as in the crystals of magnesium, which, for some still unknown reason, always employ this alternative.

Entirely different and looser ways of packing spheres or atoms, of course, may also produce cubic, octagonal, hexagonal and other forms, such as the simple salt cubes already described where each ion touches six others, the more complex iron and steel crystals where each atom meets eight neighbors and quadrilateral and octagonal forms abound, or the lovely hexagonal ice crystals where every molecule touches only four mates. Plainly, four is the minimum number of touch points in solid packing, for no lattice could keep from collapsing (melting) if all its component spheres supported one another in only three places. Yet this minimal tetrahedral packing need not be weak, as is evidenced by the diamond, hardest of natural substances, which is nothing but a four-point lattice of carbon atoms at angles of 109° 28′ from each other.

Moreover, there is plenty of mystery behind the simple beauty of a diamond, one soon discovers, for its strictly equidistant, tetrahedral structure is definitely less stable at ordinary temperatures than another somewhat similar but looser three-point packing of carbon, the nonequidistant one that forms graphite. How two such different materials as a diamond and a pencil "lead" can be made of the same latticed carbon atoms just by switching from four-point to three-point packing is difficult to see, especially as the layers of both materials show virtually identical

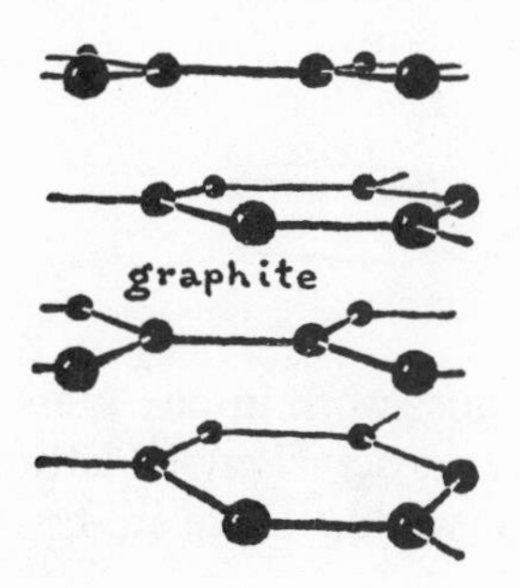

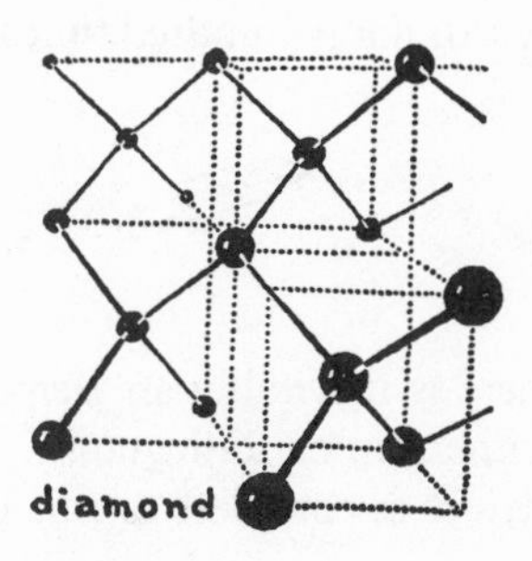

hexagonal honeycomb patterns when viewed broadside in the electron microscope. But the explanation obviously lies somewhere in the angular relationships among the carbons, which crowd closer together for diamond (density 3.51) and stretch farther apart for graphite (density 2.25). This shift is revealed by x-rays to act entirely in one direction (perpendicular to the honeycombed layers), thus inevitably separating the lattice planes of the graphite far enough to weaken seriously the atomic ties or any other causes for friction between their smooth surfaces which, once turned loose to slide easily over one another, give this solid lubricant its characteristic plumbaginous slipperiness.

It was only in February 1955, by the way, that the actual transformation of graphite into the prince of gems was proved possible for the first time when four researchers of General Electric Company finally solved this long-baffling problem in synthesis — a job that turned out to be much harder than shifting a layer of cannonballs into its alternate set of hollows, though not so different in principle. The pressure at which diamond becomes more stable than graphite had been calculated to be about twenty thousand times greater than that of sea-level atmosphere at room temperature. Yet prolonged pressures even as high as 425,000 atmospheres proved unable to budge graphite "over the hump" from its relatively unstable posture until it was heated (at a somewhat reduced pressure) to nearly 4,000° F., a point where its thermodynamic state must have approximated that in the natural diamond wombs deep under the earth, for suddenly the carbon layers snapped together into the 109° 28′ diamond configuration — at one flash turning a pencil "lead" into a jewel a third smaller but worth many hundred times more!

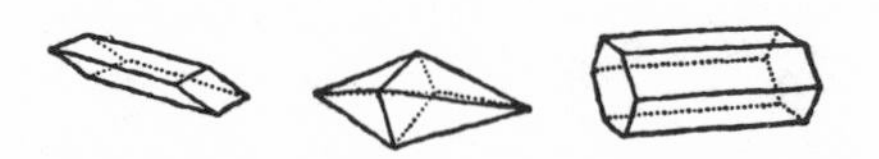

There is inevitably an element of guesswork in any pioneering experiment in crystallography, because no actual crystal lattice can be perfect or complete in this irregular world. There are bound to be holes in it somewhere or long rifts or faults where one layer shears against another, the dislocated surfaces attracting or repelling each other according to whether the areas of compression (+) in one match up with areas of tension (−) in the other or whether areas of similar "sign" clash in opposition. Such dislocations or grain boundaries are inevitably under relative stress on the average and are therefore known to crystallographers as regions of energy. Because of uneven resistance to their progress, they tend to follow spiral courses in the lattice while, collectively, their reactions upon one another cause them to align in curious invisible waves that creep forward something like dunes on the desert.

Although a single large crystal is seriously weakened and normally breaks only where there is a slipping of molecules along one plane, most crystals are compounds of many tiny semi-independent lattice systems for the same reason that marbles poured casually upon a tray will arrange themselves in tight groupings at various odd angles to one another. This is why wrought metal is tougher than cast metal, for hammering it and working it increases the number of dislocations and regions under stress — and stress means hardness or strength.

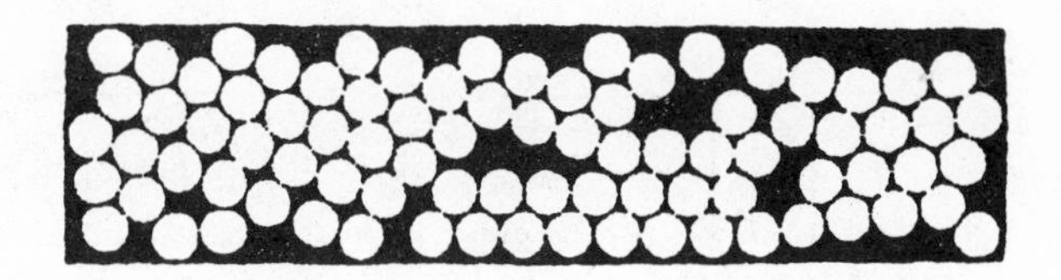

Yet there are kinds of stress, and each has its limits in each kind of lattice. The elasticity of spring steel in a safety pin obviously depends on the magnetic bonds that give its atoms their capacity to recover their original stable positions after being stretched or strained within definite limits. If forced beyond the so-called "yield point," steel will start to "creep." Elasticity will begin to be replaced by plasticity or ductility and, perhaps farther on, a fracture, depending on the exact combination of stress and material. You might not guess that a rock could be so bouncy, but by 1957 researchers had succeeded in fashioning a spring out of quartz, finer than a human hair and coiled to about an inch long, that can be stretched to arm's length, its elasticity so perfect it "never" develops fatigue. Such a spring has since been used in a gravity meter so sensitive no term is known for what it can measure.

In many compounds, elasticity is affected also by the tendency of minority atoms to coalesce like ice crystals in the sky, forming snowflakelike centers of foreign structure that, as they grow with the passing years, eventually constitute the collective weakness known as age hardening. And several kinds of evidence show that even in the densest solids atoms are far from absolutely fixed. Though most of them may be anchored most of the time, their "chains" permit them to vibrate about a mean position, while some are always shuffling from mooring to mooring, others more freely drifting or even flowing merrily along like a brook or a breeze. Under stress, minority atoms have repeatedly been observed to "float" or "sink" toward the tension or compression levels of a lattice, depending on whether they were bigger or smaller than their fellows, and in the case of quenching red-hot metal may often produce a fallout amounting to a kind of atomic hailstorm.

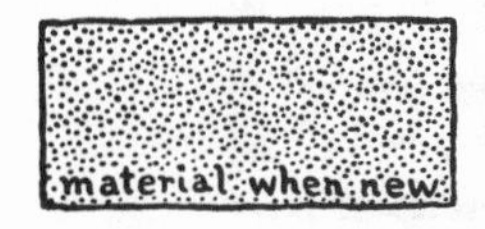

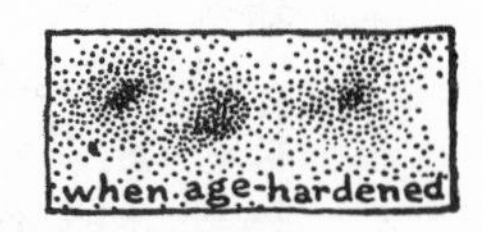

Just as perfume diffuses almost instantaneously through air and dye almost as fast through water, so do solids diffuse in their subtle way through solids. The process is surprisingly rapid when two heated metals are clamped tightly together, such as two cylinders (of gold and lead) which interfused almost completely within a few days in the famous experiment in 1896 in England conducted by Sir William Roberts-Austen, Assayer of the Mint. Even a gas will deeply pervade a solid; thus do the carbons out of methane readily creep into the lattice hollows in very hot iron to make case-hardened steel.

Solid diffusion is mechanically like putting a few black marbles into a box almost full of white ones, then shaking it. Even if the box is quite tightly packed, a little bit of play will enable some marbles to alternate their positions and the black marbles will move about and distribute themselves among the white at a diffusion rate related to the tempo of shaking which, atomically speaking, is the temperature. The tiny atoms of hydrogen thus leak through the densest materials with particular ease, making it practically impossible to purify anything else in their presence.

o·o·o·o·o·o·o·o·o
o·o·o·o·o·o·o·o·o·o
o·o·o·o·o·o·o·o·o

Purity in earthly matter is of necessity a relative term, for even so-called pure gold, the buried coin of the realm, is about .4 percent copper and other "impurities." Only recently has science discovered how to insulate and refine an element to the degree of eliminating all but one alien atom per billion, as in the case of a new metallic silicon that has doubled the heat resistance of missile transistors and as "the purest product ever made by man" (for $980 a pound) is helping to open up a new age of metals — the latest in the long sequence of metallic ages that began about 4000 B.C. with the discovery of copper, probably on the upper Nile. From bronze, which shortly afterward appeared as an alloy (90 percent copper, 10 percent tin) much harder than copper, through iron and steel, which have dominated the metals ever since the early days of swords and made possible the Industrial Revolution, the modern light-metals age of aluminum, magnesium and titanium developed into the present advent of specialized metals like zirconium, hafnium and beryllium (of great importance in atomic piles for their effect on neutrons), uranium, plutonium and thorium (as fuels), molybdenum (in jet engines), lithium, lightest of solids (as a lubricant) germanium (in transistors), gallium and so on — each individually and in combination being explored actively and continuously for new uses. A good three quarters of the elements of the Periodic Table, technically speaking, are metals — the solid underpinnings of material civilization — although many of them are too soft for usefulness while unmixed and so must be judiciously alloyed with others before they can serve humanity.

The hardness and strength of alloys as compared with elemental metals obviously derive from their compound structure: the zinc atoms lodged in the copper lattice that make it brass, the precise combination of chromium, cobalt and tungsten that creates the very hard, noncorrosive stellite, the innumerable and mysterious harmonic crystal intervals in the various resonant alloy combinations that put the ring into bells. The hardness in each case comes from the pressure of foreign matter warping or crimping the smooth lattice. It is as if one had dumped sand into an engine journal. The sliding planes have been jammed. This effect is perhaps shown most clearly by its absence in the surprisingly pliable copper-nickel alloy long used for encasing bullets, a metal mixture later revealed by x-rays to be soft because the nickel atoms added to the copper happen to match the copper atoms so closely in size that they just slip into vacancies in the copper lattice without appreciable strain.

The famous medieval swords made in India and sold to the western world from Damascus are a fine example of special alloy qualities. For one of the marks of genuine "Damascus steel" was the characteristic wavy whorl pattern which, under a modern microscope, turns out, like the Milky Way in a telescope, to be composed of curving bands of millions of independent specks. These have now been analyzed as cementite crystals, tiny clusters of molecules made up of three atoms of iron combined with one of carbon, which is the form excess carbon takes after the treated hot iron has absorbed all the loose carbon atoms it can hold in its lattice hollows.

By a long process of heating, cooling and hammering, during which the cementite crystals had their needlelike points rounded off, the swordsmith would work the steel into a state of remarkable

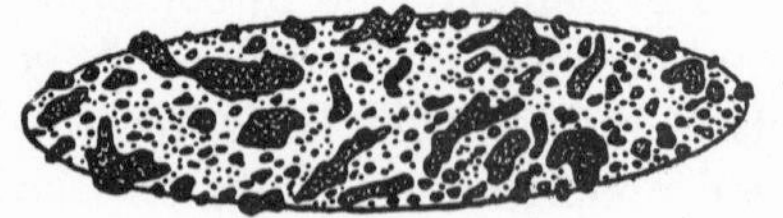

TEMPERED CEMENTITE STEEL
MAGNIFIED 1000 TIMES

strength combined with great elasticity, the edge receiving its keenness in large part from the very hard, invisible cementite particles that acted as teeth in an extremely fine saw. Such, no doubt, is the secret of the noble blades used in the famous trial of skill between Richard the Lion-Hearted and Saladin, described by Sir Walter Scott, in which the crusader, taking his sword in both hands like an axe, came down with a terrific blow upon an outstretched mace, cleaving the heavy iron in twain. Whereupon the defender of Islam threw a gossamer veil into the air and neatly sliced it by drawing his scimitar across it until the two halves fluttered separately to the ground. Veritably it was as much a contest between great qualities in the steel as between great men.

Grinding metal with a stone, of course, is just a slower way of cutting into an ordered lattice, this time with the hard crystals of silicon or quartz, whose sharp edges actually rake and hoe sizable masses of molecules away. But polishing, surprisingly enough, is a completely different thing, as it depends not on any sort of digging but on rubbing the metal of the surface until it is warm and soft enough to flow into its own pits and furrows, often literally drawing a pliant skin of material over the deeper crevices yet without quite destroying even the outer crystalline lattices.

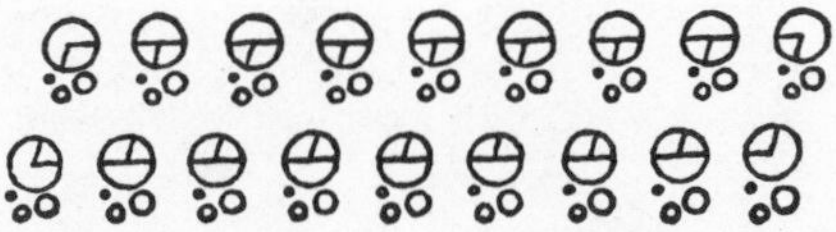

The fact that heat and electricity flow so easily through most metals is evidently due to the comparatively loose outer-shell electrons in these elements. While lattices in general may be held together by their wild-oat electrons acting as a kind of mortar, like the four shared carbon electrons of the diamond molecule, the conductive metals in particular have such lax valance electrons that their atoms can roll, swivel and slide over one another to an extraordinary degree, giving the pure state of each of these substances its characteristic ductility as well as providing innumerable easy channels for the transmission of thermal and electromagnetic energy.

Just as heat is nothing but average molecular and atomic motion (including the motion of electrons), electricity turns out to be primarily a stream of valance (outer) electrons, which may have been induced by a generator or a battery to circulate around and around a loop-shaped wire. Obviously, the generator no more creates the electrons than an engine driving a leather belt manufactures leather. It is just because each copper atom happens to have its first three shells completely filled with two, eight and eighteen electrons respectively, while that single footloose twenty-ninth electron is itching in its outer shell with practically nothing to hold it, that copper (element 29) is such a good conductor of electricity. The relationship of heat to electricity is shown by the increase in electrical resistance as a metal gets hotter until, near the melting point, atomic motion becomes so great that electricity can hardly be conducted through it at all. Conversely, electrical resistance decreases with cooling until suddenly (in the case of many substances) it ends slightly above absolute zero in the phenomenon known as superconductivity — which permits a current, once started by a generator, to keep on running (or coasting) for days after the generator has been removed.

The strange world of absolute zero, having now been virtually attained, is teaching us a good deal not only about electricity and magnetism but about several others of the deeper significances of matter, as we will see in the next chapter. Meantime, I need only mention that most solids, logically enough, take on an increasing brittleness (the opposite of moltenness) as they grow cold: a steel hammer may break into pieces if used in 50° F.-below-zero weather, while rubber at −310° F. (the temperature of liquid air) will shatter like glass, and petals crumble to the touch — yet lead, as it goes way down in temperature, becomes oddly elastic and resonant, mercury can be tied in knots and helium (superfluid at —456°) creeps out of a cup in defiance of gravity (see pages 356–57).

☊ ♀ ♌ ♂ ♄ ☿ ☽ ♆ ♃

There seems no end to the discoveries and uses of the natural materials of Earth, even though many elements have to be almost endlessly tested and tried before they achieve practical value. Sulfur, for instance, is not only expensive but extremely hard to find, coming mostly from salt domes a thousand feet underground. One could perhaps live a lifetime completely unaware of sulfur. Yet without sulfur there would be no giant industries and very possibly not enough bread or beef to feed us, for sulfuric acid, called the king of chemicals, plays some part in the manufacture of nearly everything we ever touch "from cotton diapers to bronze caskets." Land is fertilized with its aid. Oil is "cleaned" by it. Steel is "pickled" in it. Literally more than two hundreds pounds of it are consumed in the United States per year for every man, woman and child.

Have you heard of the group of fourteen elements (numbers 58 through 71) called the lanthanons or first series of rare earths? Not fitting very well into the general harmony of the Periodic Table, they form an odd little appendix in the sixth octave — yet together these metallic substances compose about .012 percent of the earth's crust and have unique magnetic properties, besides producing misch metal used in cigarette-lighter flints and as an additive in steel, while their chlorides aid the production of chrome, aluminum, silk, fertilizer and dentifrices. Most common of them, cerium (element 58) has made its mark, in oxide form, as a polish for optical glass.

Glass, incidentally, is a fused mixture of sand, alkali and lime, from which research has already developed such varied products as woolly insulation, woven fabrics, sponge-glass building blocks that look like limestone but weigh a twentieth as much and make excellent floats for life rafts, a glass "grease" that lubricates white-hot extrusion steel and a new plate glass you can drive nails through.

The Assyrians evidently were the originators of chain mail as protection against swords and spears, and chain mail has a modern parallel in fiber metallurgy, which, by using techniques borrowed from the paper industry, has now created the light, porous but strong "matted-metal" felt that bonds the linings of jet engines, reinforces jolt-proof ceramics and is easy to shape into the most complex forms.

The ancient Chinese, for their part, reputedly discovered that quilted cotton is nearly as good as chain mail at stopping sword thrusts, to say nothing of being more comfortable betweentimes — and this in turn finds its modern counterpart in the army's "bullet-proof" nylon jacket that proved its value in Korea by halving the numbers of casualties. "More efficient ballistically than any available metal," the nylon in the fused layers of this eight-pound garment will "resist a .45 bullet fired pointblank" and "virtually all grenade fragments from an explosion three feet away."

Nylon, by the way, is just one of the great and bewildering variety of "giant-molecule" synthetics that now seem to be almost exploding from the researchers' test tubes — products that include fibers, films, foams, glues, rubbers and plastics of all kinds — and which are all artificially put together out of the long chain type of molecule found naturally in oil, coal, wood, rubber or anything that is alive or has ever been alive. Without going into the organic basis of these so-called high polymers here, I must point out that study of them has at last given man his clearest concept of what causes the characteristic properties of matter and has made possible the nailable glass, bulletproof nylon, squeezable bottles, unbreakable records, waterproof sheeting, unshrinkable cloth and so on that so embellish modern civilization.

Indeed, perhaps most important of all the lessons of the giant molecules is that the degree of freedom in molecular motion is what mainly determines whether something is glassy, woody, leathery, rubbery, clayey, gluey, greasy, soupy, watery. For it is an oversimplification to say such states are just a matter of temperature, even though heat produces fluidity and the average piece of plastic will change from the glassy to the rubbery state in rising 80 degrees on the Fahrenheit scale. Nor can one say that states are dependent merely on the length of the molecular chains, even though the lengthening of chains does increase solidity — the boiling point of petroleum products, for instance, is known to rise 1 percent for every 2 percent increase in chain length. It is rather more a question of how tightly bound, how closely branched and cross-linked are the chains — in effect, how sticky or tangled the molecular spaghetti.

VARIOUS STATES OF LONG-CHAIN-MOLECULE (POLYMER) PLASTICS

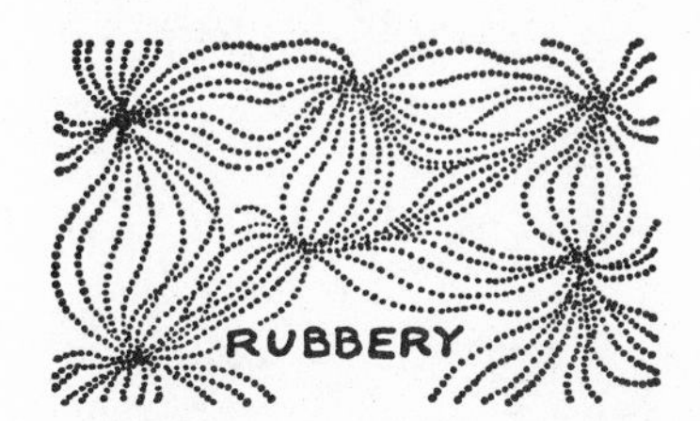

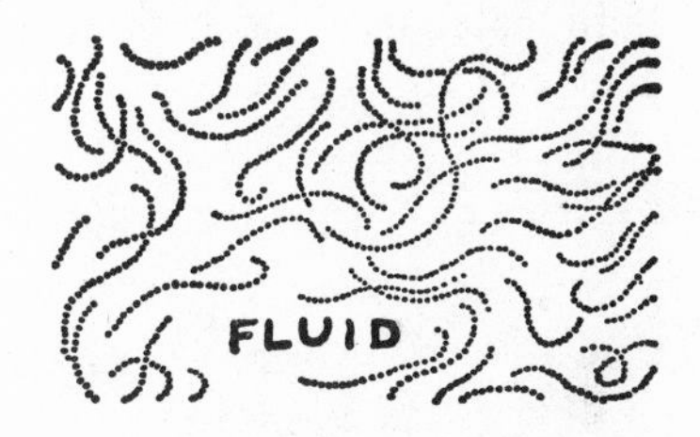

When a cool, brittle plastic is warmed and softened, its frozen molecules, which were only quivering, begin to squirm. The spaghetti comes alive. At first only short segments of it move. Perhaps a stretch of four or five atoms rolls slightly. Then a string of two dozen atoms arches its spine. What was glassy is turning leathery. And still the spaghetti is listless and not fully aroused. But as longer and longer segments start twitching and rippling in manifestation of heat, the "leather" inevitably softens. How much toughness it retains naturally depends on how tightly knotted or how frequently cross-linked are its chains. If the chains are aligned preponderantly in one direction, of course the material will have greater tensile strength in that direction. In any case, it is not long before hundreds of atoms are free to swing in unison like skip-ropes between the link points, while intermittently coiling and recoiling within the limits of their bonds. By this time, the stuff has become definitely rubbery, its elasticity depending partly on the amplitude of free coils between linkages, which, by uncoiling, permit it to stretch, and partly on the energy of its random writhing, which, by tirelessly opposing the stretching, ensures its eventual complete recovery. The molecular action in rubber can thus be compared to two girls, one at each end of a whirling skip-rope, who have a hard time pulling apart against the centrifugal force of the rope — analogous to stretching the rubber. The farther apart they go, of course, the more strongly the whirling energy in the rope resists them, until ultimately it draws them back together to their original distance — letting the rubber return to rest in its natural dimensions. Such a struggle between opposing forces well explains why a rubber band heats up when it is stretched or, similarly, why a gas heats up when it is compressed, both materials being subjected to stresses resisted by the relentless energy of normal molecular motion.

EXTENSIBILITY OF THE RUBBER MOLECULE

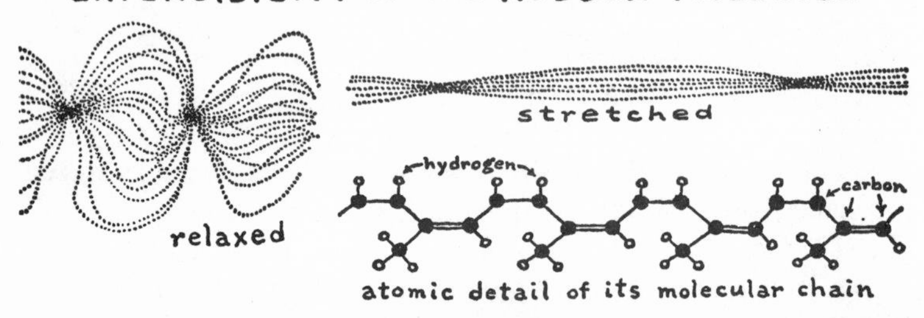

If its temperature keeps on going up, most rubbery materials must eventually yield their shape, turn gooey, and begin to ooze like honey as the link points in the spaghetti come apart, the skip-rope softens and breaks and whole molecules start sliding past one another. Stickiness is here revealed as the collective effect of friction opposing the independent motion of entire chains, or as magnetic snarls in the tangle that is writhing and struggling to be free. Plastic engineers control such viscosity mainly by regulating the chain length or degree of polymerization, cracking the molecules into shorter pieces for greater fluidity or building them longer for toughness and rigidity. A plastic's tendency to flow or creep can even be entirely stopped (up to surprisingly high temperatures) if the somewhat shifty centers of entanglement are reinforced or replaced by chemical cross-links from chain to chain.

In fact, this is exactly what Charles Goodyear accidentally discovered in the winter of 1839, when, according to one story, the bouncy little inventor got mad at the snickers of onlookers in the general store at Woburn, Massachusetts, and threw his experimental handful of sticky "gum elastic" mixed with sulfur upon the hot potbellied stove. A few moments later, recovering his composure enough to scrape it off, he was amazed to see that, "instead of melting like molasses, it had charred like leather." Quite unwittingly, he had created the first weatherproof rubber — cross-linked by vulcanization.

Yet sulfur is not the only boon to rubber. When the tire industry grew up, it was discovered that the addition of a handful of pulverized carbon black can make a 5,000-mile tire good for 40,000 miles. Although at first this was not clearly attributable to the fact that a single pound of these grimy granules of charcoal possesses more than an acre of total surface, most of which will adhere to the rubber molecules, today the surfaces of solids in general (and powders in particular) have come to be recognized as fairly bursting with energy — energy that is expendable in any form from gripping a turnpike to fueling a fire — a truth not easy to deny in the face of the all-too-frequent explosions of dusty grain elevators and ill-ventilated flour mills.

And so, we orbit onward into the mysterious fields of energy that the physical sciences, for all their experiments with ferocious boron fuels and the terrifying hydrogen plasmas (stolen from the sun), have hardly begun to explore — on into fields that await our deeper understanding of the atom, of its nucleus and, if that be possible, even of the mystic abstractions that seem to be matter's essence.

10. the netherrealm of the atom

As I look up out of the deep sky, I find it almost as easy to see atoms as stars — especially now with my mind's eye coming into fuller focus. For sometimes the electrons seem to stream past me like missiles out of the abyss. I cannot think of it as night, there being no night here. Instead, I try to remember that this black surround with the lights that streak across it is part of my native orbit, part of my world which has just begun to explore itself. And I reflect that these electron missiles are probably everywhere — moving even through the darkness of my own body — and that, as with any vehicles in the gloom, there is no knowing who or what or how many influences may be guiding them.

Nor can I seriously think of evading such ubiquitous and willful mysteries. On the contrary, they consume me. Explore them I must, while they are ingredients of myself, the grain in my world. I feel my consciousness plunging willy-nilly into their deepest depths, heeding no travail, to test their very core.

The microscope is a helpful tool, of course, but one cannot lean on it over-far. For the microscope really only divides the mystery without, in itself, solving it. Better: the ample regimen of experimenting, tooled with ordered thought, particularly abstract logic. And it needs be focused on the question area. Can energy now be treated as a substance? Or is this begrained ether within the atom a secret outcrop of some wider, subtler growth? Where next will I sink my spade, where cast my hoe?

While grubbing steadfastly at the roots of things, indeed, I find so many bugs and stones I must use a sieve for grading the soil. And I need a fresh bearing now and again — a perspective of older days to keep my brain in orbit.

What did Empedocles of Agrigentum in Sicily think the world is made of? And what keeps it from falling apart?

"The earth is a kind of meal," he said in 450 B.C., "cemented together with water."

Did any of the ancient philosophers suspect there are signs of the zodiac also in the atomic world and changing billions of times a second? It seems extremely unlikely, since there was no generally known number higher than ten thousand nor any such thing as a clock or a standardized time interval shorter than a full day and night. But in any case, it is now virtually impossible to assay the dreams of science in the day of Archimedes, when a philosopher was regarded a magician if he could lift a half-ton stone with the help of four ropes arranged in a block and tackle. Even the simple lever required deep study by no less a man than the great Aristotle, who wrote in his *Mechanics*: "As the weight moved is to the weight moving it, so, inversely, is the length of the arm bearing the weight to the length of the arm nearer to the power."

But Aristotle kept on asking fundamental questions. What is density? Is there a center of gravity? And little by little, haltingly and often incorrectly, he formulated what historians now consider one of the earliest mathematical laws of dynamics: that the velocity of a moving object is directly proportional to the "force from behind" it and inversely proportional to the resisting effect of the medium it moves through. Some of his groping passages, written on goatskin in the middle of the fourth century B.C. are remarkably close to Newton's first and third laws of motion (see page 316):

> Bodies which are at rest remain so owing to their resistance.
>
> When one is running fast, it is hard to divert the whole body from its impetus in one direction to some other movement.
>
> The force of whatever initiates motion must be made equal to the force of that which remains at rest. For . . . as there is a necessary proportion between opposite motions, so there is between absences of motion. . . . As the pusher pushes, so is the pushed pushed, and with equal force.

The thinking behind such formulas appears the more extraordinary from the fact that it had virtually no benefit of experiment or careful measurement in either time or space, and it was, along with that of Democritos, perhaps man's very first serious attempt to define the basic laws of dynamics. Manifestly, the ancient philosophers were so intent on knowing WHY that they seldom asked HOW or HOW MUCH. The latter questions seemed to them unimportant, unworthy — the kind of thing that slaves and artisans would worry about. And so, lacking the support of accurate observation, their work on the whole was lamentably lame.

By the sixteenth century A.D., however, qualitative science had begun to shift toward quantitative science. Friar Roger Bacon's cry of "Experiment, experiment!" had been heard. Tycho Brahe became a great prophet of precision, and his successor Kepler, who had learned the lesson well, summarized it for the modern world: "To measure is to know."

At the same time the new-fledged spirit of experimentation was developing the technique of postulation, of trial and error as tools of learning — a method exemplified by the reaction of the English astronomer Thomas Blundeville, who, reading *De Revolutionibus*, remarked: "Copernicus . . . affirmeth that the earth turneth about and the sun standeth still . . . by help of which false supposition he hath made truer demonstrations of the motions and revolutions of the celestial spheres than ever were made before."

Galileo, meanwhile, was taking his place as the first great experimenter and sire of modern physics. He intuitively understood that HOW is a tough enough question for this stubborn world. "Nature nothing careth," as he put it, "whether her abstruse reasons and methods of operating be or be not exposed to the capacity of men."

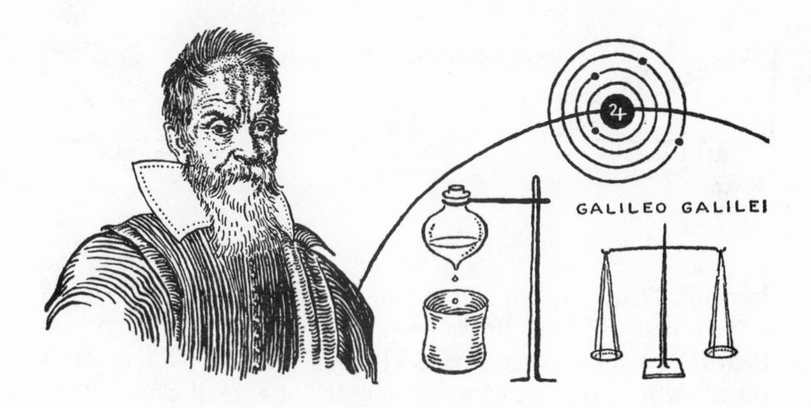

So he stalked nature with great patience, observing such simple things as a lamp swinging slowly back and forth in the cathedral of Pisa. He timed its cycles with his own pulse — he was seventeen then — until he was convinced of its absolute regularity, a realization that resulted seventy-seven years later in the building of the first pendulum clock. His meticulous experiments with the mechanics of fluids enabled him to invent the first thermometer (a glass bulb and tube filled with air and water). And his study of refraction produced the first astronomical telescope (see page 426), through which he measured the mountains of the moon, as well as his compound microscope (five feet long) through which flies appeared "like lambs, covered all over with hair, and with very pointed nails by means of which they can walk on glass while hanging feet upwards."

Less dramatic but more profound was Galileo's comprehension

of the concept of acceleration, which he defined as a change of velocity either in magnitude or direction. This was an abstract idea that no one seems to have thought much about before. And in using it to test the still accepted Aristotelian precept that a moving object requires a force to maintain it, Galileo easily demonstrated that it is not motion but rather acceleration which cannot occur without an external force. Deliberately rejecting common

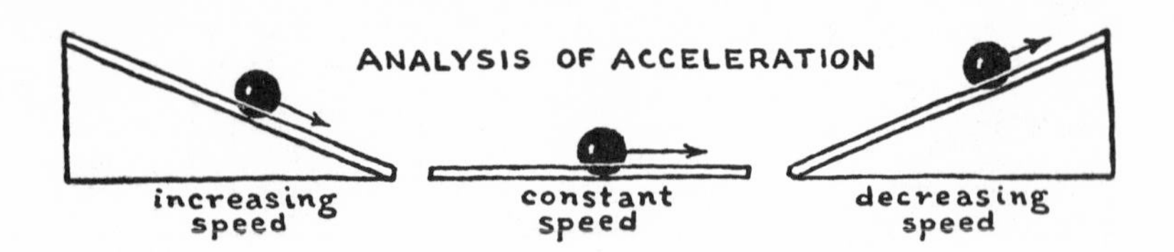

sense as a prejudiced witness, he let nature herself speak in the form of a "hard, smooth and very round bronze ball" rolling down a "very straight" ideal groove lined with polished parchment, and then rolling up another groove, clocking each roll "hundreds of times" with "a thin jet of water collected in a small glass." Thus on the humble basis of fact he showed that, while downward motion (helped by gravity force) makes speed increase and upward motion (hindered by gravity force) makes speed decrease, there is always a "boundary case" in between the two where speed ideally remains constant (without any appreciable force)—and that, by reducing friction, this boundary case can be made to approach a horizontal level where gravity has no effect.

Similarly testing the speeds of various weights dropped from various heights, he also drafted a law of falling bodies: "that the distances traversed, during equal intervals of time, by an object falling from rest, stand to one another in the same ratio as the odd numbers beginning with unity." And his beautiful analysis of a cannonball's trajectory into horizontal and vertical components of motion and acceleration was one day to be of enormous help to Isaac Newton in solving the riddle of gravity.

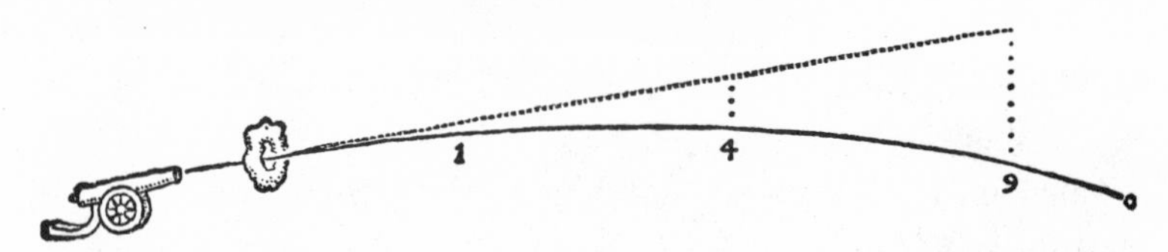

As Galileo had been born in 1564, the year Michelangelo died, so he died in 1642, the year Newton was born. Thus was the baton of scientific insight passed mysteriously from the blind, seventy-eight-year-old prisoner of the Papal Inquisition to a premature baby "tiny enough to put into a quart mug" whom no one expected to live. His father died three months before he was born and his mother soon remarried, so little Isaac was packed off to grandma's, which turned out to be an isolated farm near Grantham in the English midlands where there were no other children and little excitement save for the Cromwellian raiders who occasionally looted the "royalist" corn and animals.

Poring over medieval books on alchemy or mathematics while he was supposed to be tending sheep, the lonely boy grew up showing little promise as a farmer. But he learned to think for himself. It is told that one stormy day he wanted to find out how hard the wind was blowing. Having no weather instruments, he hit upon a unique method of his own: he jumped as far as he could against the gale and then jumped with it, carefully measuring the difference in distances, attributable to the wind during the time he was in the air, which readily gave him the wind's velocity.

At Cambridge, within three years of taking up geometry he had mastered the range of mathematics from Pythagoras to Isaac Barrow (his own teacher) so thoroughly that he had learned to generalize the so-called "infinitesimal geometry" into a new kind of analysis he called "fluxions," meaning roughly "rates of change" or "velocities." From Galileo's dynamics he naturally thought of a moving point as describing a curve. And, using Descartes' analytic geometry, he confidently placed the curve on a graph between coordinate axes, calling the vertical and horizontal components of the point's velocity the "fluxions" of x and y, denoted as "$\dot{x}$" and "$\dot{y}$." In logical consequence, the "fluxion" of the "fluxion

$\dot{x}$" became "$\ddot{x}$," a change of a change (motion) or an acceleration — and so on. This form of mathematics has since become known as the (differential) calculus.

Newton had to leave Cambridge in the fall of 1665 when the university closed on account of the great plague then raging in London, which eventually accounted for 68,000 deaths. So he went back to live in the dreary little house in Woolthorpe where he was born and there continued his amazing solitary research which included experiments with prisms into the nature of light and, most portentous of all, an effort to coordinate and complete the great investigations of Kepler and Galileo.

Since it had become accepted knowledge that the earth is round with real antipodes where Chinese and Bushmen live upside down and where things fall upward in relation to Europe, there were a number of theories of gravitation already under serious discussion in university circles and in books — but evidently few of these envisioned the pervasive tendencies of falling objects as extending continuously beyond the region of the earth. Aristotle had maintained that the heavenly bodies, being composed of ethereal substance, could have "neither gravity nor levity." Copernicus, on the other hand, had speculated that "gravity is just a natural inclination, bestowed on the parts of bodies by the Creator . . . and we may believe this property present even in the sun, moon and planets." Some earlier philosophers had long since rationalized a law of symmetry as the motivating force of this ubiquitous hunger under which all things crave to attain the center of the universe, but Copernicus' demonstration that the earth may no longer be assumed to be in the central position had recently begun to undermine their argument. Descartes, with characteristic imagination, had visualized the whole heavens as a high-pressure

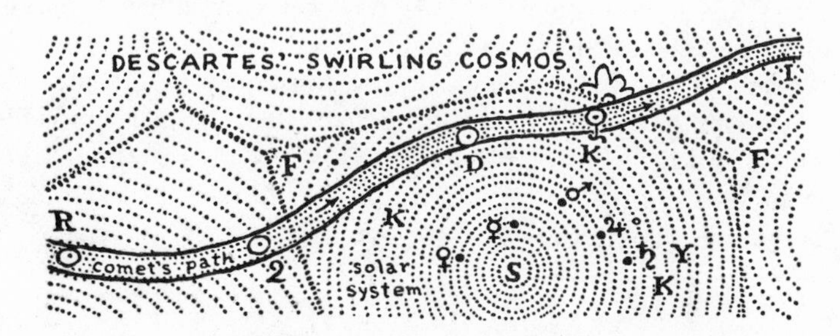

vortex of invisible atoms elevated by their swift circling motion and relentlessly pressing all heavy objects toward the center of the earth. Galileo had circumspectly contented himself with writing that "we do not really understand what principle of virtue moves a stone downward any more than we know what moves it upward when it is separated from the thrower, or what moves the moon round, except possibly that word (which more particularly we have assigned to falling), namely, gravity."

It remained solely for Kepler, most intuitive of all, to make the bold deduction that not only does a falling stone approach the earth, but also the earth approaches the stone. "If two stones," he wrote in his *Astronomia Nova* in 1609, "were removed to some place in the universe, close to each other but outside the sphere of force of any third cognate body, the two stones, like magnetic bodies, would come together at some intermediate place, each approaching the other through a distance proportional to the other's mass [*moles*]." And in his *Tertius Interveniens* (1610): "The planets are magnets and are driven around by the sun with magnetic force." Whereupon he predicted (correctly) that the sun must rotate and pictured its rays as giant arms directing the planets along their appropriate orbits.

But all these theories (and a score of others) seemed somehow weak and unconvincing, if not evasive, to Newton. And what did they add up to mathematically? He instinctively knew there must be a better explanation somewhere. He pondered intently upon the extent and mystical steadfastness of gravity, which, he felt, must be the manifestation of some simple, logical and general law. Why should gravity stop a few miles or a few thousand miles above the earth? Why should it stop anywhere?

"I keep the subject of my inquiry constantly before me," he was to write later, "and wait till the first dawning opens gradually, by little and little, into a full and clear light." It was about when the apples were ripe in the fall of '66 that he "began to think of gravity extending to the orb of the moon."

The ways of minds, of course, are inscrutable, and Newton could not later recall just how it happened (if, indeed, he was ever seriously interested in doing so), but a falling apple could well have been the means. After all, an apple in the air a few feet away appears about the same size as the moon. And both are objects that move naturally in our sight. Was there any reason to suppose the moon is excused from obeying the same law that applies to apples? Has any object ever been proved to be too big or too remote to fall? What about shooting stars, which Newton must have seen and wondered about? They come from far beyond the earth and yet apparently fall even faster than apples.

Is it not quite likely also that Newton had read in the introduction to *Astronomia Nova* Kepler's specific declaration that the earth is attracted by the moon as well as the moon by the earth? And might he not have heard of Kepler's original speculation that the intensity of this yearning is inversely proportional to the square of the distance of separation — though Kepler later emasculated his conjecture by unsquaring the distance, probably in the interest of simplification? Could some such hint have suggested a mathematical formulation of the all-pervasive influence that leaps from star to star?

Whatever put the far-fetched idea into his head, twenty-three-year-old Isaac suddenly saw the moon as something falling and, in his own words, "compared the force requisite to keep the moon in her orb [orbit] with the force of gravity at the surface of the earth, and found them to answer pretty nearly." From Kepler's established harmonic law that "the squares of the years are as the cubes of the orbits" he logically "deduced that the forces which keep the planets in their orbs must be reciprocally as the square of their distances from the centers about which they revolve." Since the moon was well known to be 60 times as far from the earth's center as we are at its surface, the earth's pull thus ought to be 60 squared, or 3,600 times as strong here as out where the moon is. And since things here fall sixteen feet toward the earth's center in the first second, out there things (including the moon) ought to fall only $1/3{,}600$ as far, about $1/20$ of an inch, in the same time—which tiny bit, to Newton's delight, turned out to be just what was needed to keep the moon (traveling $5/8$ of a mile a second) from flying off at a tangent out of her orbit.

In this manner did Newton derive his great law of the inverse square, which, however, he did not publish for more than twenty years, presumably because of doubt about certain accepted but crudely measured astronomical "facts," or perhaps an apprehension of controversy and probably neglect under the pressure of concentration upon such rival interests as alchemy and the solution of various challenging and cryptic riddles of medieval tradition: the elements of "the philosopher's stone," the elixir of life, the dimensions of Solomon's temple . . . Also, there is evidence that, although he knew intuitively from the beginning that "you could treat a solid sphere as though all its mass was concentrated at the centre," he did not hit on any mathematical proof of it until 1686. The time lag here can explain something of how Newton voyaged through his "strange seas of thought alone." Great ideas came to him first as flashes of vision. Only later could he build logical foundations under them. A perennial bachelor who often forgot to eat or change his clothes and has been called "profoundly neurotic," his deepest instincts were occult — so much so that John Maynard Keynes called him "not the first of the age of reason" but "the last of the magicians."

Yet Newton did not pretend to know the cause of gravity but only its behavior. "I do not deal in conjectures," he once remarked. And the wonderful generalizations of his laws of motion and gravitation, published in his *Philosophiae Naturalis Principia Mathematica* in 1687, won acceptance through their beautiful simplicity as practical mathematics rather than as abstract philosophy. Behold the universality of his dynamic declarations, which are the first and only criteria by which the Copernican doctrine can be tested by observation.

Law of Inertia: A body at rest will remain at rest, or a body in motion will continue in motion in a straight line with constant speed, unless constrained to change that state by the action of an exterior force.

Law of Acceleration: Change of a body's motion is proportional to any force acting upon it, and in the exact direction of that force.

Law of Reaction: Every acting force is always opposed by an equal and opposite reacting force.

And again, his famous Law of Gravitation: Every particle of matter in the universe attracts every other particle with a force proportional to the product of their masses and varying as the inverse square of the distance between their centers.

Of course, the strangest thing about this last law, which was the first to raise mass to a basic concept and the first to encompass the whole material universe with a measurable force amounting to a kind of inorganic love, was not that the sun pulls the earth but that (as Kepler divined) the earth pulls the sun just as hard and that every sparrow, fish, snowflake and molecule influences not alone the whole earth but also the moon, sun and all the stars. Indeed, that the entire universe is full of falling bodies, every mass-point pulling every other mass-point inevitably and forever from the descending mists to the rising tides and even unto the slow precession of the equinoxes.

But Newton was well aware of forces besides gravity in the world and distinctly foreshadowed modern atomic chemistry when he wrote:

> And now we might add something concerning a certain most subtle spirit which pervades and lies hid in all gross bodies, by the force and action of which spirit the particles of bodies attract one another at near distances and cohere, if contiguous; . . . and there may be others which reach to so small distances as hitherto escape observation . . . and electric bodies operate to greater distances, as well repelling as attracting the neighboring corpuscles; and light is emitted, reflected, refracted, inflected, and heats bodies; and all sensation is excited and . . . propagated along the solid filaments of the nerves.

Newton must have wondered whether this "subtle spirit" binding matter together could be related to gravitation — whether the atoms can be tied mathematically or harmonically to the stars. But finding a common equation for electrons and galaxies certainly cannot be easy, for many physicists, including Einstein, have since sought just such a unified formula, and there is no clear assurance that anyone has yet come anywhere near success. And now the whole idea of force in the world is being deeply questioned, with the modern view tending toward replacing it with the more definitive concept of interaction.

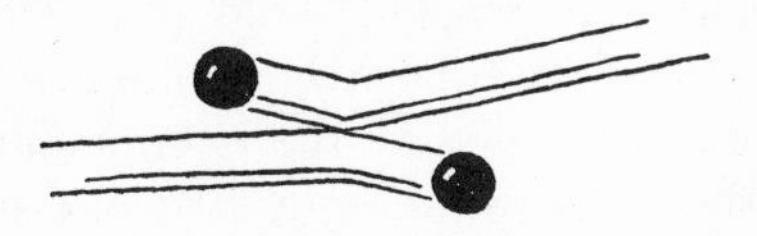

Since the only known place where force can be measured is where something material acts upon some other material, it may be interesting to consider now just what "material contact" really is. Is the familiar "force" of friction what it seems to be and was assumed to be until this century: an actual grinding of particles against particles with nothing in between them?

Of course, we know now that the answer is no. Physical impact is far from what it appears, and the phenomenon grows in complexity as it is studied more closely. When two billiard balls click together, they bounce in the way a comet "bounces" off the solar system, guided in its curving path by mysterious and seemingly remote influences. Or you might visualize the colliding atoms as two nations at war for, although the territories may consist largely of empty fields and woods with a hundred feet of "nothing" between one soldier and the next, still the two armies are firmly held apart.

The first two classic laws of friction were formulated by Leonardo da Vinci, who said that when solids rub against each other, the frictional force is (1) proportional to the load or pressure of one upon the other and (2) independent of the area of contact. These observations still "generally hold true, with no more than ten percent deviation," and they seem quite reasonable under the old view of friction as caused by the intermeshing of protrusions in adjacent surfaces. But surprisingly, modern experiments of high precision have proved that both of Leonardo's laws are wrong and, in fact, that they should be reversed. Even though they seem correct because the contact area is usually roughly proportional to the load, the frictional force has turned out to be really proportional to the area of contact and independent of the load. Indeed, the intermeshing of protrusions has recently been exposed as the great delusion of friction, since all the tests show that smoothing out roughness eases friction only in the coarse stages and that there is definitely a limit beyond which polishing a surface *in*creases rather than *de*creases resistance.

This is because friction comes almost entirely from molecular cohesion rather than from the grinding of humps — which explains why the best bearing metals have proven to be alloys with various large atoms that make lattices too coarse and irregular to form neatly interlocking chemical bonds. In fact, it is practically certain that the real function of teeth in a friction hacksaw is to carry enough oxygen from the air into the cut to maintain the microscopic "flames" essential to the proper working of the saw — and this despite oxygen's being such a good lubricant that it amply accounts for surfaces sliding more easily in air than in a vacuum.

Another recent finding is that the increase of friction with the slowing down of sliding surfaces reaches a maximum while the surfaces are still moving and, below this very slow speed, friction decreases again — probably because of creep. As we saw in the last chapter, all solids, including lattice crystals, creep or slowly change shape even though the forces acting upon them may be very slight, the creep speed being proportional both to the frictional resistance (inversely) and to the softness of the material. The creep of hard steel, for example, is so slow it cannot be observed, while soft lead may creep up to a millionth of a centimeter a second, a foot per year.

A much faster kind of friction is the sudden impact of two colliding bodies, which may either bounce apart without damage or break to some degree depending on their structure. According to one calculation, if a golf ball had as high a "coefficient of restitution" (bounciness) as have some precision crystals, it would, after 440,000 bounces off a crystal pavement in a vacuum, still be rebounding to half its original height. In the case of less resilient materials, however, much of the energy upon colliding goes into plastic deformation. When two cannonballs hit, for instance, an "ellipse of contact" is created around the momentarily compressed surface where they touch. And to a depth of about a third of the mean diameter of the ellipse below the surface in each ball, the crystal breakdown keeps pace with the ellipse, its outer boundary spreading in the form of twin ellipsoidal fronts inside the ellipse — fronts whose advance cannot be halted until the compression energy (momentarily stored in the crushed molecules) is sufficient to make the balls stop approaching each other and begin to separate.

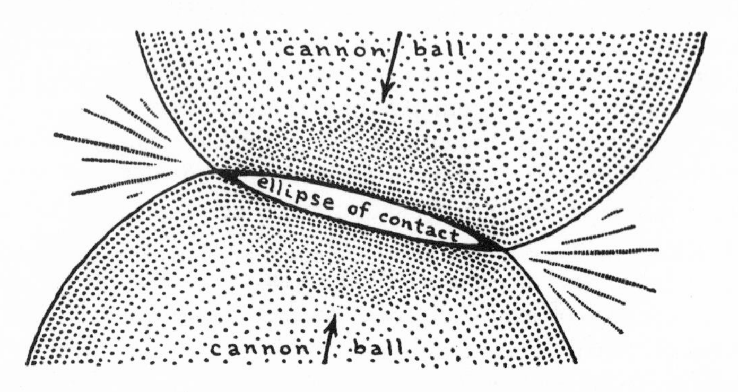

According to the law of entropy (second law of thermodynamics), all these kinds of friction convert energy from a more orderly form (as in a regular lattice) to a less orderly form (as in a broken lattice or the random molecular motion of heat). But the process does not stop with breakage or heat, for it is now known that most chemical reactions either absorb or release energy in the form of light or sound or electricity.

A tiny crystal cube one millimeter on a side, for example, has some 100,000,000,000,000,000,000 energy "levels" occupied by the valence electrons of all its constituent atoms — levels so closely spaced that, for most practical purposes, they form a band or continuum within the crystal. Such a grain of sand can be made to glow, sizzle or emit sparks, perhaps discharging energy in many forms at once — energy which, we shall see presently, is actually always "granulated" or quantized into distinct parcels in much the same way that matter is made up of separate, measurable atoms, and for the very good reason that it and matter are basically the same thing.

This surprising two-in-oneness of energy and matter, first clearly demonstrated by Albert Einstein in 1905, explains how the mass of an atom's nucleus can be less than the sum of the masses of all its nuclear components, the difference being the "packing loss" or binding energy that holds the nucleus together. A strange kind of addition, you think? But such relationships will become clearer when we have had a chance to look into Einstein's famous equation $E = mc^2$, which says, with beautiful simplicity, that energy (E) equals mass (m) times the square of celeritas (c), symbol of the velocity of light.

To comprehend this, in turn, we need to get better acquainted with the atom as it is now known, so we can visualize its many and peculiar parts and at least build up some sort of mental picture of what this weird inner frontier of knowledge is like. There is, of course, hardly a need to point out the difficulties physicists have had in discovering the atom's so-called elementary particles, since these "objects" are much too small to be seen by any known kind of microscope, and they must be deduced only from the evidence of their tracks and deeds in a bubble chamber, photographic plate or in one of the great new particle accelerators whose increasing dimensions and importance are making them the modern architectural counterparts of the pyramids of Egypt, of the Gothic cathedrals of Europe. Such indirect techniques have been said to make the task of discovery at least as hard as figuring out the path and caliber of a bullet fired at a flock of birds just by observing the birds.

Yet at the present writing, more than 100 subatomic particles have already been discovered and at such a rate that there has not been enough time yet to complete the tougher job of thinking up theories to explain them. As these particles or "states," as they are coming to be called, evidently inhabit the atoms of all elements when conditions favor them, they constitute a Table of Particles that amounts to a kind of inner family of elements of the second order within the classical Table of Elements described in the last chapter. As such they are a big new step toward the ultimate "partless parts" sought by Leucippos in Abdera — if, indeed, there should ever turn out to be anything anywhere small enough to be really indivisible.

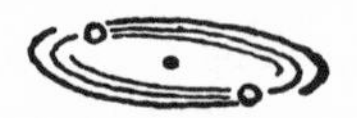

As recently as the 1930's there were only four "known" subatomic particles: the electron, proton, neutron and photon. The first three, discussed in the last chapter, are the main building blocks of atoms: positively charged protons and uncharged neutrons in the central nucleus and negative electrons revolving through the vast firmament around it. The photon, on the other hand, is the modern name for the quantum unit of radiation first vaguely postulated by Newton as a "corpuscle of light" three centuries ago but not generally accepted until 1905, when Einstein convincingly demonstrated that it could literally knock electrons from metal surfaces and formulated his Photoelectric Law to establish it as "the building block of the electromagnetic field." More elusive than the electron, as we will learn in Chapter 12, the photon exists only while moving at the velocity of light, 186,282 miles a second. It is a particle that can never be at rest, for its mass and energy are created by its motion in the proportion $E = mc^2$. Don't be dismayed if this strikes you as mysterious, because it is mysterious. It was mysterious to Einstein too, for he knew enough, in this dawning era of knowledge, to ask only HOW of the world. Honoring the unknown as a kind of sacred manifestation, he once said, "The most beautiful and most profound emotion we can experience is the sensation of the mystical."

As the years went on, physicists, experimenting humbly with the four primary particles, found that they needed new concepts to satisfy their equations — that something other than mass, density, electric charge and velocity must distinguish one particle from another. What was this difference? Could it be described or explained?

In testing various ideas, it turned out that the concept of spin untangled several mathematical snarls. For the evidence is that all these particles spin on their axes like planets and asteroids and, if they are electrically charged, the spin makes them tiny magnets. This can hardly be explained except to say it is part of a natural electromagnetic law under which electric generators and motors operate all over the earth. But the spin fits in with quantum theory as an integral and measurable trait of each particle, the electron, proton and neutron being considered to have a spin of ½ while the faster spin of the photon is 1.

How actual the spin of these particles is is hard to say, but there is no doubt that they behave as if they are really spinning. They are certainly doing something that has rhythm to it. And those that are magnets are responsive to the magnetic fields they happen to be in, abiding by the quantum restrictions that permit a particle with spin ½ only two positions: its axis pointing either with or against the field. A particle having spin 1, on the other hand, enjoys a choice of three positions: its axis with, against or perpendicular to the field.

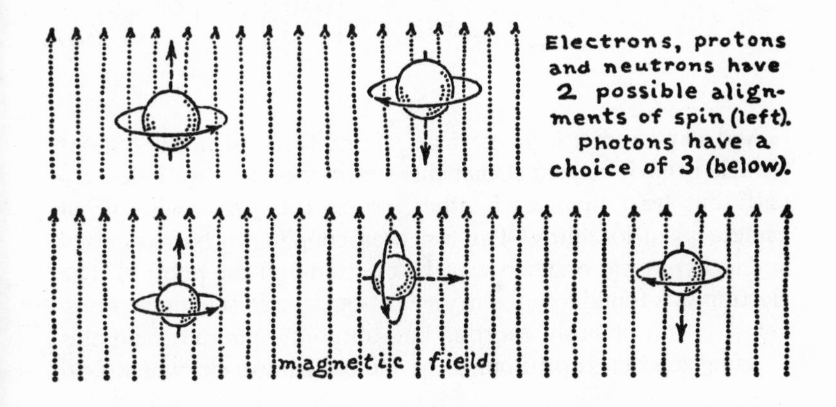

Another new and helpful concept is known as the exclusion principle. This famous law, formulated by Wolfgang Pauli of Austria, applies to all particles with spin ½ and says in mathematical language that inside the atom, just as outside, no two objects can fill the same space simultaneously or, more specifically, that only one particle of a kind can occupy a given quantum "state." Thus only one electron at a time can be spinning in a particular direction and revolving in a specific orbit around a given nucleus — which goes a long way toward explaining why there can be no more than a fixed number of electrons in each shell of an atom, and this in turn (as we have seen) is a key to the harmony of the Periodic Table.

Of course, it would be only natural at this point to ask why particles spin and why electrons obey an exclusion law. But we must try to be patient and not revert to the primitiveness of the ancient Greeks even if the atom world now seems to make little sense. For there are still plenty of HOWs that need answering first and, as Galileo discovered, the HOWs not only get answered much more easily but their answers are apt to throw more light on the WHYs than the WHYs' own answers do — if indeed the WHYs get answered at all.

And so we come to the next peculiarity that was noticed about particle behavior: the fact that these entities do not live independently but react upon each other just as definitely as a pair of gamecocks or lovebirds. The first such coupling to be recognized and analyzed was one between the electron and the photon. The electron was found to pulse in an electromagnetic field, unaccountably but continuously emitting and absorbing photons, and this amazing reaction turned out to be the basic means by which field and electron exert force upon each other — a vital, physical relationship that soon established itself as the cornerstone of quantum electrodynamics and, incidentally, produced the first apparent violation of the law of conservation of energy.

By violation I mean the photon's capricious appearance and disappearance. For how can a photon containing energy be spontaneously emitted or absorbed by a stable electron without suddenly altering the total of energy in the system? This was the baffling question to which quantum theory answered that "the photon is emitted and reabsorbed so fast that the gain in energy cannot be detected, even in principle." As a result, the phenomenon has come to be called a "virtual process" to distinguish it from an actual one. But the phrase "virtual process" is not meant to imply an unreal process, for the fact that the photon's energy is not detectable in this case does not prove it is nonexistent. It only testifies that it does not affect the conservation of energy, on the ground that "quantum laws deal only with observable quantities."

By such down-to-earth reasoning is the first law of thermodynamics neatly spared its exception. Yet the elusive photon exists as something empirically tangible — and this can readily be tested by the addition of enough energy from outside (accelerating the electron, for instance) to convert the photon from a virtual to an actual particle.

If the four basic particles as thus presented seemed to come close to explaining the atom in 1930, the scientific world would soon realize that they had not come close enough. One of the best-known though paradoxical fruits of the quantum theory was that fundamental particles not only behave like particles, they also behave in some ways like waves. And the solution to the electron's wave equation yielded a puzzling negative frequency as well as a positive one, in much the same way that $\sqrt{4}$ equals -2 as well as 2. But since frequency is proportional to energy in quantum mechanics, this in turn suggested the idea of negative energy — whatever that might be. Should such an absurdity be taken seriously or was it just something extraneous like a shadow, a forgotten dream or the lingering grin of the Cheshire cat? Could there really be an antipode of energy?

Physicists did not agree at first as to how to handle "negative energy," but presently one of the best and youngest among them, Dirac of England, found a mathematical way of proving that this weird concept has a "physical significance" — that if an ordinary, negatively charged electron could lose its normal energy and pass into a state of negative energy, it must reverse its charge and become a positively charged electron — which would be something like a hole or bubble in a sea of electrons. Not only that, implied Dirac's theory, but "if a positive electron collided with a negative electron, they would annihilate each other and their mass would be converted into photons with an equivalent amount of energy. Conversely, if enough energy could be concentrated in a small volume, as in a high-speed collision between two particles, a positive and a negative electron could be created."

Although this fantastic implication was not uttered as a valid prophecy, it could as well have been. For it was fulfilled with promptitude and precision when Carl D. Anderson of the California Institute of Technology discovered in 1932 an actual positive electron, soon to be named the positron. Not only did it have the mass of an electron and exactly one unit of positive electric charge but, when an electron and positron were induced to meet, they annihilated each other. Yet still they could be (and literally were) re-created by certain "energetic collisions."

Does it not seem preposterous that a mathematical equation, a purely abstract relationship not based on experience, should thus show us what is going on inside the atom? But it has happened again and again in physics, chemistry and equivalently in many other branches of science. The method actually works, and one cannot avoid realizing that it must express a profound truth about the nature of our world. Bertrand Russell has gone so far as to define matter as merely "what satisfies the equations of physics." And more and more, we see, as we delve deeper into the nature of things, that the abstract relationships of elementary entities are what really constitutes matter — that the particular harmony of each combination is the true essence of what it is made of. In fact this idea, far from new, was strongly implied by Pythagoras and developed quite remarkably in Plato's famous intuition that the four Greek elements (fire, earth, air and water) were essentially four abstract shapings of empty space into pure geometric form: the tetrahedron being the basic atomic shape of fire, the cube of earth, the octohedron of air and the icosahedron of water.

Anyhow, the abstraction of material symmetry that developed out of Dirac's theory came to fruit not only in the positron, which is now considered the antiparticle of the electron (which it cancels), but also in the general concept of antiparticles that can cancel or counter all particles. Thus new equations for the proton and neutron soon brought to theoretical light the antiproton (of reverse charge) and the antineutron (magnetically reversed), which were actually produced and detected twenty-odd years later in 1955 and 1956 with the help of the giant Bevatron at the University of California, which was built especially for the purpose. And even the photon now has its antiparticle, mathematically speaking, though here "the two solutions to the equation can be interpreted in the same way and the photon and antiphoton are indistinguishable" — or, as one physicist put it, "the photon is its own antiparticle."

If anyone seriously thought that antiparticles (theoretically combinable into what is termed antimatter) could be the final word in subdivisions of the atom, however, he did not think so for long. For even the theory of the antiparticle left little room for doubt that encounters between particles and antiparticles would sometimes produce brief exchanges of material analogous to a lightning stroke between a positive cloud and a negative mountain. I am not referring now to anything like a head-on collision between a proton and an antiproton, which would completely destroy both, but to a near miss, when they shave close enough to affect each other drastically but incompletely. Here theory indicates that the antiproton would flip a quantum of negative charge to the proton, leaving both particles electrically neutral: the proton would become a neutron and the antiproton an antineutron, but the pair of them would still be symmetrical and therefore able to annihilate each other if they ever collided. The obvious point in this instance, however, is the quantum of electrical charge flipped from the antiproton. Could this brief shot of energy (if it really happened) be considered another elementary particle?

The Japanese physicist Hideki Yukawa thought so and called it the meson. He also calculated that it must be emitted and absorbed again all in 1/100,000,000 of a second and, since it lived in the very heavy nucleus of an atom (the mysterious central habitat of protons and neutrons) it must be quite heavy itself (between 200 and 300 times as massive as an electron, he figured) and might well contain the powerful energy that holds the nucleus together — something that had remained completely unexplained up to that time. It took researchers about twelve more years to find Yukawa's particle, but when it showed up at last in 1947, weighing some 260 electron masses and in three forms (positive, negative and neutral), it fully confirmed his brilliant intuition. It is now called the pi meson or pion and is considered to be the unit of a virtually continuous emission and absorption from the nuclear core, serving collectively as a kind of potent, mystic ligature binding the nucleus together.

Yet another discrepancy appears whenever a neutron for any reason has been knocked out of its natural place in the nucleus, for the normally stable neutron seems to have a streak of delinquency in it, becoming hopelessly unstable as soon as it gets out of sight of home. In fact, about eighteen minutes after its departure from the nucleus on the average, it spontaneously ejects an electron and turns into a proton. Such sorcery, of course, makes sense only if these "particles" are all basically made of the same primordial stuff which can somehow "flow" from one to another. Yet the proton and electron added together are some 1.5 electron masses lighter than was their mother neutron so, according to the law of conservation of energy, 1.5 electron masses or its equivalent in energy (780,000 electron volts) must have escaped somehow in the decay and birth.

Where could it have gone? The Italian physicist Enrico Fermi, with the help of Pauli, suggested that some almost undetectable unknown particle must also have been emitted like a ghost's placenta when the neutron broke up, and that it took off with the missing energy. He gave the shy little whatever-it-might-be the appropriate name of neutrino, but for some twenty years this hypothetical will-o'-the-wisp looked to many theorists like just a bookkeeper's symbol for a slight discrepancy in the accounts. All agreed that it was the epitome of elusiveness, with zero charge and virtually zero magnetic moment, to say nothing of being as swift and tiny and intangible as a photon, only vastly more unstoppable and completely invisible. Study of radioactivity in the sun, however, eventually showed that neutrinos must be produced so abundantly there that "some 100,000,000,000 of them pass through each square centimeter of the earth's surface per second" and so subtly that all but a freak few continue on straight through the molten interior of our planet and out the other side like a bullet through a cloud without hitting a single electron or any other particle on the way. To make them even more incredible, physicist George Gamow has declared that "a beam of *neutrinos* would go without much difficulty through the thickness of several light-years of lead!"

When a long series of tests culminated in actual detection of neutrinos in the famous Los Alamos experiment of 1956, physicists sighed with conscious relief that finally this "last of the theoretical parts of the atom" had been proven a reality. Could anything have more perfectly fulfilled Lao Tzu's ancient intuition: "The softest of stuff in the world penetrates quickly the hardest; insubstantial, it enters where no room is"? The neutrino even completed the so-called "dozen-particle theory of matter" in which all of the particles and antiparticles so far described were logically sorted into four groups: (1) the heavy particles (proton, neutron, antiproton, antineutron); (2) the short-lived middleweight mesons (positive, negative and neutral pions); (3) the light particles (electron, positron, neutrino and antineutrino); (4) the photon. It was a neat arrangement with the groups nicely interconnected by the three basic reactions already mentioned: the Yukawa process connecting heavy particles with mesons, the Fermi process connecting heavy particles with light particles (neutrino, etc.) and the Dirac process connecting light particles (electrons) with pho-

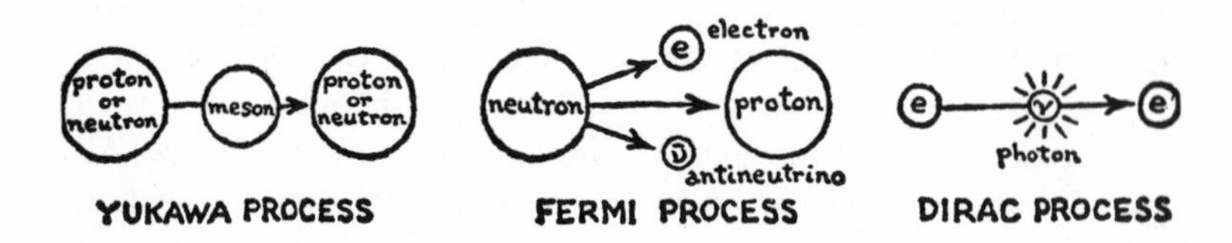

tons. With the known general laws of physics, such as the conservation of energy, of momentum and of electric charge, physicists now had a kind of atomic algebra for writing equations and solving problems about particles — a method which might seem to an outsider like just shuffling symbols around but which actually could, and usually did, yield a reasonably exact prediction of what would happen in a new experiment, how long it would take and so on. The fact that both atomic bombs and atomic generators became realities on little more than the "dozen-particle theory" suggests how much can be accomplished with such a simplified concept of matter.

And then came the mu meson or muon! There was no theoretical justification at all for this surprise morsel of mystery (discovered by Anderson as early as 1937), which turned out to be one of the decay products of the charged pion — which, if you remember, lives only about a hundred-millionth of a second and theoretically should decay into a positron and a neutrino. Yet this unwanted baby on the doorstep of science was but the first gentle hint that the age of atomic innocence had ended on Earth, for around 1950 a whole procession of new ephemeral particles appeared — all utterly unexpected and unexplainable by previous theory. Strange V-shaped patterns were noticed on lead plates struck by "cosmic rays" inside cloud chambers, mysterious splittings whose careful analysis established the existence of what are

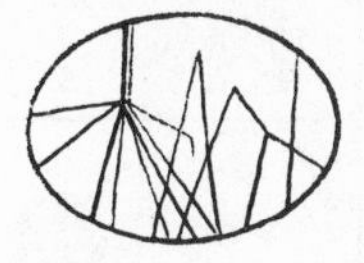

now known as K mesons or kayons and also the very baffling lambda particle, which is even heavier than the neutron. Then came the still more massive and stranger sigma particles and xi's — and in the 1960's an avalanche of new ones, almost all very massive, being discovered at the rate of one a month. Nearly all the particles as heavy as protons or heavier are now classified as baryons and live in the nucleus but are somehow abstract, composite and divisible: not truly elementary. They interact through the potent short-range nuclear force, each kind helping to spawn others (which in turn spawn it) into a spectrum of energy levels that reveal it to be, like music, fundamentally harmonic.

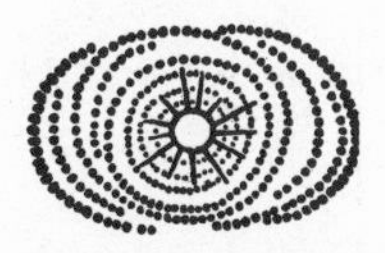

Of course, it is pretty generally admitted among physicists that there are now already far too many particles within the atom. Indeed, the present era of atomic knowledge is reminiscent of the long Ptolemaic age in astronomy when the confusion of

TABLE OF SUBATOMIC PARTICLES KNOWN IN 1960

Weight class	Mass in units of 9.1085 X 10 −28 grams	Particle symbols: matter		Particle symbols: antimatter	Name	Generic name
No weight	0		γ		photon	LEPTONS
	0	ν°		ν°	neutrino, antineutrino	
Light weight	1	e^{-}		e^{+}	electron, positron	
Middle weight	206.9	μ^{-}		μ^{+}	muons	MESONS
	264.5		π°		pions	
	273.3	π^{-}		π^{+}		
	966	κ^{+} κ°		κ° κ^{-}	kayons	
Heavy weight	1,835	p^{+}		p^{-}	proton, antiproton	NUCLEONS (BARYONS)
	1,837	n°		n°	neutron, antineutron	
	2,181	Λ°		Λ°	lambda	HYPERONS (BARYONS)
	2,327	Σ^{+} Σ° Σ^{-}		Σ^{+} Σ° Σ^{-}	sigma	
	2,583	Ξ° Ξ^{-}		Ξ^{+} Ξ°	xi	

"obviously needed" planetary spheres, cycles and epicycles blinded astronomers to the simple but unobvious truth of sun-centered motion. Many physicists are therefore trying to boil the hundred odd particles back down to a dozen or less — but it is not easy in the face of the tangible tracks actually made by these errant objects on photographic plates and in bubble chambers with accumulating evidence of their individual masses, spins, charges and other characteristics. At the very least they must be something more than figments of mathematics. Even a geiger counter can readily record some kinds of single particles, and anyone may literally see the so-called alpha particles (helium nuclei) by looking at the luminous numerals of a watch in the dark with a magnifying glass — the light being composed of tiny sparklets that correspond to sunshine on the ocean, each speck of luminescence separately shot out by a radioactive atom which, in doing so, transforms itself into a different kind of atom. And lately, not content with merely observing particles, physicists have actually begun to create or transmute them artificially in the great accelerators — even to manufacture whole atoms and fairly large molecules out of carefully combined parts.

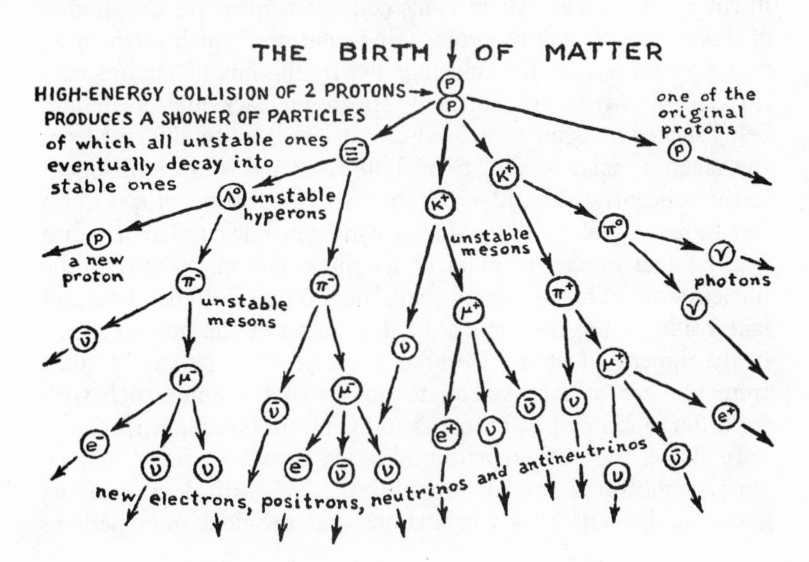

The frontier of understanding in the atom is meanwhile advancing inexorably deeper and deeper into the nucleus, which not only contains much more than 99 percent of all matter but is so small that the whole of it occupies only a thousandth of a millionth of a millionth of the "vast emptiness" enclosed by the outer electrons. This, of course, makes the density of the nucleus a thousand million million (10^{15}) times that of ordinary materials. Yet the evidence shows that the nucleus is itself far from solid, containing room for free motion and having a still heavier core at its center made of stuff so dense that a normal-sized pill of it would weigh two million tons.

Whether that core in turn may be more or less "fluid" again, with a still smaller and denser kernel somewhere within it — and how far such a sequence might continue — are questions still apparently far beyond our means of investigation. Even the density of the nuclear core is somewhat speculative, but the nucleus as a whole is becoming well established under the mounting data of modern nuclear research.

Perhaps the most accurate picture of the nucleus is the "liquid drop model" devised by Niels Bohr and John Wheeler after the discovery of nuclear fission. This concept emphasizes the fluidity of this region of moiling protons and neutrons, which seem to be somehow lubricated yet held together by the flux of mesons continuously bursting out of them, streaming back and forth and being sucked in again. Obviously, a nucleus is quite different from the vacuous "solar system" of a whole atom and is more analogous to the concentrated body of our sun. Its comparative crowdedness can be judged by the fact that a hydrogen nucleus has a radius nearly equal to the diameter of its single proton, so that if the nucleus were as big as a tennis ball, the proton would be like a golf ball inside it. On the same scale, the nuclei of the heavier atoms in the upper half of the Periodic Table of elements would range from the size of a basketball to double that volume, each with from 100 to 275 golf-ball protons and neutrons buzzing within.

In shape, the nucleus naturally tends toward a sphere, and research indicates it has a kind of fuzzy "skin" with density fading away on the outside — a mysterious and nebulous aura perhaps knit together by something on the order of surface tension which, however, for a reason unknown, never seems to make it stiff. And since it stays always flexible, the nucleus does not have much constancy in its round shape. Instead, like a falling raindrop, it evidently oscillates from egg to bun forms and may sometimes (like the earth) be pear-shaped. In various ways it quivers, sending rapid but measurable waves around and around its surface, waves which in turn affect (or participate in) the tumbling motions and complex orbits of the particles inside it. For all we know, the so-called particles (protons, neutrons, mesons . . .) inside it all merge together like drops of milk in a spoon. Or are they really more like puffs of smoke in a bag — or thoughts in a mind?

THEORETICAL MODELS OF ATOMIC NUCLEI
AND THEIR PARTS

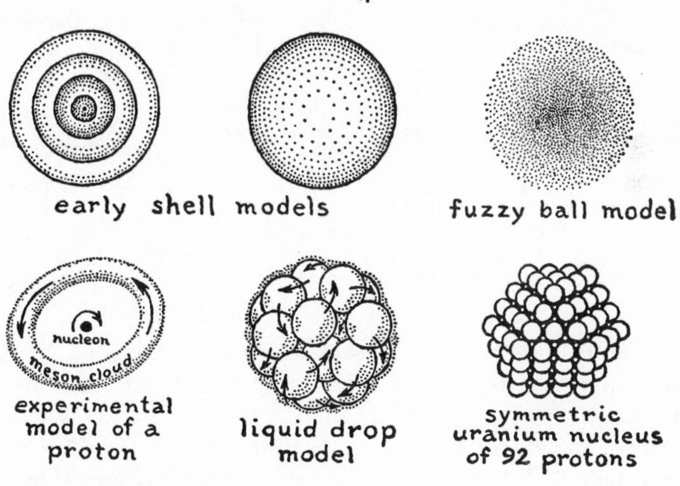

An even more troublesome thing about the nucleus, though, is the nature of the force that binds it together — that holds the positively charged protons in their close brotherly embrace despite the familiar electromagnetic law that poles of similar charge must repel each other and with increasing strength as they draw closer together. Nuclear physicists have long since concluded that the nuclear binding force must be as distinct from electromagnetic force as electromagnetic force is distinct from gravitation. Although all three are undoubtedly related and perhaps are just three aspects of the same influence, each has its own properties. Electromagnetic force, for example, is stronger than gravity within its own range but, unlike gravity whose pull increases with nearness, its symmetrical positive and negative forces tend to cancel each other to create the balanced neutral atom of ordinary matter. But the nuclear force, a lot more powerful than either of the others in its unimaginably tiny precincts, evidently feels no polarity and is centrally cumulative like a kind of supermicroscopic gravity of the second order, increasing with nearness at a much faster rate than the inverse square of the distance up to an extremely close range, when it not only begins to decrease but ultimately reverses its direction and becomes repulsive.

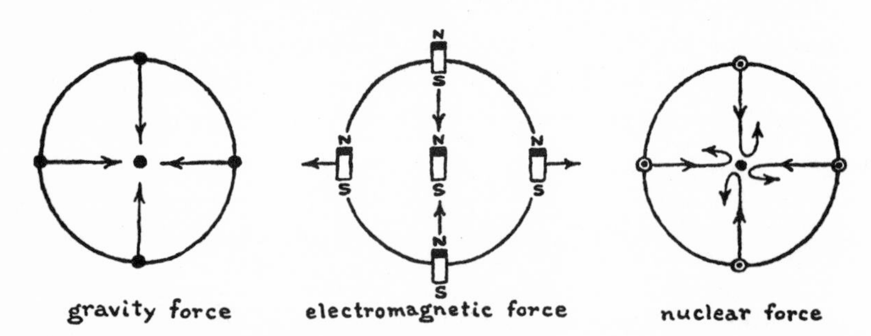

Besides these three best-known forces or interactions, I should mention that a fourth has recently been accepted, intermediate in strength between electromagnetism and gravitation, governing particle decay (p. 333). And a fifth, weaker than gravity, is suspected. Yet beyond such garnered scraps of knowledge, the crowded complexity of the nucleus — more like a bunch of cats fighting in a sack than an organized dance of entities — has made it very difficult for nuclear physicists to make head or tail out of the influences involved. For while a planet's interaction with the sun can easily be treated as a simple two-body problem since other gravitating bodies are much too far away most of the time to have an appreciable effect, the tight-milling nucleons in any atom heavier than hydrogen inevitably pull one another with nearly equal but shifting strength in several varying directions at once.

All such problems notwithstanding, the nuclear researchers are probing their way forward through the night, shooting streams of electrons and neutrons into abstract darkness like volleys of ping-pong balls, some percentage of which must bounce off an unseen stone at significant angles, trying to analyze nucleons as complexes of interrelated two-body reactions, measuring what proportion of its time the average meson spends "inside" other "particles" — ever seeking better ways of penetrating the incredible smallness of the atom, a smallness most of us find only too easy to forget. By that I mean it is not enough to realize that ten million atoms arranged in one straight line would barely reach across the head of a pin. For each of these atoms must be enlarged to the size of a house to bring its nucleus up to pinhead dimensions, which at the same time would make the ten million house-sized atoms extend to the moon. Thus one would have to imagine a real pinhead at least as big as the whole earth to make an atom's nucleus just visible on the same scale — and this nucleus is the speck that science weighs with an accuracy of one part in a million and whose "skin" and component particles are now being measured separately and are found to have deducible thicknesses and densities and all sorts of characteristics that are turning out to be what J. B. S. Haldane once called "not only queerer than we suppose but queerer than we can suppose."

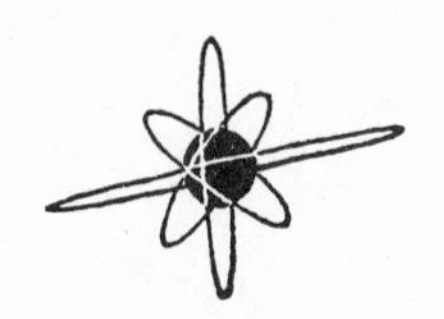

To look into this strange netherrealm of the atom in a slightly different way, let us now go back into the early 1920's and see how it all appeared to Niels Bohr, then the leading atomic physicist. You remember from the last chapter that the famous Bohr theory of the atom, published in 1913, pictured the sunlike nucleus as surrounded by planetlike electrons revolving in strictly defined shells around it. Bohr, who worked for years in Ernest Rutherford's laboratory in Manchester, had developed this idea from Rutherford's discovery that the atom possessed a nucleus and from J. J. Thomson's earlier beanpot concept of the atom as a swirling flow of electrons like beans, each element of the Periodic Table being composed solely of atoms with a particular number of electron beans.

But Bohr went much further than either Rutherford or Thomson and tied his atom neatly to the new quantum theory by showing that not only did all three of Kepler's laws for planets hold in the case of the electron revolving around the nucleus of hydrogen, but another law — a curious quantum law which was not noticeable among bodies as big as planets yet was truly fundamental — also applied. This was the law that electrons can move only in certain orbits and in no others. It is a harmonic law akin to the rule that restricts Saturn's rings to precise dimensions and to the abstract principle under which there are just 32 symmetry groupings of crystals and exactly 230 distributions of identical objects in space (see page 291).

The Bohr atom was thus elaborated into seven shells, roughly analogous to the seven notes of the scale or the seven planets of antiquity, the inmost shell having room for exactly two electrons, the second eight, the third eighteen, and so on, according to the curious rule of doubled squares described in the last chapter. But there turned out to be a lot more harmonic order than that, for these shells are themselves sized according to a strangely simple rule of single squares. Thus the first shell of the hydrogen atom is one angstrom unit (1Å) in diameter, the second is 4Å, the third 9Å, the fourth 16Å, etc., which is mathematically reducible to the series 1^2, 2^2, 3^2, 4^2 . . . Å.

If you should wonder how a hydrogen atom with only one

electron can have so many shells, it is because any electron may jump from one shell to another and will do so (on the average) as soon as it has the required energy. Moreover when a shell is not occupied, it is still there in abstract truth, as is shown by the fact that an electron can always find it. For the electron never orbits between shells: it always jumps the whole way or not at all. That is how the law of quantum mechanics has been found to work within the atom.

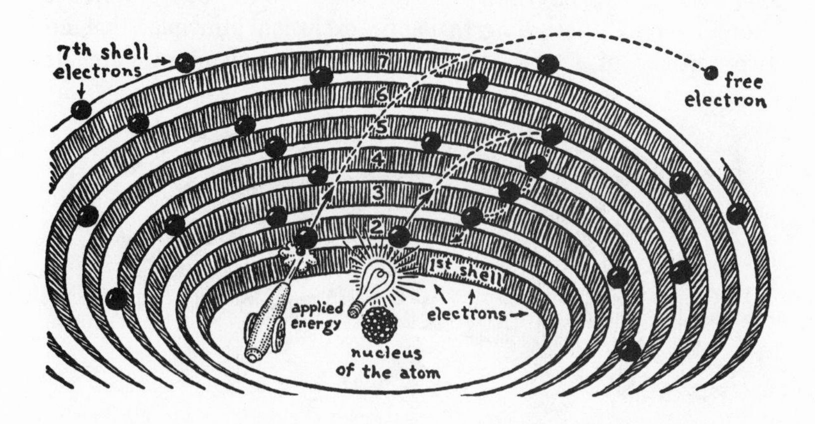

Until we look closer at what the electron is made of in the next chapter, it is perhaps enough to say now that the atomic shell structure is something like a set of invisible terraces built of nothing but abstract law. If enough light, heat or some other form of energy is applied to an atom, its electrons cannot help but jump to a "higher" terrace — farther away from the nucleus. One quantum of energy received makes one electron jump one terrace farther up, and if the electron receives enough energy to go beyond the seventh terrace, then it is completely free and the atom has become ionized (missing an electron). Likewise, an electron may jump down the same series of shells or terraces, giving back to the world its quanta of energy in discrete photons of radiation — evidently strictly fulfilling the law of conservation of energy without neglecting its quantum method of doing so. It is a strange performance, this — hard to get used to after our years of extravagant, if not always carefree, irregularity in the macroscopic life — but it is entirely law-abiding and plain as plumbing once you catch on to it.

Each terrace, according to one analogy, contains a room or apartment in which electrons may dwell in pairs of opposite spin like married couples. The lowest and smallest terrace (inner shell) is the most exclusive and popular, but its room holds only one couple: two electrons who, of course, must be of different "sexes" (spins), according to Pauli's exclusion principle: that no two particles of a kind can occupy the same quantum state. If

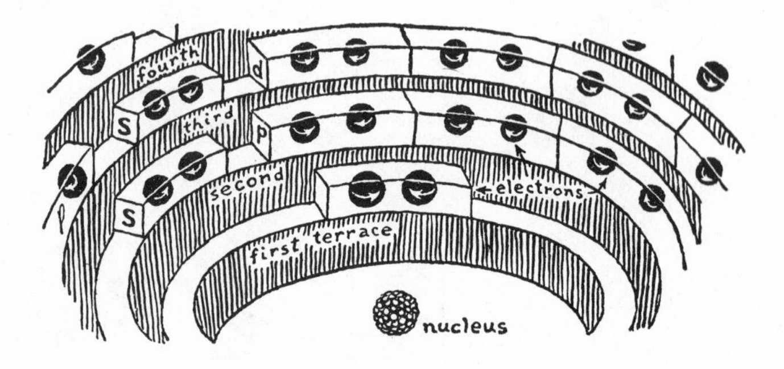

there happens to be only one electron in this "bridal chamber" (as in the case of a single hydrogen atom) that one will inevitably be lonesome and any footloose electron in the neighborhood who can contrive to fill the vacancy will eagerly do so. The second terrace shell has an apartment of four rooms, one room being slightly smaller than the other three. This room will hold an electron couple with "zero angular momentum" and the other three will take couples of "one unit of angular momentum," the smaller room being technically subshell *s* (for *sharp*, a spectroscopic term) and the three larger ones constituting subshell *p* (for *principal*).

And so is constructed the Bohr atom, shell by shell, subshell by subshell. The third shell, if you remember, has room for eighteen electrons — distributed among three subshells: *s* holding its usual 2 electrons, *p* holding its 6 electrons, and subshell *d* (for *diffuse*) holding 10 electrons "of angular momentum two." The fourth shell now adds a fourth subshell *f* (for *fundamental*) with room for 14 electrons ("ang. mom. 3"), making its total 32. Thus the physicist has a neat way of designating the exact arrangement of electrons in any part of an atom: $3d^{10}$ refers to the third shell's subshell *d* containing 10 electrons; $6p^5$ means the sixth shell's subshell *p* with 5 electrons. The designation $6p^5$ also indicates that this shell is part of element 85, astatine (At), since that is the only element with a subshell *p* of 5 electrons in its sixth shell. And as all subshells *p* have room for six electrons, $6p^5$ clearly further reveals that astatine must be chemically receptive to any element such as hydrogen, lithium, boron or sodium with a single fancy-free electron in its outer subshell.

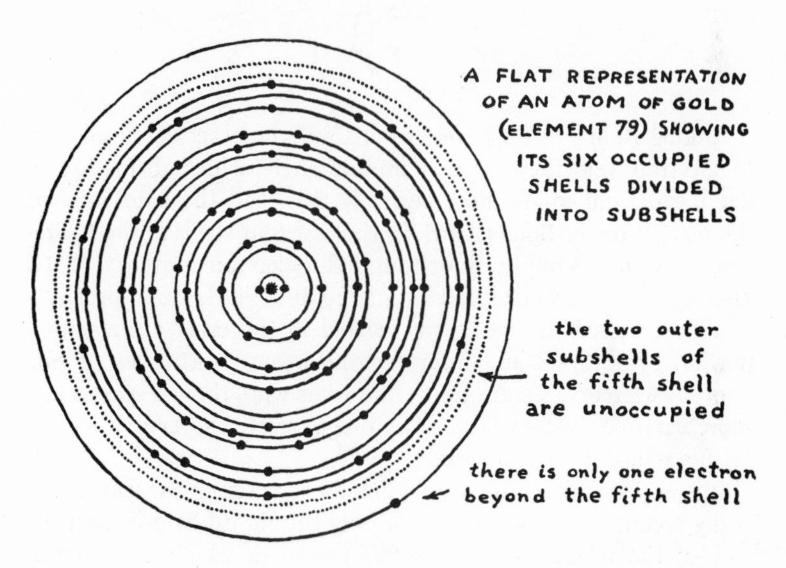

Inevitably, all this must strike an innocent pedestrian as pretty complicated, but Bohr's model has nevertheless proved enormously clarifying to chemists and in many ways fruitful. Fact is, God's world, as seen by science, is not turning out to be the essence of simplicity — certainly not in the usual sense of that term. On the contrary, complexity seems to be one of its surest ingredients if not a basic criterion of its value — for the same reason that complexity is an accepted indication of quality in a watch. Bohr faced this realization squarely, and his extraordinary vision of the atom did not arrive simply or blossom without a deep struggle. His friends describe him as having hammered out his beautiful electron shells partly in speculative conversations with many a skeptical fellow theorist in England, Germany and at the Institute of Theoretical Physics in Copenhagen during World War I and in the early twenties — pacing intently around and around his table like a symbol of one of his own electrons, arguing profoundly, distractedly lighting his pipe again and again while absent-mindedly strewing the matches like photons behind him.

Among Bohr's first concrete results was the exact determination of electron velocities in each atomic shell, which, being quantized, came out much more perfectly than had the harmonics of the planets under Bode's Law. In the first four shells of the hydrogen atom, for example, Bohr found the electron moved at 2,160 kilometers per second, 1,080 k.p.s., 720 and 540 k.p.s. respectively — almost incredible speeds, yet related in the exact abstract proportion of 12:6:4:3 and following the inverse-square law as precisely as Mercury, Venus, Earth and Mars who, though spaced in a more carefree manner, also orbit progressively slower as they are farther removed from the sun.

But why was there this subtle difference between the way of the stiffly regimented atom, in which only certain orbits exist and the way of the relaxed, carefree solar system, in which any orbit is possible? This curious but fundamental disparity in an otherwise harmonious world seemed unaccountable and was very disturbing to physicists of the day. It figuratively shook what they had assumed to be bedrock under their principles — and it shook it in tiny, regular, quantum jerks.

What could this mean? How could one unified vehicle be a track-bound trolley car one minute and a free-wheeling bus the next? How could the quantum world of electrons operate on one principle and the classical world of stars (full of electrons) operate on another? Was there any dividing line between the two? Did they overlap, or did each fade into a mysterious no-man's-land of unknown dimensions somewhere in the middle?

Bohr boldly led the way out of the woods by showing mathematically that, although the quantum theory is never exactly equivalent to the classical, there is always a correspondence between them that grows more and more exact as dimensions increase and the incredible vibrations of the microcosm relax into the lower frequencies of our familiar macroworld. This correspondence principle, in fact, became very important in the relentless efforts of physicists to understand what matter really is.

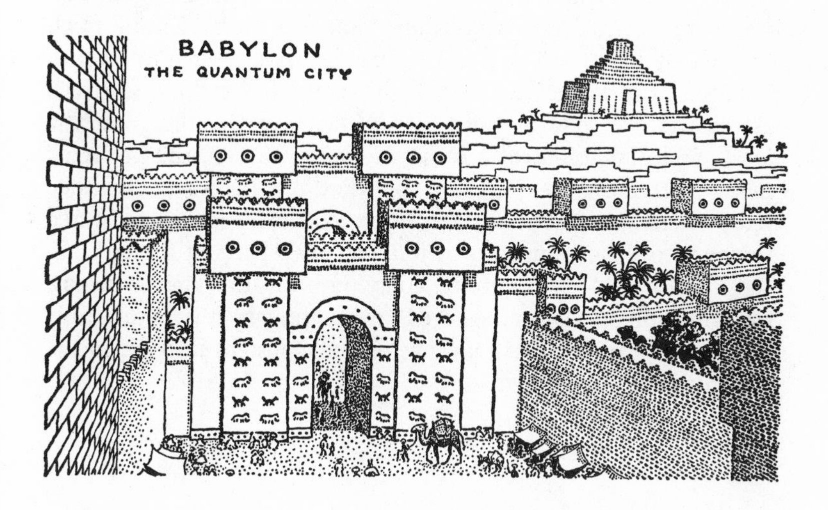

For a good analogy of the principle, imagine ancient Babylon in the seventh century B.C. with its straight brick-paved avenues and vaulted sewers extending over several square miles of land. And let's assume that the bricks are all identical and laid in one continuous system throughout, exactly fitted end to end and side to side, and that the same bricks form the foundations of all the houses from the Ishtar Gate to the famous Tower of Babel.

Then any mason building a house in Babylon will naturally lay his first course of bricks neatly upon those already there in an exact relation, according to the sacred Marduk Building Code which none dares defy. Thus every block of buildings will be incorporated into the same system, spaced in integral multiples of the Babel brick — leaving, say, exactly 22 bricks across a street from wall to wall, or 69, but never 46½ or 33⅝ or 81.44327+. You could call this a quantum law, the brick being the quantum unit or minim than which there is nothing smaller.

But of course, brick dimensions need concern only the brick makers and masons who actually handle them, and even these handlers probably have no awareness of their bricks' over-all implications. For every landowner feels free to build his house whereever he chooses and people are not conscious of any quantum restrictions. Still less need King Nebuchadnezzar's Lord High Planner of Royal Works, overlooking Babylon from the seventh temple of the Tower (290 feet high), consider anything so trivial as the width or position of a single brick when working out the locale and dimensions of a future Hanging Garden, the projected rebuilding of the great Temple of Marduk (1,500 feet by 1,800 feet) or the construction of the unprecedented 3,000-foot covered bridge across the Euphrates to unite the two parts of the city. In the macroworld of ordinary consciousness, the quantum law can thus practically be ignored, even though it is universally obeyed — a paradox that may give us a new insight into the meaning of liberty.

While Niels Bohr plumbed the atom in Europe, others were sounding it from various angles in other continents and running up against the same abstruse quantum limitations — even in the realm of electromagnetism. As early as 1909, Robert Andrews Millikan in America successfully measured the charge of an individual electron in one of the most elegant of the great classic experiments. This involved watching a laboratory cloud of microscopic oil droplets through a special optical instrument as the droplets floated in air illumined by a beam of light. Shining like asteroids in sunshine, hundreds of these tiny orbs were made to drift between two exactly horizontal metal plates so equipped that Millikan could regulate the rate of falling or rising of any visible droplet by means of a set of very sensitive controls that changed the electric potential of the plates to whatever positive or negative charge was required. Naturally, the oil droplets had been subjected to friction when sprayed by atomizer beforehand, and this had rubbed extra electrons upon most of them, giving them a negative charge. So the attraction of the positively charged upper plate and the repulsion of the negative lower plate tended to counteract the slight downward influence of gravity, a circumstance that en-

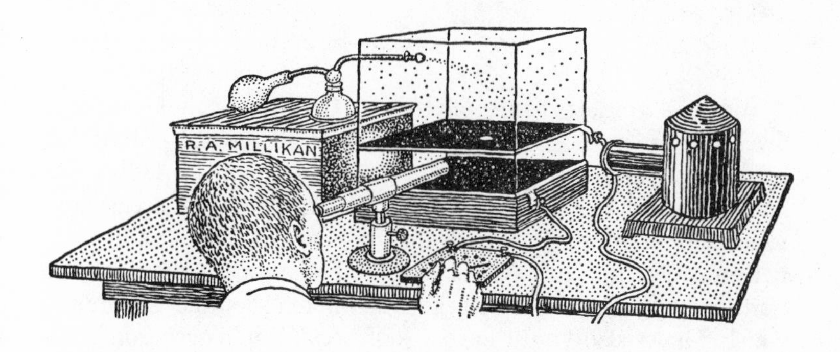

abled Millikan to balance any convenient droplet in suspension between the plates and to hold it stationary long enough to record the strength of the electric field around it — a measured quantity which, combined with the droplet's observed rate of fall when both plates were electrically neutral, made possible a calculation of the droplet's exact charge. This droplet charge, in turn, always came out as an integral multiple of a certain fixed minim charge, called "*e*" because it is the charge of an electron — which finally confirmed Faraday's famous intuition that any atomic charge must be a multiple of some basic electromagnetic quantum (see page 251).

So electricity had been revealed to be made of something like Babylonian bricks, too. And in the process the electron had been awarded a new dimension. Calling an electrical charge a dimension, of course, may be taking liberty with a plinth of our material universe, for electromagnetism seems to be one of those ubiquitous underlying abstractions that can be defined only in terms of its behavior. Like heat, it is a statistical function of motion. But where heat is random motion or the collective effect of the speeds of countless atoms or molecules, electricity is nonrandom or polarized motion, the collective effect of the velocities of many electrons flowing in a particular direction. And if heat can propagate itself through an intangible radiation made of photons moving at the speed of light, so has electricity a corresponding aspect of propagation by electromagnetic radiation, likewise made of photons.

We shall go deeper into photons in Chapter 12, but here let's dwell briefly on electromagnetic behavior in general. Electricity is named after *elektron*, Greek for amber, a substance known at least as early as 600 B.C. to "attract mustard seeds" and sometimes emit sparks after being rubbed with wool. And magnetism is named for the Greek coastal district of Magnesia on the northwestern arm of the Aegean Sea where lodestone was found, which Thales said "must have a soul because it moves iron." Although the Chinese seem to have known a similar magnetic iron oxide (Fe_3O_4) even earlier (some say in 2700 B.C.), calling it "the stone that loves" and fashioning floating compasses of it that steadfastly pointed toward the lodestar, no one seems to have proved any definite connection between electricity and magnetism until the last century, when an Italian jurist named Grandomenico Romagnosi accidentally discovered in 1802 that an electric current flowing through a wire will deflect a magnetic needle in its vicinity.

This discovery, duplicated and properly announced by Hans Christian Oërsted, the Danish scientist, in 1820, set many scientists to experimenting in the now combined field of electromagnetism.

And progress accelerated, soon culminating in the discovery by Joseph Henry in America in 1830 and Michael Faraday in England in 1831 that a moving magnet, in turn, will induce an electric current to flow in a wire.

This great revelation of inductance, soon developed by Faraday and later James Clerk Maxwell of Scotland into the strange new theory of the "field," was, of course, the key to the dynamo, which led directly to the large-scale generation of electricity that is now considered indispensable to modern civilization. But, beyond its obviously practical side, the integration of electricity and magnetism sounded the knell of the 200-year-old mechanistic theory of matter as an association of mass-points by exposing the incongruity of any kind of material point as a physical reality and replacing it with the pervasive concept of continuous fields of energy — a switch that the unborn Albert Einstein would one day applaud as "the most profound and fruitful one that has come to physics since Newton."

And what did Faraday and Maxwell mean by "the field"? Clearly, they meant a mathematically definable region extending through space and containing measurable influences: the gravitational field of the sun, the magnetic field of a magnet and so on. In this way they neatly got around the old Newtonian puzzle of what transmits "action at a distance" by eliminating the distance — so that instead of reaching out unaccountably and instantaneously across a void of nothing, action simply propagated itself at a finite speed (that of light) and permeated space with the familiar inverse-square intensity in all directions. Electromagnetic forces were admittedly more complicated than those of gravity but that was natural, for they were evidently more fundamental and probably included gravitation as "a special case."

Meantime, the many ingenious experiments by Faraday, Henry and others had clearly demonstrated the working differences between electricity and magnetism, including the particularly striking contrast in their field orientations. Lines of electric force, for example, form a field that is everywhere *perpendicular* to the surfaces of conductors (wires, tubes, etc.), while lines of magnetic force compose a contrary field that is always *parallel* to magnetic

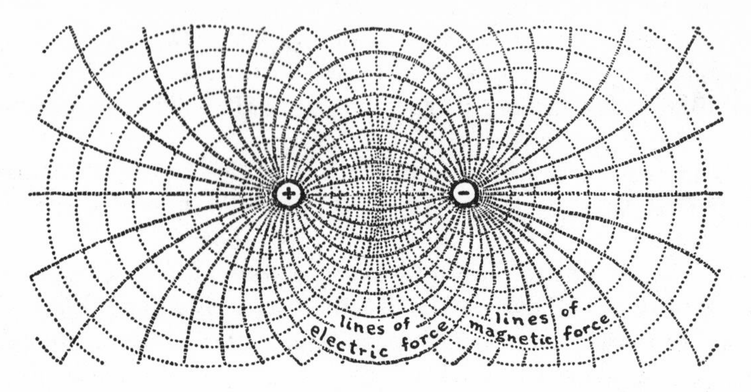

surfaces. And these two sets of force lines are respectively analogous, in familiar gravitational terms, to the lines down which falling objects drop *perpendicular* to a surface (as of the earth) and contrariwise to the path lines of moons or satellites orbiting *parallel* to the same surface. Such rather mysterious yet graceful correlations illuminating the warp and woof of the electromagnetic tapestry were eventually compiled, along with all other experimental data, into laws governing basic electrical and magnetic behavior which the mathematical genius of Maxwell found a way to express in four beautiful relations — which have served as the cornerstone of electromagnetic theory ever since.

Because of Maxwell's untimely death at the age of forty-eight in 1879, before the electron was discovered, he could hardly even have guessed what was going on inside an electric conductor or a magnet. But that did not really matter much, since atomic particles are virtually beyond visualizing anyway. And besides, Maxwell's equations proved so conclusive he could accurately predict almost any electromagnetic behavior with them and often did — including such abstruse things as radio waves, which, in his day, were considered — if they were considered at all — as mere metaphysical fantasy.

By the time the electron was discovered to be the natural unit of electrical charge and current, the electric industry was growing up fast, gaining experience and new insight daily. Its confident engineers accepted in their stride the new knowledge that the so-called positive charge possessed by glass after being rubbed with silk really meant it had lost electrons, while the antipodean negative charge found on hard rubber brushed with fur indicated a corresponding gain in electrons. The theoretical physicists had already extended the early analogy between electromagnetism and gravity to the idea that currents in wires tend to persist for the same basic reason that the motions of large masses tend to persist under Newton's law of inertia. For do not electrons, after all, have mass and therefore momentum when in motion — a momentum that turns out to be strongest when the wire carrying the current is wound in a coil of many turns, creating the maximum magnetic flux? And from here, what could be more natural than to compare the flow of current through such a coil to the easy sliding of a material body upon a slippery surface — and the electrical resistance of the coil to the gentle friction between body and surface which steadily resists its motion?

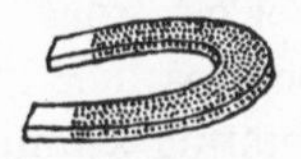

If a mathematical structure built up from a sequence of such analogies has helped engineers to harness intangible electricity into modern electronics, a somewhat similar technique at the same time advanced the even more mysterious lore of magnetism to its own remarkable present state. Indeed despite a continuing failure to find the "elementary particle" of magnetism once proposed by Dirac, researchers nevertheless discovered something almost as useful in the so-called gyromagnetic ratio between the magnetic force and the spin of subatomic particles. For the stability of this relationship is strong evidence that the "desire" of the whirling top to keep its axis upright is in essence none other than the "love" of "the stone that loves" — that the gyroscopic and magnetic compasses of modern navigation both work, basically, on the same principle — indeed, that the almost emotional yearning exhibited upon these disparate macroscopic and microscopic scales may even be a component of the primal stirring of life itself.

In any event, every spinning electron is now accepted by physics as "a tiny permanent magnet": its strength designated as one magneton. Even protons and the uncharged neutron have definite "magnetic moments" attributable to spinning, as has the rotating nucleus as a whole. But the entire atom may or may not add up to a magnet, depending on how many of its magnetic parts (electrons, etc.) neutralize one another. This is because electrons, protons and neutrons tend to go in pairs, spinning in opposite directions like the propellers of a twin-motored airplane geared to cancel each other's torque. The atom as a whole will be a magnet only while there is an imbalance of spins, which must happen when the number of electrons or other particles is odd or when most members of some particle group spin one way instead of pairing off in counterrotating symmetry. The iron atom, for instance, is bound to be magnetic because its third electron shell has an incomplete subshell *d* with five electrons spinning in one direction and only one electron the other way, producing a net magnetism of four magnetons.

From this you may wonder why every piece of iron is not a powerful magnet. It is only because magnetic atoms in their turn tend to neutralize each other through being knocked about in random directions by their natural thermal energy. In fact, no material can become magnetic until some force stronger than thermal agitation aligns a majority of its magnetic atoms in one direction. This force could be a bolt of lightning striking ferrous oxide on Mount Olympus, instantaneously turning whole cliffs of the ore into magnetite or lodestone. Or it could be just the natural sorting process of crystallization organizing atoms as they cool down and stack themselves into a lattice. In the latter case at least, there is little doubt that random atomic and molecular motion (ordinary heat) is the mortal enemy of magnetism, for all magnetic materials lose their magnetism when heated to the point where the forces of thermal disorder begin to overpower magnetic discipline. In iron this critical temperature is 1,420° F.

But obviously coldness alone is not enough to make metal magnetic. This is because the single unified crystals of magnetization, called magnetic domains, are usually very small (numbering a few million or billion atoms), and their magnetic axes of polarity ordinarily have a choice of at least two directions (six in the case of iron), paired parallel but opposite each other in conformance with neighboring polarities, which results in most of them being neutralized almost as completely as paired electrons. Thus not only must a piece of metal, to be a magnet, be made of magnetic atoms like multiengined planes with more propellers spinning one way than the other, and these arranged in magnetic domains like huge aircraft carriers with most of their thousands of planes heading, say, forward, but the domains in turn must be given a preponderant orientation as if vast fleets of the carriers were ordered to steer most of the time southeastward, seldom northward or westward.

The domain phase of magnetization has been actually seen under the microscope in the past decade, by smearing magnetic metal with fine iron-oxide powder in colloidal suspension to outline the domains as they bodily shift their axes step by step under magnetic field pressure — often accompanied by faint audible clicking sounds known as the Barkhausen effect. Perceptible transition regions called walls separate the domains by a few millionths of an inch and, during the process of magnetization, advance fairly steadily in waves, showing a kind of zipper action compounded of millions of atoms and molecules successively swinging into line. Thus do the individual domains grow wherever their constituent crystal axes are most nearly parallel to the field, converting new lattice members, rank by rank, row on row, in surging tides at the expense of less favorably oriented domains next to them — measurably changing the dimensions of hard chunks of iron, cobalt, nickel, gadolinium or other magnetic alloys as their straining lattices snap and contort in response to the mystic "love" force of gyrating magnetism.

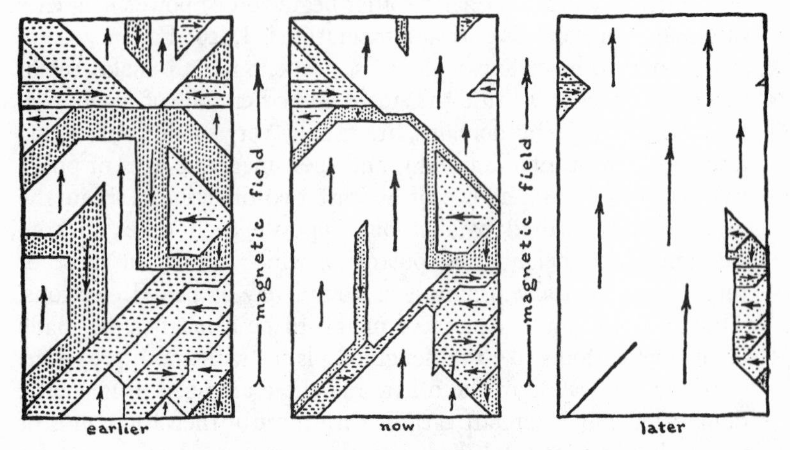

GROWTH OF MAGNETIC DOMAINS
IN METAL PLACED IN A STRONG MAGNETIC FIELD

If this new-discovered wonder of nature has not been made crystal-clear by this presentation, it is not surprising. For magnetism is still on the frontier of knowledge, and plenty of deep mystery remains. Yet the researchers and engineers have already made astonishing progress in their current struggle to develop stronger magnetic fields. They measure these in gauss, a unit of strength named for Karl Friedrich Gauss, the great German mathematician, the potency of which can be judged by the fact that the earth's magnetic field is about a third of one gauss at New York City. The strongest permanent magnet yet made has a field of about 10,000 gauss, while an electromagnet, consisting of a coil of wire around a metal core that amplifies the field generated by the current in the coil, can approach 60,000 gauss. Although this seems to be the limit of the magnetic capacity of metals, the most powerful modern magnets operate without cores in air or a vacuum and have attained to several million gauss, a concentration of energy able to melt steel like butter or explode it with the violence of TNT, producing a pressure of approximately a thousand tons per square inch.

Today's theorists think of a magnetic field as something like a gas, which can be intensified by compression, by crowding its lines of force. And the fact that its density and energy increase as the square of the field strength gives enormous potentiality to this form of power — a power whose curious natural manifestations have ranged from turning the ammonia molecule (NH_3) inside out to unaccountably reversing the polarity of giant stars in the course of a few days.

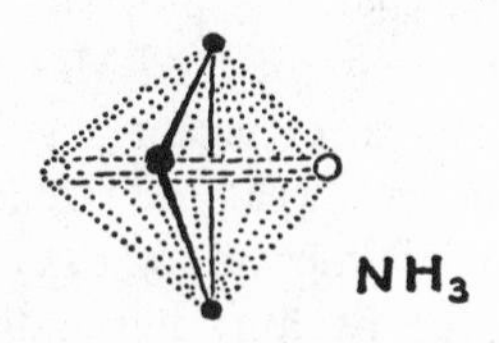

And don't forget that, as we saw was true of heat, magnetism in large perspective is a *relative* property whose magnitude depends ultimately on the frame or viewpoint it is measured from. Which is a way of saying that, because an airplane lands "hotter" on the deck of a carrier moving in the opposite direction, so must a great fleet of coordinated carriers loaded with buzzing planes (representing a magnetic field) be stronger to an observer moving or spinning *against* rather than *with* their prevailing direction. This can be considered as but basic groundwork for grasping relativity — which we will come to in Chapter 13 — and clearly suggests one of its aspects in the beautiful lines of force that define the electromagnetic field, flowing abstractly on and on without beginning or ending anywhen or where.

If electromagnetism is but one form of energy in the world — and thunderous evidence from the nuclear front denies it is the greatest one — we might do well to consider now the strange genie lurking among protons and neutrons whose muscle is stronger than his weight by the incredible amount of the speed of light multiplied by itself!

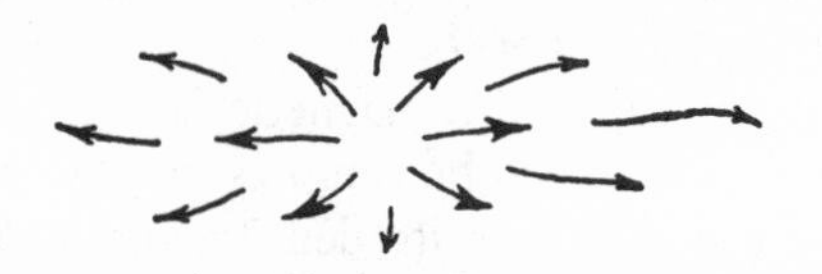

Appropriately, it was none other than Einstein who explained that the only reason men had not noticed the nuclear monster before our century was that this being kept his real life strictly to himself like an overcautious miser who had never been known to spend or give away a penny. Nevertheless, one of the miser's moneybags eventually began to split, showing a tiny hole from which a keen Polish researcher named Marie Curie and her French husband Pierre managed to extract a peculiar glowing substance. This they refined, after two years of strenuous work, into something worth a thousand times more than gold: a new element they named radium for its amazing radioactivity — and which made Marie the most famous woman scientist in history.

In such fashion did the dawning twentieth century reveal the awful potency of the atom's center, whose binding force there easily counteracts the electromagnetic repulsion of proton for proton, + for +, but whose steep decline in power with radial distance rapidly diminishes its grip on the bustling outer fringes of the bloated nuclei of the heavier elements. This is the secret of radioactivity, the disintegration or fission of heavy matter and the tendency that explains the instability of every atom heavier than lead (element 82) in the sixth octave of the Periodic Table, particularly seething radium (88), exuberant uranium (92) and all the transuranic synthetic elements. Some of these turgid substances spontaneously transmute into lighter elements in so few minutes that it is almost impossible to find out what they are — or were — made of. And of course, present atomic power is little more than a controlled exploitation of uranium's natural instability, based on the discovery in Germany in 1939 that blasting an atom of the uranium isotope U^{238} (92 protons, 146 neutrons) with a fast neutron or two was enough to break it apart into an atom of barium (56 protons, 81 neutrons) and an atom of krypton (36 protons, 46 neutrons), plus the splash of a few neutrons left over and, most significant of all, about 200 mev (million electron volts) of energy in the form of light, heat and radiation. At the same time, if the cast-off neutrons could immediately be made to strike more uranium, a chain reaction would be produced. This was the principle of the famous Hiroshima A-bomb.

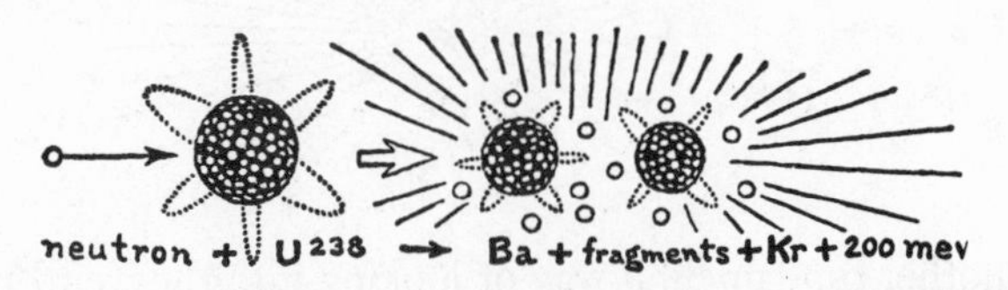

Although fission is thus an efficient power releaser from big atoms that lose mass and energy whenever they split apart, the opposite is true of small atoms which, being stable, must absorb a great deal of energy before they will crack, spending none in the process. Small atoms instead shed mass-energy ("the packing loss") when forced to fuse together against electromagnetic repulsion. It is a fusion requiring a lot of initial energy from outside, because protons with identical (positive) charges must be hurled extremely close to each other before the nuclear binding force will grab and lock them, but it pays handsomely in the end and is the key to the mighty H-bomb, in which two deuterium (H^2) atoms are knocked and welded into a single helium (He^4) atom plus 24 mev of radiant energy. In comparing this with the output of uranium fission, it is apparent that fission produces more energy per nucleus, yet, on the more practical "pound for pound" basis, fusion exceeds fission by "about seven times," not to mention using a "fuel" that is vastly more abundant and inexpensive.

There are many ways of setting up both fission and fusion, of course, and undoubtedly some of each will be important sources of power in the future, along with direct energy from the sun, cosmic "rays" and surely other developments now still undreamed. Whatever comes, it can hardly be more revolutionary than the domestication by fission and fusion of the atom's nucleus, which is certainly our profoundest technological accomplishment up to now. For the atom has just given man an impressive hint of how he may eventually escape the solar sovereignty and become something more than a mere son of the sun — how ultimately, perhaps even in a literal physical sense, he may graduate into a citizen of the universe.

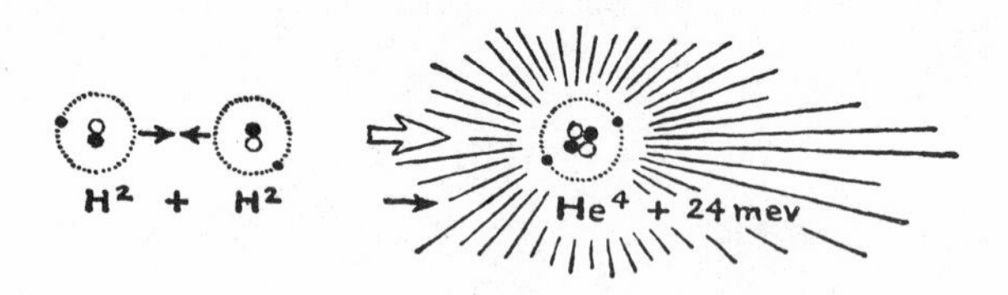

Still another fundamental way of looking into the weird netherworld of the atom is through the gateway to absolute zero. Here men have already reached to within much less than a hundred thousandth of a degree of what is defined as the complete absence of heat. One of the most successful techniques for approaching this unearthly purity of negation has been to line up the nuclei of very cold copper atoms in a magnetic field, then remove the field, letting the nuclei revert back to a random arrangement. The energy expended in this reversion leaves the nuclei weaker and quieter, which is the same thing as cooler. Theoretically, after many repetitions of the process, they should attain absolute zero when their disorder (entropy) is complete, but their own slight nuclear magnetism in real life does not quite permit this totality of relaxation.

Certain metals, however, cooled even to within a few whole degrees of absolute zero, have revealed the phenomenon of superconductivity (described on page 300), when "all" electrical resistance vanishes and surplus electrons flow onward through atom after atom as freely as the moon flows through "empty" space. This seems to be related to the equally fantastic superfluidity of liquid helium at 4° above absolute zero, a component of which flows in the form of "whole atoms completely without friction," slipping along as a film a few millionths of an inch thick that tends to cling to solid surfaces yet moves at a foot per second, even siphoning itself spontaneously out of cups. Physicists theorize that such a superfluid can exist only after the last traces of heat have somehow "separated" themselves from it to vibrate independently "relative to the superfluid background" of absolute zero. If this turns out to be more than an abstract interpretation, it seems to indicate not only that absolute zero is attainable on Earth but that friction is a function or attribute of atomic motion (heat) and that, in the astronomical perspective, it may be something peculiar to certain kinds of vibrant atomic worlds like ours, while other more elemental, perhaps more innocent and generalized, matter moves and spins somewhere somewhen with as little restraint as the headlong flight of the galaxies.

There is little doubt, you see, that heat is a more restricted property than we usually remember, for already physicists are working to discover the laws that govern "a temperature range" that reaches downward from absolute zero — where the motions that still exist must soon be given a separate collective name. These motions seem to consist primarily of the spinning of nuclei, which continues after all motion of whole atoms, and perhaps even of electrons, has died away. This subabsolute state, of course, is very hard to explore, but we already know that some spins have more energy than others and that spins (in their magnetic essence) react upon one another, with the result that high-energy spin can spread through matter in much the same way heat does. Even low-energy spin or the lack of spin seems to spread like a "wave of cold" where the energy of spinning has dissipated, permitting a substance to go on losing energy below absolute zero.

Whatever the nature of this nadir world beyond the ordinary laws of thermodynamics, its recent discovery is sure testimony of the fantastic potential of nature in every direction. From out of the sun and stars, for example, and from galaxies, probably particularly from quasars, come the assorted particles called cosmic "rays" that are so dense in the high atmosphere that black mice exposed in a balloon 20 miles high have turned white in a few hours. Although many of these particles are known to be the nuclei of heavy elements, including gold, lead and (appropriately) the rare earths, a few have been measured to have the almost impossible energy of many quintillion electron volts — of which one eminent physicist remarked, "To accelerate a flea to an equivalent speed would require all the energy released by a million hydrogen bombs."

How many the lessons to be learned from outer space! And how endless the prospects from within, where curious momentary elements completely separate from the Periodic Table keep revealing themselves: positronium, the fleeting partnership of an electron and a positron with no nucleus at all; superdense mumesic hydrogen where a negative muon takes the place of the hydrogen electron; and the flash-lived kayon^{+} electron^{-} or pion^{+} electron^{-} "atom" — glimpses of grist for the dream mills of some latter-day Leucippos.

A Chinese poem written by Lao Tzu more than two thousand five hundred years ago says that the best knots are tied without rope. This hoary wisdom may be interpreted as meaning that relationships are more real than things. It is a conclusion we can reasonably have come to in this chapter on the physical depths of the atom, for, plumb as we will the nuclear core, we cannot fathom matter entirely as matter. A factor of unpredictability amounting to a subtle willfulness begins to appear in that impalpable postulate, the electron, even while its behavior explains gross properties (such as tension and electric charge) to mathematical perfection.

Come to think of it, what do we mean by perfection anyhow? Is perfection more abstract than the song man sings in this paradoxical world of known form but unknown content? Could total perfection be already here unbeknownst — its immeasurable virtue throbbing in the atom, its axes but willingly spinning out their little quanta of warmth and love, the very compend of goodness?

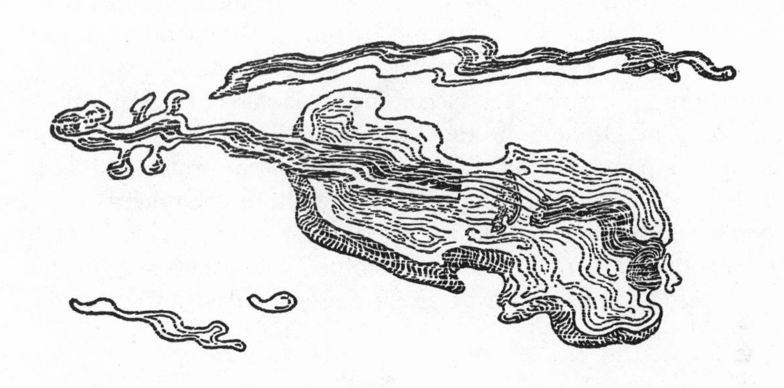

11. of waves and music

WE HAVE A SMALL CABINET marked SPACE LIBRARY in this station and, as anyone might have guessed, its four dozen books include a Bible. Although I am not much given to reading Scripture, I was idly thumbing through it after my last watch when what should my eye light upon but the ancient Mosaic commandment: "Thou shalt not seethe a kid in his mother's milk."

At first thought, this hardly seems to apply to us here. Nor, I might add, does it strike one as a reasonable restriction even for terrestrial nomads or goatherds who traditionally stew their meat in milk and who may have few nannies with any to spare. Yet there are certain mysterious discords in life — even in space life — that instinctively repel us, and one must admit that boiling any creature in a symbol of maternal love must inevitably be among them.

On the other hand, the sympathetic frequencies and deep harmonies in this universe, which can be intuitively felt but hardly explained, may well be a source of more than matter or life or mind — as I have long believed and hope presently to show. At least the harmonic nature of the world is known to have been sensed as far back as the days of Pythagoras, when the great star Vega, according to legend, first shone upon the harp of Orpheus in the little constellation now known as Lyra. In popular belief Vega's celestial harp strings had been tuned by the Sun himself with the result that Orpheus' music was so enchanting to the trees that they bent to listen, while the most savage of beasts were soothed into gentleness by the bewitching strains and even rivers ceased to flow lest they miss a single note.

As for Pythagoras, the manifest perfection of the intervals of the spheres (see page 67) was to him but one aspect of an omniscient harmony — for had he not observed and defined the exact musical-scale intervals on a harp string and had he not recognized fundamentally similar abstract relationships throughout all creation — even to the essence of matter itself? According to the books I've read, the Ionic lyre had been perfected early in the seventh century B.C. by old Terpandros of Lesbos, "father of Greek music," who increased its strings to seven and "canonized the heptachord." No one knows for how many tens of thousands of years before that men blew on reeds or twanged strings to make music, but by Pythagoras' day every competent phorminxist or citharist must have known that shortening or fretting a string changed its note, and no doubt many had learned by improvisation how to play octaves, fifths, fourths, thirds or other pleasing intervals. This was both natural and expectable, for children and birds often sing in fifths and fourths and a simple horn (such as the bugle) will produce many such exact intervals just by being blown with differing intensities.

But Pythagoras seems to have applied the philosopher's WHY to all this and to have noticed that the extremely sympathetic relationship we call an octave comes from exactly doubling or halving the string length, that is, in a 1:2 proportion, while the harmonious fifth has a 2:3 ratio and the fourth 3:4. He or his followers may even have explored the 4:5 interval of the third or still less obvious consonances.

In any case, the evidence is that these discoveries led to deep contemplation of the abstract ratio of numbers and geometric figures and particularly to a new mathematical relation: the harmonic mean. The harmonic mean expresses a pitch ratio be-

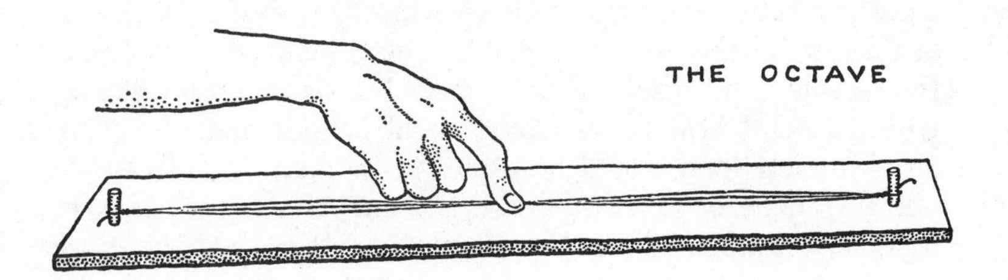

THE OCTAVE

tween neighboring musical notes that is a good deal more subtle than the arithmetic mean, which merely averages them, or than the geometric mean, which equally tempers their proportions. Thus if we take, for example, the numbers 6 and 12, their harmonic mean is 8 (which exceeds 6 by one third of 6 and is exceeded by 12 by one third of 12), their arithmetic mean is 9 (which differs from both 6 and 12 by the same number, 3) and their geometric mean is approximately 8.486 (such that $6 : 8.486 = 8.486 : 12$).

It was the fortune of the harmonic mean that it came to appear to the Pythagoreans as one of the most divine endowments of nature, not only in music and the heavens but in flowers and hills, in moving animals and waves of the sea. Even the abstract cube was held sacred because its eight corners form the harmonic mean between its six faces and its twelve edges. And the other known means and proportions and harmonies all had their special significances as symbols of the integral order in the universe — a concept that was almost wholly intuitive, since no one in those days knew how to analyze a flower or measure a moving wave or count the wingbeats of a sparrow or the musical vibrations of a lyre, nor was there such a thing as a second or a minute to time them by.

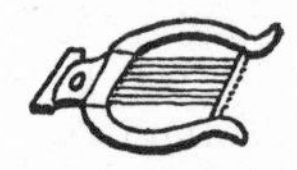

It is, of course, hardly possible now to recover the vague nebulosity of that primordial world in which neither maps nor calendars agreed with each other or were considered important and when it was almost unthinkable to make a major decision without consulting an oracle in some spooky cavern or offering a sacrifice upon the altar of Apollo. There were no laws of nature — only tendencies. Stones tended toward earth. Smoke tended toward heaven. Anything could happen and often did. So if musical tones could be expressed in numbers, why could not any shape or sense or quality? Some of the Pythagoreans regarded the number 4 as the essence of justice, perhaps as we think of square (4-sided) dealing, while number 7 signified the "right time." And they theorized about the harmonic relation of heat and cold, of wetness and dryness, light and darkness, maleness and femaleness . . . indeed about all balances and symmetries they could imagine.

That is what the members of the brotherhood probably meant by their doctrine that "all things are numbers" and that every structural form is, in essence, a piece of "frozen music." Explaining it long afterwards, Aristotle wrote that "the elements of numbers are the elements of things and therefore things are numbers." Of course, in Pythagoras' lifetime this was only an unwritten concept, for those were the days before writing was used for anything much except royal monuments, religious inscriptions, legal records, maybe a few business tallies. They were the days of the great lyric poets — Pindar, Simonides, Bacchylides — when prosody and melody were a single art and the ode a public celebration. Most teaching then consisted of stories of the great deeds of heroes and gods in far places and times long gone, which came from the lips of bards, rhapsodists and minstrels. Priests, in turn, relied heavily on the emotional effect of chants and ritual dances, for it was obvious that music could control the human heart, raise the courage to fighting pitch, quench the fires of anger, even cool erotic yearning. Specifically, the *Iliad* tells how the heroes of the Trojan War kept their wives faithful at home by leaving them to the care of the right musicians. And it was accepted among Ionian physicians that an oboe played in the Phrygian mode was a sure cure for sciatica, particularly when aimed at the affected part. All in all, it was an appropriate era in which to name music the common denominator of the universe. There were even occasions when a choice of melody directly affected the outcome of an important battle, and I can find little reason to doubt that one particular change of tune, attributed to Pythagoras himself, actually sealed the doom of Sybaris, then the greatest city in Europe. It is a significant tale.

Sybaris, you may recall, was the famous home of the Sybarites in the arch of the foot of Italy, renowned for its extravagance and luxury. Founded in 720 B.C. under a liberal charter that allowed "any person of Hellenic speech" to become its citizen and backed by the merchant princes of great Miletos in Asia Minor, whose last king was Croesus, it had grown so rapidly that in a little over a century (according to the historian Timaios), it attained a population of 300,000 free persons, while its city walls (in Strabo's account) were more than five miles around. Even if caution should lead one to halve the first of these figures, Sybaris was still a city without peer in Europe: not only the largest but by all odds the richest, and "overflowing with finely woven woolen materials, painted Ionian and Corinthian earthenware, Oriental jewelry and silver plate, unguents, medicines, spices, glass and carved ivory," all towed to its warehouses up the short canal from the sea. And there also, it is said, were unloaded "cargoes of Egyptian cats," an exotic novelty to Greeks, who had formerly made pets only of dogs and ferrets.

So plush and decadent did the Sybarites eventually become that it is claimed their patricians "never walked" but, even to go two houses down the street, "took a chariot." They were the first Europeans to have steam baths and elaborate plumbing and had earlier invented the chamber pot as a between-course accommodation to be passed around at great banquets. For an indication of how seriously they took their eating, Herodotus describes a certain "Smindyrides of Sybaris who went to a wedding at Sicyon with his private fleet which included a thousand cooks, fishermen and huntsmen." And Sybarite hedonism at its culmination relaxed sexual inhibitions to such a degree that "their women had complete liberty to do as they pleased and lay alongside the men at their licentious parties, waited on by naked youths and girls." From this, one hardly need ask why there was no known class of Sybarite courtesans. Chorus girls and models there were aplenty, however, and mannequin parades were often a feature of the more formal feasts where the quality of the gowns exhibited may be judged from Aristotle's report that a particular one containing no gold or silver thread was sold for twenty talents (about $15,000).

Outdoor processions were more imposing still. The dashing Sybarite cavalry, "five thousand strong and recruited of young men from the best families," often rode through the city wearing saffron robes over their breastplates. Their proudest stunt was the coordinated prancing of the magnificent matched horses which had been painstakingly trained to dance to the music of flutes, an accomplishment so marvelous, so unsurpassed that no one saw any flaw in its perfection. No one, that is, except Pythagoras, who held sway over his humble, barefoot devotees upon a beautiful cape near the neighboring city of Croton.

Pythagoras' observation might never have become known to the world but for the fact that when Croton at last grew big enough to rival Sybaris toward the end of the sixth century, serious differences arose between them, culminating in war in 511 B.C. Aristotle described the principal engagement of this strife as a charge by the entire Sybarite cavalry upon the smaller Crotonite army. Few of the Sybarites seemed to doubt that they would overwhelm their enemies in the first rush, but the Crotonites were better prepared than they knew. For Crotonite scouts, on Pythagoras' advice, had learned the music of the Sybarite cavalry band. Furthermore, Milon, the Crotonite general, already a legendary hero for having been six times the champion wrestler in the Olympic games and six times in the Pythian ones, was dressed in a lion skin and brandished a great club like Hercules. He inspired supreme confidence in his troops and well-distributed musicians, who calmly awaited the advancing enemy. And just before the clash, Milon gave a sign to his flutists who piped up such a tune that all the Sybarite horses started dancing and prancing, permitting the Crotonite spearmen to close in with deadly effect. Within a few minutes the whole Sybarite army was in flight, fiercely pursued by the Crotonites, who swarmed over the bridges into the city and flooded and destroyed it — so completely that it soon vanished and was forgotten, forgotten so thoroughly, in fact, that modern archeologists, who well know its general location, have scarcely yet found a trace of it.

To what degree this classic victory of the flute may be literally true is at present impossible to tell. But certainly it is close to the tradition of Pindar, who grew up when Pythagoras was an old man and was wont to limn the victories of the spirit in lilting Hellenic phrases — as in his Pythian ode: "Short is the space of time in which the happiness of mortal men groweth up . . . Creatures of a day, what is any one? What is he not? Man is just a dream of a shadow; but when a gleam of sunshine cometh as a gift of heaven a radiant light resteth on men."

Without a doubt, music has long held charms for the human ear, but not until men learned how to count the vibrations of a harp string and compare them with those of reed pipes (as Aristotle and Aristoxenos tried to do) could they begin to understand the true nature of pitch or harmony or melody or even realize that sound is wave motion. But men did eventually analyze pressure and vibrations enough to discover sound waves and, along with sound, they studied waves in general, which turned out to be important not only for comprehending music but, surprisingly, for grasping the very fundamentals of matter itself. This development would probably have astonished Pythagoras — yet it ties in so beautifully with his visionary harmonic order of creation that the most advanced of modern physicists are becoming in effect Pythagoreans again, and it looks as if his celestial music will long be heard in the laboratories of basic research.

The idea of a wave, of course, is very ancient, for not only the sea but all sizable bodies of liquid, and some of solid, reveal surface waves. Fields of grain form obvious waves in the

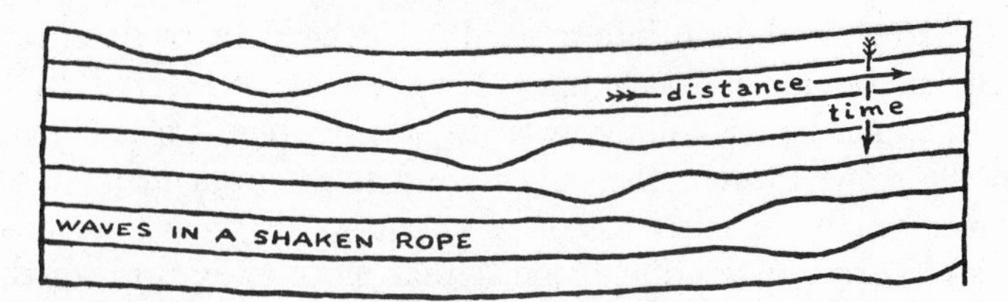

wind, and waves of curvature can be made to flow along a rope — even waves of quavering flesh upon a portly thigh. Yet there is much more to a wave than at first appears. Its depths contain vast mysteries that are still unplumbed by the greatest philosophers and among which must be included no doubt a key to the ubiquitous melodies of life.

What is a wave made of? A child gazing at the ocean would be most likely to assume it to be made of the ocean — in other words: salt water. This impression would be almost unavoidable if the wave were viewed from so far away that its details of motion were lost. But of course, a closer look soon reveals that the ocean wave does not carry water along with it any more than a wave of wheat transports the wheat. The wave but momentarily shapes and uses the water it passes, being itself composed of something less palpable, less constant, more abstract. For if a wave that marches forward a thousand feet can be said to be still the same wave (an arguable point), the water in it cannot possibly be defined as the same water. Like the spindrift and bubbles, the H_2O molecules inevitably drop behind. Furthermore, if you follow any ocean wave as a single identity, you will discover that it fades away entirely in a very short time — to be replaced by another or other waves. In some cases this replacement may appear as a dividing into two waves, like an amoeba giving birth, yet it always occurs somehow sooner or later, being a basic manifestation of the nonmaterial essence of the wave.

So of what, then, is this discarnate monster made? Is a wave really part of the objective world? Yes — it is made of energy — energy in as pure and palpable a form as we can expect to know it anywhere: pure energy in motion, perhaps the best example of ethereal might in the corporeal universe! To illustrate:

when a stone is thrown into a pond, what happens to its energy which, as we have heard, must somehow be conserved? Obviously, much of it is flung away in splash, some carried down with the stone to dent the bottom and some given off as heat. But most of the remainder is dispersed outward through waves — waves of pressure vibration (sound) that move away through the water, air and solid earth faster than the eye can see them, while some also goes into the familiar slow surface waves that live where air and liquid meet, spreading out in the beautiful rings that are too well known to need description.

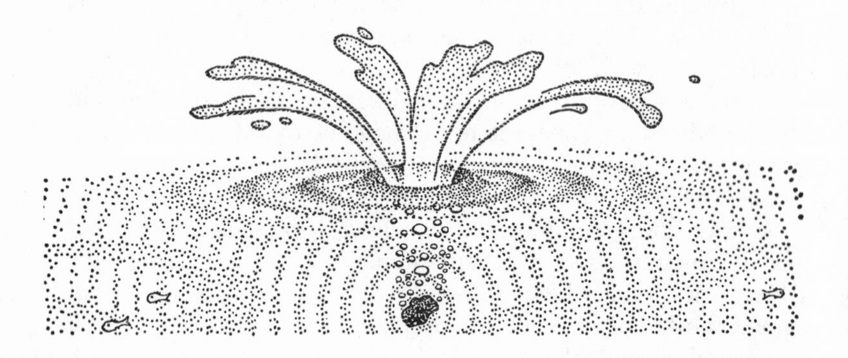

These latter, being among the most observable and best-known carriers of energy, are a subject we cannot afford to overlook in our study of reality. They are the essence of the common ocean waves whose own energy, however, comes mainly from the wind.

The three important dimensions of a wave are (1) amplitude or height from trough to crest; (2) length or distance between one crest and the next in their line of motion; (3) frequency or the number of them passing a given point in a given time which, multiplied by the length, gives the speed.

One would think that virtually everything knowable about ocean waves would have been discovered by the sailors of the earth during their thousands of years of experience upon the sea — but waves are immensely and deceptively complex. Their shapes, for instance, which tend to gravitate from the wind-tossed trochoidal toward the gentler sinusoidal, vary so much in the complicated interplay of forces at sea that hardly any rule applies. Neither wave height nor size are necessarily proportional to wind velocity, which may change abruptly while the waves coast steadfastly on in remarkable verification of the conservation principle. Furthermore, the wind pulls the waves as well as pushing them in the same way it tows sails by lee suction or lifts wings by the vacuum above them. And the wind is a "two-edged sword" that can propel and level a wave in one action, raising up peaks of water often just to thrash them into spray. That is why the record-sized waves, around 110 feet high, rarely seen and almost impossible to measure from a heaving ship, generally appear during brief lulls (as in a hurricane's eye) when the wind is not there to topple them. Such giants also undoubtedly pack the combined energy of all the lesser waves that blended to form them, a phenomenon corresponding to what is termed in acoustics the looping of sympathetic vibration.

In their attempt to understand the waves, mariners naturally have collected a few rules of thumb, such as "a wave's height in feet equals half the wind speed in miles an hour," according to which a 50-mile gale should create 25-foot waves. Yet this is far from always true, because a wave's height increases also with the length of time the wind has been blowing and with the fetch or distance over which the wave has been building up, so that a 50-mile squall on a small lake may yield only five- or ten-foot waves, while an equally strong gale in mid-ocean has been known to produce heights of 35, perhaps 40, feet. The comparatively recent fetch law states, for example, that the fetch is always proportional to the square of amplitude (other factors remaining the same). Thus a doubling of wave height implies a quadrupling of fetch if nothing else changes. In an actual case, of course, all the factors are constantly changing and are so numerous that predicting the shape or dimensions of sea waves is about as tricky as forecasting weather — and for a similar reason.

More basic to wave nature than its relationship to the wind is a certain interdimensional mechanism within the wave itself.

Have you ever considered the strange evolution of steep, white-capped combers into long, gentle swells? Sea waves as a collective phenomenon eventually pass beyond a storm area, and their height diminishes while their length and speed increase, the conserved energy apparently dispersing itself until each wave has flattened gradually into a swell. Deceptively mild in height and more or less disguised with overlying waves or ripples, the mature swell is nevertheless long, broad and fast, for it still possesses (in spreading form) most of the total power that went into its creation. This basic fact, which will be news to some sailors, can be expressed more specifically by saying that as the frequency decreases or period (time elapsed in one swell's passing) increases, the latter being the reciprocal of the former, the speed correspondingly increases and the wave (swell) length increases faster still. Or in terse, quantitative terms: a wave's velocity is directly proportional to its period, and its length varies as the square of either. A general law for all surface-water waves larger than a small ripple, this can be checked at sea through one of its practical formulas: a wave's speed in knots (nautical miles per hour) equals three times its period in seconds, and its length in feet is 5⅛ times the square of the period. Thus if a patch of foam takes 11.4 seconds on the average to move from one watery summit to the next, the swells should be running at 34 knots and measure 666 feet apart.

The obvious exception to this hydrodynamic rule is where the water becomes shallow, as near a beach, for shallowness is naturally a drag on wave motion, since it imposes a limit on its subsurface components. Under any deep-water wave, all the individual molecules of liquid have been found to move in circular orbits like planets, swinging up and forward with the crests, down and backward with the troughs. While the radii of these orbits decrease rapidly with depth so that, half-a-wave-length down, the motion is only 1/23 of what it is at the surface, still the movement extends in some slight degree all the way to the bottom of the "incompressible" sea, enabling any wave to "feel" its approach to the shore. The "feeling," of course, comes through the molecular orbits, which, having no room for a vertical component at the ocean floor, must reconcile themselves even at the surface with varying elliptical shapes that revolve flatter and flatter with depth until they are nothing but back-and-forth shuttle motions on the bottom. It is this orbital drag of diminish-

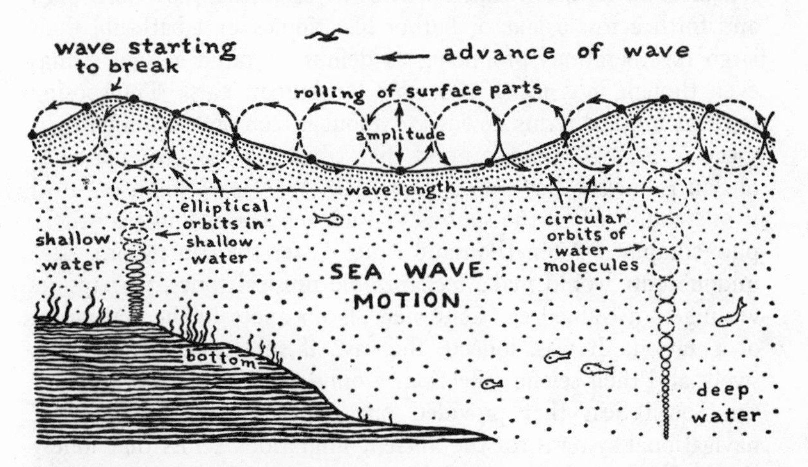

ing space, in fact, that breaks (and brakes) the arriving wave at the beach by holding back its roots — the wave velocity varying, according to Lagrange's law, as the square root of the sea depth — until its crest leans too far forward, curls and thins, finally toppling as surf when the depth no longer exceeds the wave height. And some similar thing undoubtedly happens to sound waves inside musical instruments — indeed, to any kind of waves anywhere that are constrained by the pinch of space.

If the vertical cramping of shallowness thus imperiously cuts down the amplitude of a wave, the horizontal convergence of shore lines has almost as dramatic an effect through the waves' collective reflections, particularly when the shore is steep enough to rebound the waves without much loss in amplitude. Such

SEICHE REFLECTIONS IN A HARBOR

reflected waves, often called seiches by scientists, may slosh back and forth across a lake or harbor like ripples in a bathtub, their largo reverberations producing as definite a pitch as viol strings even though it is much too slow for human ears. Transposing familiar musical terms downward about fifteen octaves, one might speak of a small soprano pond that vibrates every 28 seconds or of San Francisco Bay's deep bass seiche rate of 43 minutes. And if one could record the sounds of such reflections and play them phonographically ten thousand times faster than in nature, they undoubtedly would make recognizable musical tones, perhaps revealing to a trained ear some new significances in the waterways of a nation. It was, indeed, the vast, if subtle, patterns of the swells and their seiche reflections around the islands and atolls of the South Seas that provided one of the most important of navigational systems for the ancient migrations across that lonely third of our world — a kind of tuning in upon the ocean's harmony that has scarcely begun to be understood.

Great as these wind swells can grow on the Pacific, however — and some are known to be half a mile long after a 7,000-mile fetch — they are as nothing compared to the occasional earthquake swells or tsunamis (often misleadingly called tidal waves) which have been estimated up to six hundred miles in length. These seldom-seen phenomena are really a kind of heart beat of the aging earth, for they are created by the periodic jerks or shifts of its crust, perhaps sometimes by volcanic outbursts or rock avalanches deep under the sea. One such seismic sea wave struck the Bay of Bengal without warning on October 7, 1737, and within a few minutes had swept over hundreds of coastal villages and towns, wrecking nearly 20,000 boats and killing some 300,000 people in what was probably the most destructive deluge since the time of Noah.

You may remember the more usual tsunami that hit Hawaii on April 1, 1946, five hours after an earthquake had domed up the sea near Unimak Island in the Aleutians 2,300 miles to the north. As the resulting swells were 90 miles from crest to crest, they were far too big to be noticed at sea, and ships rose gradually over them totally unaware of the deadly power racing southward at some 450 miles an hour to lash its fury upon drowsy Hilo with a series of waves more than twenty feet high and several minutes apart that smashed houses, bridges and railways, accounting, all told, for $25,000,000 in property damage and 159 lives.

At first thought, the approach of a swell 90 miles long and only a few feet high might seem too gradual for such violence — and it is true that tsunamis really are comparatively gentle upon a steep shore. But where they meet a broad coastal shelf or a long, shallow inlet, the sea's depth becoming much less than half the waves' length, Lagrange's law inevitably has a drastic effect on their shape. Their advancing lower slopes are retarded, while their crests continue full speed, overtaking and overreaching from behind, and so the steepening wave front may eventually curl upward and forward into a towering wall to break and crash upon a beach town with catastrophic force. This is presumably the reason why the 1946 tsunami attained its greatest heights at flat Pololu Valley, the third and fourth waves reaching an estimated 55 feet above normal sea level, while the eerie ebbs between them withdrew almost as much below. And the famous Krakatoa swells of 1883 were credited with heights well above a hundred feet, combined with a length that grew to hundreds of miles and a speed no less than that of sound in air which, when augmented by air blast, produced in some parts of the Pacific a particularly strong impression by arriving at the same time as the distant thunder of exploding Krakatoa itself.

The ghostlike tendency of waves or swells to fade away individually while being replaced by others is, of course, what induced scientists to start dealing with them collectively. Hence the designation of the "group wave." An example of a group wave might be the band of breakers moving in upon a beach where, although single waves keep entering and leaving it, the group as a whole remains nearly the same in size, position and average wave velocity.

Another kind of group wave is the rather prim pattern that forms around a moving ship in calm waters, an enclosing series of billows of almost crystalline rigidity that I used to think must be as much a function of the H_2O molecule as is the hexagonal design of snow. I thought so because I had read that the bow wave has always been measured to slant outward at an angle of 19°28′ from a ship's keel line, regardless of the vessel's speed or size. But when I discovered that the bow wave maintains that exact same 19°28′, not only when the vessel is moving through deep water but also when it is traveling in comparable bodies of oil, glycerin, mercury or any other liquid on earth, I realized that the angle cannot be attributable to water alone but rather must be a function of the gravitational force of the whole earth, as the now well-established bow-wave equation ($Vw = \sqrt{g\lambda/2\pi}$) shows. From which one may deduce that space travelers in future centuries, if they find themselves voyaging on the methane or ammonia seas of some planet they don't remember the name of, may simply measure their bow-wave angle and look up the name in a pocket planetary guide (every planet having its own mass and proportionate angle), thus identifying it uniquely as a world.

Another curious characteristic of the bow wave, no matter in what material world it may be, is that, taken in the singular, it becomes concave forward due to acceleration, yet, as its faint

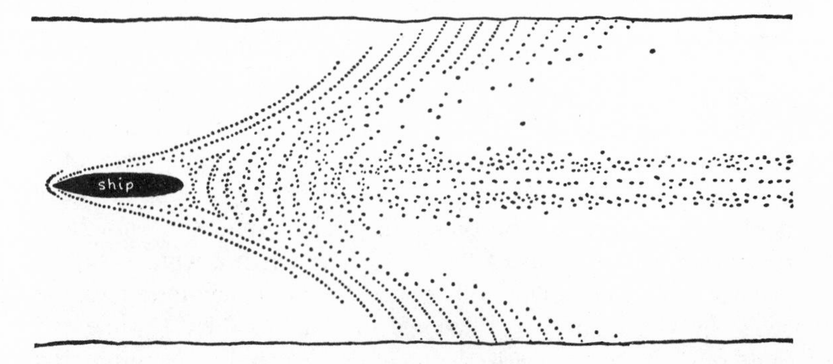

ends keep speeding up and individually fading out, the steady grouping of their maximum phase proceeds in the plural at the same speed as the ship. Likewise do stern waves accelerate in concave-backward form without apparent heed to their equally stable group confederation whose individual members come and go like metabolizing cells through a body.

But the life of every wave is relative not only to any group of which it is a part but also to the angle of its motion. Hence, if it is waning in its march perpendicular to its own axis, it may at the same time be waxing if followed parallel to the same axis — which explains how a slightly oblique comber can break and plunge to its death at your feet on a beach while its continuation in another dimension (sideways) may roll on and on down the length of the coast as successive parts of it reach land in a kind of itinerant lateral immortality.

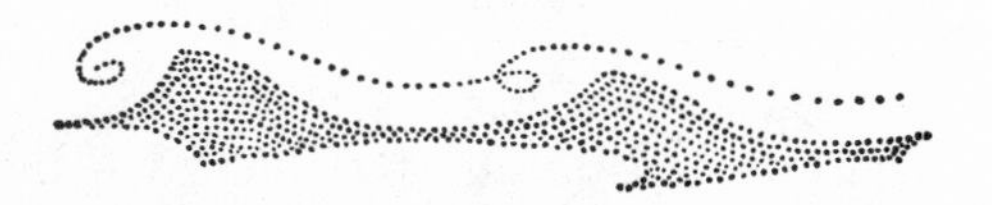

What is true of ocean waves is, of course, generally true of waves upon other surfaces and in other mediums, such as the unseen submarine undulations where fresh river water meets the sea or the river-bottom ripples of sand which commonly parade up or downstream at speeds as high as an inch a second depending on the complex interplay of current and granular dimensions.

Have you ever watched the serried terraces on the up-wind slopes of snowdrifts give way to the abrasive pressure of a gust only to regain more than their lost substance from a dense flurry of flakes in the succeeding lull? Or have you considered the dynamics of the sand dunes which flow so slowly but inexorably across the world's deserts and beach coasts under their own kind of wave law? The rhythm of the dunes has been found to come from the flow of the billions of sand grains which the wind naturally sorts by size and weight, the coarser grains being tossed to the crests by the main jets (blowing slightly spiralwise) while the finer sand floats off with the gentler side eddies, forming the familiar crescent-shaped dunes known as barchans in Africa and Asia or as medaños in Peru. Along the Nile south of Cairo, small barchans march 30 feet apart at an average pace of about a foot a day, as similar in interrelationship as swells at sea or cloud ranks in a mackerel sky, though lacking the momentum that probably gives liquid waves their acceleration. The dunes' height, limited by the wind, remains one eighteenth their wave length,

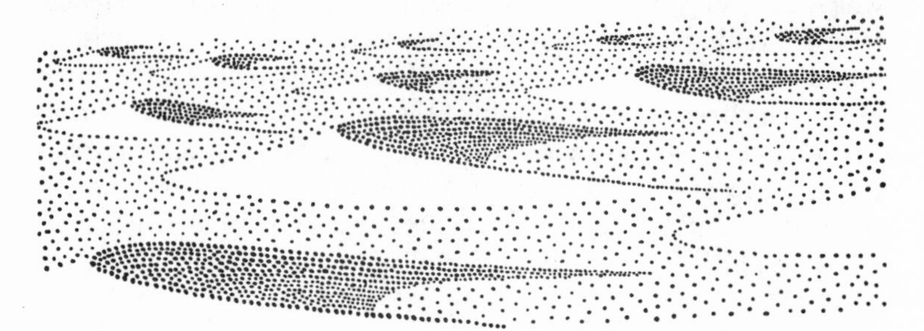

while they are laced all over like sea billows with tiny moving ripples of height one fourteenth their length — the whole a wondrous creeping organism sculped and driven in nearly every detail by the action of air upon flying grains of sand, a corporeal entity whose balance, feedback control and metabolism may be one of nature's most persuasive demonstrations of a prime ingredient of life.

If one were really to analyze the action of any of the kinds of waves mentioned, one would have to separate the complex movements of sand, water, snow, wheat or other substance into

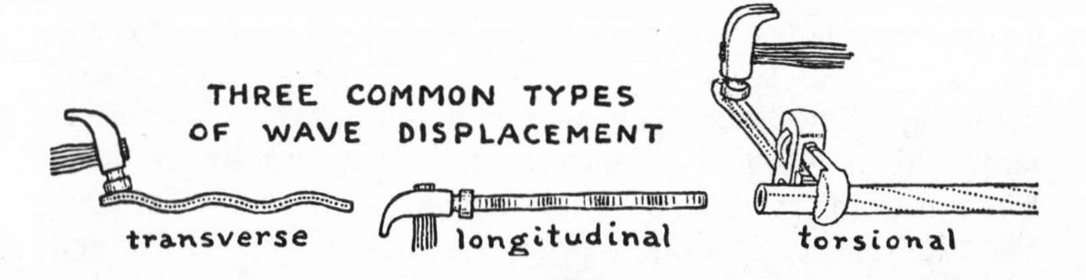

their simpler components known to physics as *transverse, longitudinal* and *torsional* displacements. The crosswise waving of a flag in the wind is largely a *transverse* displacement of cloth, the pulse of steam compression sent lengthwise through a locomotive cylinder by a piston moving in the same direction displaces the vapor *longitudinally,* a twist of a screw driver handle imparts a wave of *torsional* displacement to the screw at the other end. There is no end, however, to combinations of such motions among waves of many other sorts, from magma rhythms in the earth's core that move radially outward in expanding spheres of compression to others that oscillate bodies from stars to raindrops prolate-oblate-prolatewise (alternating football-doorknob-football shapes) and even to the eerie electromagnetic melodies of the human brain.

o o 0 o o o o o o o 0 o o o o o o

Going beyond the physical altogether, it might be appropriate here to mention even such an abstraction as the "group velocity" of waves, a mathematical concept used in theorizing

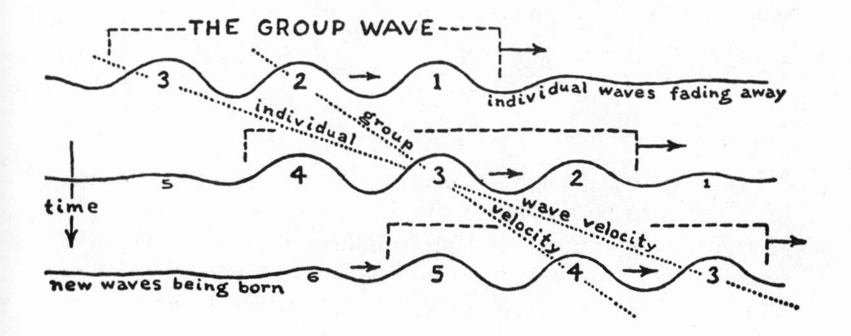

about elementary particles. To visualize this, you could imagine ocean swells of a particular wave length, period and velocity occurring progressively outward from a storm area at a constant rate. This rate is their group velocity. Group velocity is therefore something more remote from tangibility than waves or even than the velocity of waves, for it is the velocity of a velocity and one that must continuously advance from waves to waves. Yet in deep water, surprisingly, the group velocity has been found to remain constantly equal to half the wave velocity, a discovery as simple and beautiful in its way (if you follow me) as the so-called Pythagorean theorem. And, in the case of electromagnetic waves in a plasma, the group velocity equals the square of the speed of light divided by the individual wave (or phase) velocity—which likewise has a meaning that (to a mathematician) seems to sing.

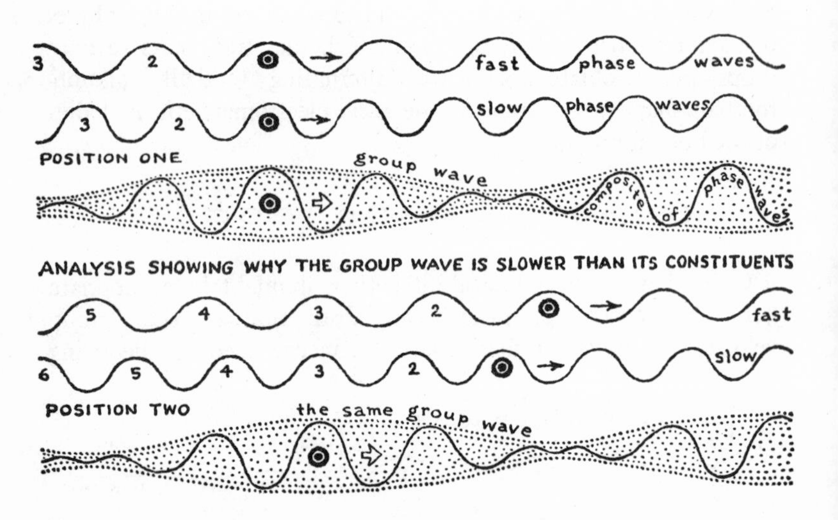

I could go on about waves of waves of waves, of course—and there seems to be no end to the little-known ripples and swells of radiation, gravitation and other influences we are constantly recording up here in space. But the kind of waves easiest to understand harmonically are the longitudinal variety, which include the rapid waves that our ears are almost continuously bringing to our attention as sound. So let us consider the nature of these invisible impulses that mystified the world before modern times — that caused the ancients, according to Leonardo, to wonder whether a bell might be entirely consumed from too much ringing.

Was it not the same Leonardo, incidentally, who sized up sound as a kind of "percussion of the air," whatever that meant? And Newton two hundred years later, who deduced sound's transmission as alternate waves of compression and rarefaction in any material?

Indeed, we now know that sound vibrates all molecules longitudinally: directly toward as well as away from the sound's source. In fact, it is these waves of molecular motion that *are* the sound. And as the natural vibration rate of materials increases with temperature and density, sound as naturally travels faster in warmer or more solid mediums: 4 miles a second along a steel railroad track, 1 mile a second through water and about ⅕ mile per second in the air — depending on how hot it is. Sound is thus only a milder form of the shock waves made by explosion blasts and moves at the same rate. In a still more general sense, it is one of numerous ways in which concentrated energy diffuses itself about the world, as we noted in the case of the stone thrown into the pond.

To be heard and understood, of course, a sound normally needs to have a duration of at least some plural number of waves — waves do not remain in the single state anyway — and such repeated waves add up to a vibration of whatever substance they are in. This vibration, moreover, is on a very different scale from the elementary, submicroscopic molecular vibrations (see page 270); it is a majestic and sophisticated motion by comparison, being a manifestation of a passing energy pattern in the outer macrocosmic world.

A classic example of such a vibration is the twang of a bow string, a sound at least seventy thousand-odd years old on Earth (as evidenced in cave paintings) and very probably ancient enough to have been the beginning of all stringed music. It is again very different from, and newer than, say, the noise of a tree crashing in the forest or of a hoofbeat upon gravel, for it is a regular tone of one frequency or pitch rather than a medley of many irregular, shifting vibrations. In short, it is a single musical note of the simplest kind which, when first heard, must have seemed like purest magic.

Just to be sure its basic nature is clearly understood, may I explain that waving one's finger once in the air will send a slight shock wave of pressure out in all directions at a fifth of a mile a second. But this is neither sound nor music and it makes no sensation in the ear, which cannot detect a single wave. Even shaking the finger back and forth as fast as is humanly possible will not break the silence. But if one could wave it fifty times a second, a faint deep humming note would begin to permeate the air, something like that made by a hummingbird's wings, rising in pitch with any increase in the waving rate.

That, actually, is what pitch is — and all that it is. Pitch is sound-wave frequency. It is to sound what temperature is to molecules or what color is to light (see page 451). Run a stick along a picket fence, slowly at first, then faster and faster, and you will notice the same thing: the accelerating taps of the stick will blend into a rising tattoo that becomes a musical tone. Pour liquid out of a jug and you can hear its gurgling melody descending the tonal scale as the incoming air takes longer and longer to reflect its pressure waves back and forth across the expanding cavity within, thus lowering their frequency — the principle of the bagpipes, trombone, oboe and other wind instruments. And the humming string of the bow or the harp is no different in essence for, waving back and forth, it hits the air each time, sending out regular impulses of pressure that collectively amount to a musical note of steady pitch.

No doubt, the string was particularly inspiring to Pythagoras because it is so easy to observe in action, so simple and amenable to handling and experimenting, and perhaps also because it is so close to the border (if there is one) between the concrete and the abstract. The latter consideration, which Pythagoras seems to have felt intuitively, has since been made much clearer in a modern laboratory. Fixing a needle to a vibrating tuning fork (similar in principle to a string) so that it continuously scratches a wavy line on a long piece of smoked glass passing steadily under it, scientists have caught a permanent graphic record of every motion the fork made — actually a sort of basic phonograph record — revealing the beautiful sine curve of "simple harmonic" sound. Not only is this curve mathematically pure in its simplicity but it has an amazing regularity, repeating itself hundreds of times a second (depending on the pitch), with each wave maintaining an identical length and the same characteristic sine shape even while the amplitude gradually diminishes as the fork's energy runs down.

By such experiments it was discovered that the note called "middle C" on the piano, for example, consists of about 262 complete pressure waves a second, that any sort of regular vibrations at the rate of 262 per second, whether they come from a vacuum cleaner's motor or a bumblebee's wings, will play middle C — confirmation, if any were still needed, that middle C, like every note, is not made of any particular kind of sound but is purely a *rate* of impulses, a *number* — in other words, an *abstraction*.

Pointing up the incorporeal side of harmonic motion still further, the modern general theory of vibrations shows that all materials oscillate in somewhat the same basic way that celestial bodies swing around their orbits: not necessarily describing complete ellipses, however, but commonly alternating their directions in some partial orbit or folded ellipse like a pendulum, or perhaps in the rocking motion of a librating moon. This happens, generally speaking, because every material possesses at least one position in which it can theoretically remain at rest — otherwise it would be a "perpetual-motion machine." And if or when it is in such a position of equilibrium, all the forces influencing each particle of it must (by definition) be exactly balanced. But of course, any little disturbance such as a whisper or a ray of light arriving from outside will move the structure at least a little out of equilibrium into some new and unstable position in which each particle of it will feel a "restoring force" trying to return it to balance again. By the time this force succeeds in getting it back to equilibrium, however, the structure is virtually certain to have picked up such momentum that it will overshoot the position and coast beyond it until another (and opposite) restoring force reverses it once more. Thus is an oscillation set up: the natural harmonic motion called vibration when its path is short but which, on the larger scale, keeps clocks on time, holds stars to their courses and is inextricably involved in the origins of waves and music.

Kepler would have been particularly delighted with the implications of this now well-proved theory, and it seems a pity his enthusiastic letters to Galileo (beginning in 1597) did not elicit more response or some serious collaboration. For Galileo's experiments with oscillating pendulums might have given Kepler just the appreciation of the ellipse he needed to solve the orbit of Mars a long, hard decade sooner. Galileo must have known before 1600 that while a single blow upon a motionless pendulum will start it swinging back and forth over a straight line marked on the ground beneath it, this being the simple harmonic motion of the plucked string, any additional blow or blows at different angles (each also imparting simple harmonic motion) will send it into an ellipse, this being the compound of all the simple wave motions. And if an ellipse could thus be the natural path of a suspended object swinging under complex influences on Earth, why in heaven shouldn't a similar law hold for the paths of the celestial spheres suspended in complex space? Indeed, Kepler might literally have transposed the actual "pitch" of the six known planets to his famous oracular music, for Galileo had firmly established that the pendulum's frequency (pitch) is rigidly dependent upon its length, corresponding in the sky to an orbit's invisible radius.

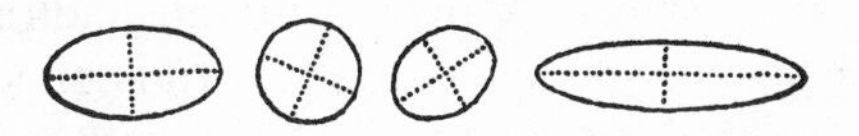

Understandably, Pythagoras, standing seven times farther back in history, could hardly be presumed to have had more than a vague inkling of the pitch of his wandering stars, yet his observations on the natural intervals of strings were mathematically exact so far as they went — and founded our modern science of harmonics. Without counting vibration frequencies, he could naturally hear pitch and see the nodes that accompanied the elementary overtones. He could see that the pitch is raised to the sympathetic octave or "eighth note" by momentarily holding a finger on the exact mid-point of the string while plucking the center of either half, leaving the whole then vibrating freely in two equal, seesawing parts separated by a node of immobility at the fulcrum. Or raised to the further interval of a twelfth by making the string hum in three equal parts with two nodes.

SIMPLE HARMONICS OF THE PLUCKED STRING

fundamental octave twelfth double octave

Or to two full octaves with four parts and three nodes. Or to any of several other congenial intervals (fifth, fourth, etc.) by shortening the string in simple or aliquot proportions.

This is not to say that Pythagoras knew that when the string is divided in two, by a node or a fret, it naturally vibrates twice as fast — or, divided in three, three times as fast. Nor is there any evidence that he clearly realized the effect of the string's weight or tension on its pitch. Still less could he possibly have understood that the string's outstanding musical quality derives from its free vibration pattern matching its natural harmonics so perfectly (in frequency) that every time its fundamental tone is set going, its harmonics respond as well. More probably, Pythagoras was conscious only of the mysterious relationships between pitch and length of vibrating string segments, and of the mathematical implications of combining two or more notes into a harmonious chord which somehow in the addition — or was it multiplication? — became more beautiful and perhaps more divine.

If it is still hard to accept the obscurity of such musical recesses in the ancient mind, there remains, at least, the sober fact that twenty-one and a half centuries were to pass before the French mathematician Marin Mersenne would at last discover the true basic law of pitch for strings in 1636, stating that the frequency of a string's vibrations varies directly with the square root of its tension but inversely with its length and the square root of its weight. This remarkable three-way summary, formulated just after Kepler's death and during the old age of Galileo, came none too soon for teaching the clavichord pioneers how to avoid having to make their bass strings a hundred times as long as their treble ones: by greatly increasing both the weight of the basses and the tension of the trebles.

Then soon afterward, in the days of Newton and Hooke, while a young man named Johann Sebastian Bach was still experimenting with new keyboards in Thuringia, scientists began to realize that if pitch could be a manifestation of the length of sound waves, loudness or volume of tone must logically be determined by the same waves' amplitude. And also that it would be reasonable to expect that the detail of the wave curve, which could be very complex, might somehow account for its timbre or sound quality.

Of course, it was by no means yet obvious that the hard-to-visualize sound wave was normally a composite thing, blended of many simpler waves crossing, passing and overlapping each other like ripples upon waves upon swells at sea. For the analysis of wave shapes would inevitably be a long and painstaking job requiring the utmost of many dedicated mathematicians — a work that continues to this day with increasingly elaborate equipment but without foreseeable end.

Probably one of the first experiments in compound-wave analysis consisted of hitting one prong end of a vibrating tuning fork with a hammer. This makes a "clang tone" about two and a half octaves above the fundamental note of the fork, because of new and very rapid vibrations that ride upon the fundamental ones like ripples on an ocean wave. But the clang tone alone actually is composed of just as simple a harmonic motion as the fundamental tone, though aligned in a different direction and with a frequency some six and a quarter times higher. It is only the combination of the two simple motions (as shown in the illustration) that becomes complex or double motion. The addition

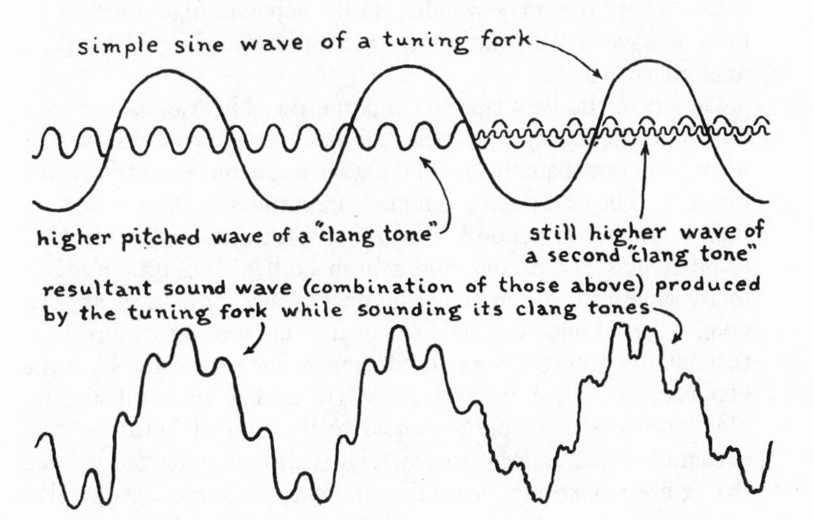

of another clang tone, an octave and a half higher than the first, may then increase the complexity to triple motion — something that happens very commonly. In fact, it is almost impossible to start a tuning fork humming so purely that no admixture of such overtones can be heard.

In the case of a bell — an instrument usually designed for volume and beauty rather than purity — the overtones are actually considered vital to the character of the ring. For it is undeniable that the simple purity of tone sometimes obtainable with a "perfect" tuning fork grows dull in the ear, being in the long run no more of an improvement over a random noise than the architecture of the barracks is an improvement over that of the junk yard. Which goes a good way toward explaining why the world's accepted music has come to be a blend of tones, a deliberate confluence of pitches, volumes and timbres that overlap in every imaginable way.

But what, you may wonder, really happens in a mixture of musical waves? Why are some combinations so delightful, others so excruciating?

Let's take the very simplest mixture possible: two waves with the same frequency and phase, which means two waves that not only endure equally long but always begin and end at the same instants. They may have different amplitudes or shapes but, as their frequency or period is equal, they must certainly be identical in pitch like, say, a gong and a horn both hitting F♯. Furthermore, since they are in the same phase, they must meet at least upon every node and while moving in the same direction, thereby lining up crest over crest, trough under trough. Thus the two waves will reinforce each other, giving their resultant or combination wave an amplitude equal to the sum of both constituent amplitudes. Indeed, so powerful is this compounded volume that it exerts a strong retroactive influence back upon the weaker

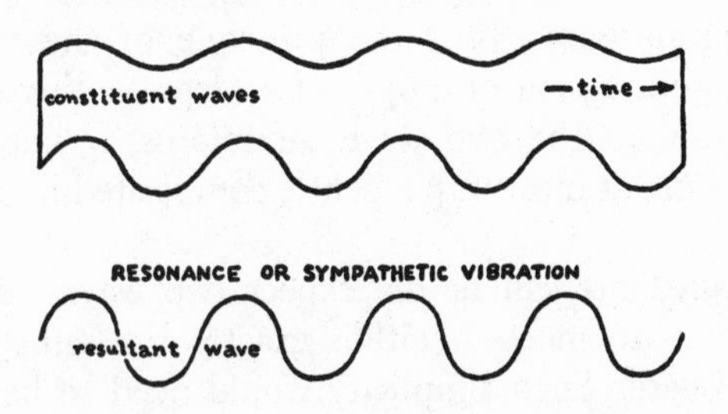

of its constituents and even upon completely external or inactive wave capacities within its range, often rousing them from silent potentiality to loudly energized motion. This is the phenomenon called sympathetic vibration or resonance, and it is what makes an unstruck tuning fork hum in sympathy with another set to vibrating in the same natural key. It also explains the gaps in Saturn's rings (see page 97) and the exaggerated bounciness of certain washboard roads where the natural spring frequency of your car happens to match the wave frequency of the passing undulations.

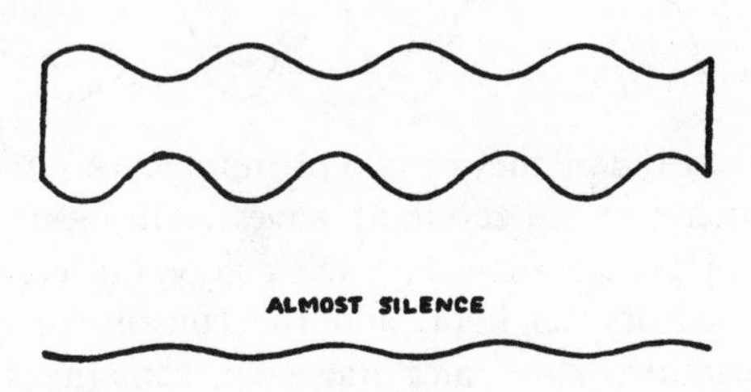

The next wave mixture in order of simplicity is two waves of the same frequency (pitch) but opposite phase, similar to the previous combination except that crest will come over trough and trough under crest. This makes the constituent vibrations pull in opposite directions, partially neutralizing each other, since the resultant amplitude is now the difference (not the sum) of their single amplitudes. It is the principle of the silencer, for obviously the subtraction of amplitudes (representing energy and loudness) combines the two wave mountains, if not into complete cancellation, at most into a gentle composite hill of comparative quietude.

But in nature, one can hardly expect two waves of the same pitch to happen to meet in either exactly the same or exactly the opposite phase. Such simplicity would need to be contrived, as it is rather an exception to the normal independence of wave systems. In ordinary random life, the crests of one vibration will come over neither the crests nor the troughs of another vibration but somewhere in between, creating a resultant (see below) that is, however, in the case of equal frequencies, still simple harmonic motion and still obtained by adding or subtracting amplitudes at each point along its length.

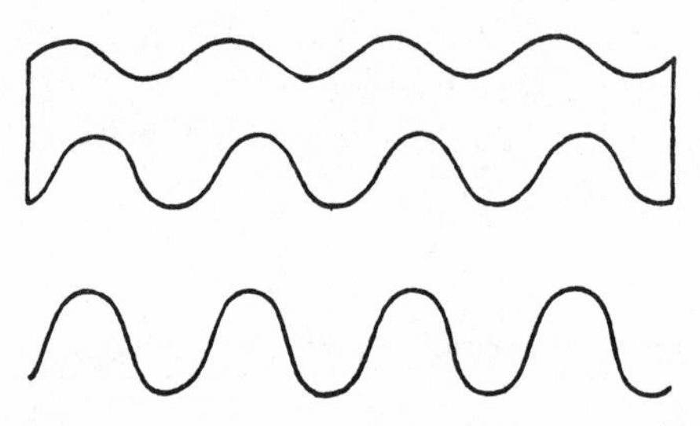

But it is when the pitches of two or more wave systems are different that their blended resultant waves really begin to take on harmonic complexities, for then each compound crest or trough ceases to be like its neighbor, and the constituent crests come sometimes together, sometimes staggered, sometimes nearly opposite. If two waves of equal amplitude, for instance, differ in pitch by only two or three vibrations a second, their phase relationship when they are combined must change constantly (see top of next page), repeating itself likewise two or three times a second, a repetition that puts a kind of throb, called a beat, into the resultant tone. Such beats are really audible group waves, and when they are created by two nearly equal frequencies, they can readily be heard up to a rate of six or seven a

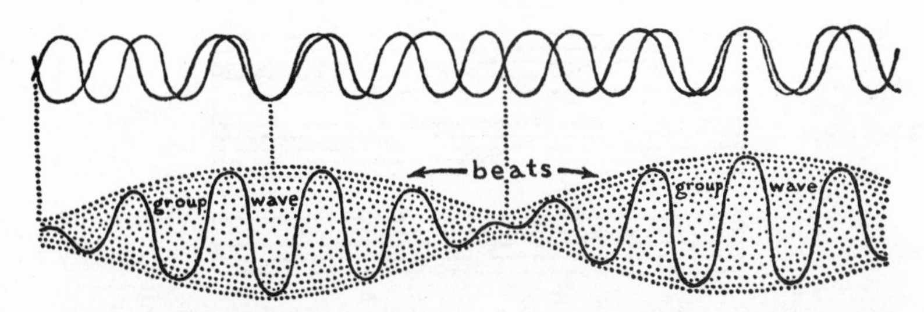

second. Piano tuners commonly listen for them between the paired or trebled strings of supposedly equal pitch. Beyond seven a second, however, the beat frequency merges into sensations of dissonance or consonance, depending on the vibration ratio of the combined notes. The unpleasant clash of two pitches that differ by something between about 5 and 25 percent (in frequency number) is usually considered the main belt of dissonance, beyond which begin the consonant intervals: thirds, fourths, fifths . . . (named for the frequency spans to the third, fourth, fifth . . . notes of the scale), followed by the integral harmonics: the octave, the twelfth, the double octave . . . and so on, with sundry dissonant zones interspersed among them.

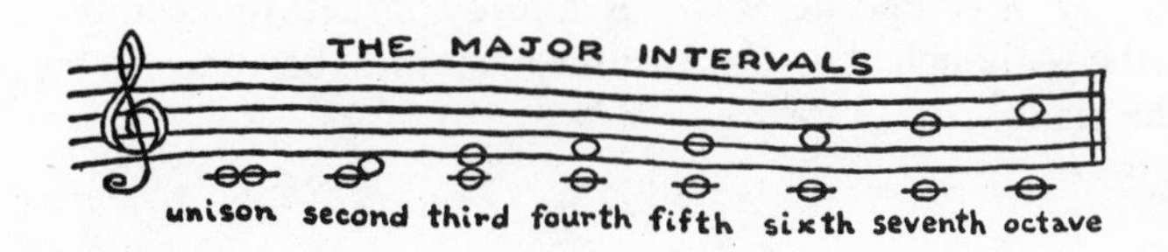

It is these consonant intervals that were measured precisely by Pythagoras and that are exceptional among combined notes of different pitch for their simplicty of ratio: the octave or 1:2 ratio being a combination of a low note whose vibrations might be expressed as "bang . . . bang . . . bang . . . bang . . ." and a high note of double the frequency, "bang bang bang bang bang bang bang bang," which, put together, come out—

bang bang
bang bang bang
bang bang bang
bang bang or bang BANG bang BANG bang BANG bang. Played fast enough to make the in-

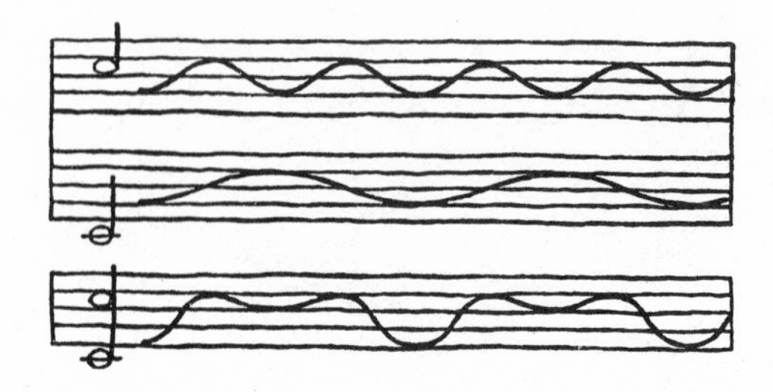

dividual vibrations blend into a continuous tone, it sounds like an octave — as indeed it should. For it *is* an octave.

Actually, a real compound tone is not *quite* so simple as this, for the changing phases subtract as well as add. And in the 2:3 ratio of the interval called a fifth, the combined "ba-ang — ba-ang" (2 syllables) and "bang bang bang" (3 syllables) come out something like "BANGebang ma-bang MANG, BANGebang ma-bang MANG, BANGebang ma-bang MANG . . ." (6 syllables), which, at speed, turns out to be the congenial do-sol (C – G) dyad. Then, adding only a middle note, we have the handsome triad of C major, the do-mi-sol (C – E – G) chord, which, as a composite wave, is already rather too complex to describe — though it is the easy first chord a beginner is taught on the piano.

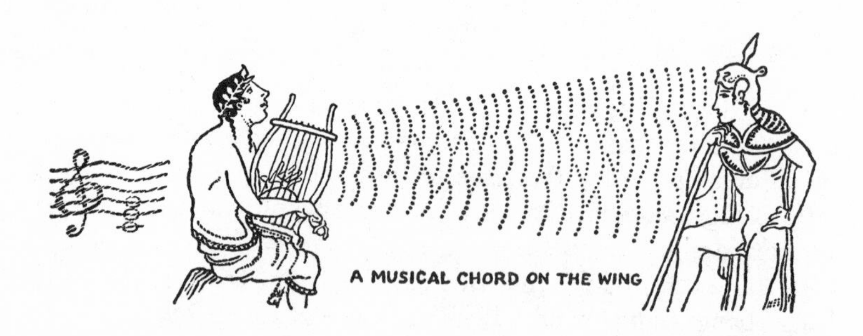

A MUSICAL CHORD ON THE WING

Beyond here, it would seem better to resort to the almost mystic generalizations of Fourier's theorem of harmonic analysis than to try to itemize any more of the vagaries of the musical wave curve, composed of its unending combinations of intervals and accordances, consonances and dissonances, chords and discords. For Jean B. J. Fourier, the French mathematician, discovered early in the nineteenth century that not only can an infinite number of resultant waves be created by mixing these ingredients — just like putting chemicals into test tubes — but, amazingly, any wave curve, no matter how complicated, can be sorted out again into its constituent parts. At first blink, this seems about as feasible as unscrambling eggs, yet there really is an easy technique for analyzing any wave into nothing but simple harmonic curves, and Niels Bohr actually used it in working out his "correspondence principle" (see pages 342–43). I will not take space to pursue the method here, but I must just say that a little of its surprising capacity for revealing the ultimate nature of things can be surmised from the fact that decomposition of waves and curves, like disintegration of bodies, turns out to be possible in any of an infinite number of ways — just as there is no limit to the ways you can wreck a ship or a house or tear up a piece of paper or blow smoke into the sky. And if you are tempted to reject the compositeness of a wave as something mathematical but unreal, just remember that waves themselves, presumably without thinking, combine and sort their own shapes to perfection — passing straight through each other like geometric monsters to emerge completely restored on the other side (see illustration on next page). And this is true of waves ranging all the way from the ponderous surges of galactic momentum to the incredible negative ripples of intramolecular magnetism.

If natural wave behavior is sometimes hard to swallow, however, the artificial wave combinations tried out by the early pioneers of musical science can be about equally unexpected. The very oldest-known scale in the world evidently progressed downward in pitch instead of up and may be described approximately by the note sequence E – C – B′ –B – A –

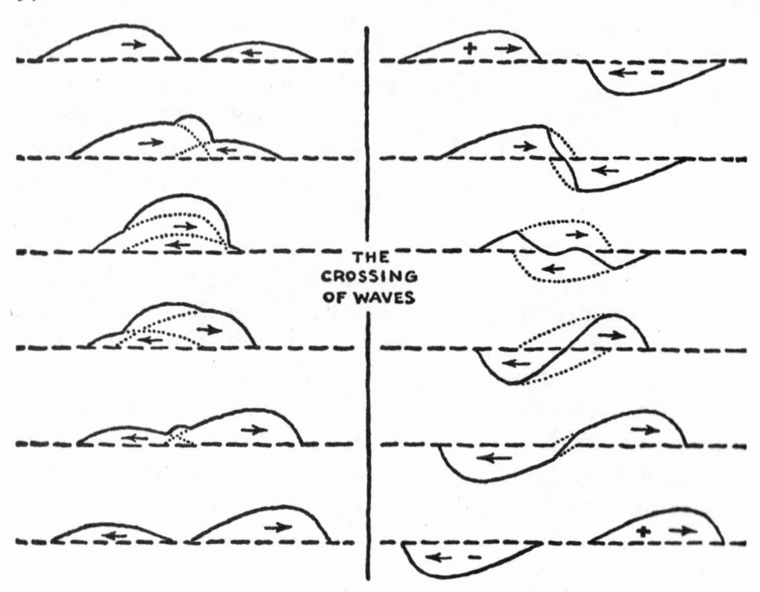

F – E′ – E, in which B′ and E′ represent tones between sharp ♯ and natural ♮. Curiously, the slowly vibrating bass notes in that dim era were reputedly considered of "high" pitch, while young girls sang in what were called the "low" registers. And on just such weird quarter-tone modes and half-, full- and double-tone ones were hung the earliest plain songs of temple and church, those haunting chants of which some are said to have been handed down uncorrupted from the psalms of David or the Song of Solomon. Of such may have sprung Aristophanes' beautiful chant of the Initiated, his lays of Lysistrata. Of such the wild piping of the Corybantic priest — perhaps even the primeval paeans of Sargon.

Meantime, the Hindus promoted their unbridled scale of 22 alternative notes of variable sequence and the Arabs various modes in quarter tones, while other peoples of the earth tried third-, sixth-, eighth-, twelfth- and even sixteenth-tone music. Yet such difficult tonal parings never could become generally appreciated, and even the relatively simple third tones remained unrecognizable, if not physically painful, to occidental ears.

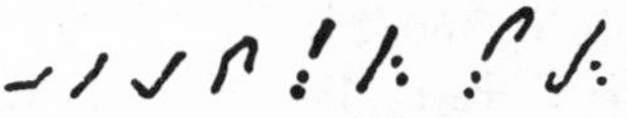

It was Aristoxenos of Tarentum, the great theorist of music in the late fourth century B.C., who seems to have done the most to develop the Pythagorean intervals of the so-called Lydian mode into a complete diatonic scale. In his one surviving book, *Elements of Harmony*, he defined the pitch difference between the fourth and fifth (intervals) as the unit tone of his scale. Then, needing a subunit, he divided the tone arithmetically into two semitones, and rather arbitrarily placed five semitones in the fourth, seven in the fifth, and twelve in the octave. The fact that these semitones did not fit into their places exactly but only approximately, like planet orbits in their "ideal" spheres, must have disillusioned Aristoxenos, leaving him with what the Greeks called a *leimma* or residue — but it also led him to an interesting (if inconclusive) "calculus by logarithms" that made his work one of the masterpieces of Hellenic thought and established music as a major Greek science alongside arithmetic, geometry and astronomy. Indeed, these four constituted the *quadrivium* of higher classical learning until the end of medieval times.

In A.D. 384, many of the ancient "tones," including Aristoxenos' and some variations by Ptolemy, were reviewed and set in order by Saint Ambrose to be later developed in the Gregorian tradition, and by the Visigoths in Spain and the Byzantines in the East. And gradually, among others, there evolved the five-note melodies known as pentatonic modes, like "Auld Lang Syne" and the Japanese national anthem, both of which were built around the interval of the downward fourth (as from C to G) and can be played entirely on the five black keys of a piano.

And by the eighteenth century, the diatonic scale had pretty well crystallized into its modern form — its notes taking the numerical relationship C 4 – D 4½ – E 5 – F 5⅓ – G 6 – A 6⅔ – B 7½ – C′ 8. Thus C and C′ make a perfect octave (4:8 = 1:2), C and G a perfect fifth (4:6 = 2:3), C and F a perfect fourth (4:5⅓ = 3:4) and so on. And the appealing seventh chord C – E – G – – B♭ – C′ was discovered to derive its magic from the simple exactitude of its proportions 4:5:6:7:8! Just why this strangely insistent musical "call" should find its most satisfying harmonic "answer" in a chord exactly a fourth higher, F – A – C′ – F′, is still not well understood. I think it will undoubtedly be explained eventually by wave symmetry but, in the meantime, the seventh chord seems certainly one of the most beautiful mysteries in nature.

The diatonic scale has thus served its purpose ideally within its own limits, even offering a few perfect intervals and triads in other keys than its own, as the accompanying table shows. Aristoxenos would probably have been impressed by the fact that all its tonal differences in frequency are multiples of the prime number 11. And Pythagoras would surely applaud its plurality of means: E the *arithmetic* mean between C and G, F the *harmonic* mean of C and C′, and G the *geometric* mean of C and D′!

Yet, as musicians well know, the diatonic scale failed utterly as a practical all-around tool of music — for the same reason that Kepler's circumscription of the spheres failed as a tool of astronomy. It just would not quite fit! It could do very nicely at certain points or within particular limitations but the mere act of pressing it into place in one region would always and inevitably force it out of line in another. Ill-fitting notes were so common, in fact, in the early clavichords and pianos that they were given the name of "wolves" — they howled so. If a major diatonic scale were constructed starting at D instead of C, it could not possibly match up with all the notes of the C scale since several would be at the wrong pitch intervals. Instead, it would need four new notes for its diatonic perfection. And, to provide for all the twelve musical keys, every octave would have to have 72 notes!

FREQUENCY RELATIONS IN THE MAJOR DIATONIC AND EQUALLY TEMPERED SCALES

	Do	Re	Mi	Fa	Sol	La		Ti	Do′	Re′	Mi′	Fa′
Frequency relations	Middle C	D	E	F	G	A	B♭	B	C′	D′	E′	F′
Octave, key of C	1								2			
Major triad and octave, key of C	4	(4½)	5	(5⅓)	6	(6⅔)	(7)	(7½)	8	(9)	(10)	
Major triad and octave, key of F	(3)			4		5			6			8
Major triad and octave, key of G		3			4			5		6		
Vibrations, diatonic scale, key of C	264	297	330	352	396	440		495	528	594	660	704
Diatonic vibration differences		33	33	22	44	44		55	33	66	66	44
Diatonic interval ratios		9/8 whole	10/9 whole	16/15 half	9/8 whole	10/9 whole		9/8 whole	16/15 half	9/8 whole	10/9 whole	16/15 half
Equally tempered (geometric) scale, suitable for music in all keys	261.6	293.7	329.6	349.2	392.0	440		493.9	523.3	587.4	659.3	698.4
Tempered interval ratios		$\sqrt[6]{2}$ whole	$\sqrt[6]{2}$ whole	$\sqrt[12]{2}$ half	$\sqrt[6]{2}$ whole	$\sqrt[6]{2}$ whole		$\sqrt[6]{2}$ whole	$\sqrt[12]{2}$ half	$\sqrt[6]{2}$ whole	$\sqrt[6]{2}$ whole	$\sqrt[12]{2}$ half

The mathematical reason for this discrepancy (in case you're interested) is that the perfect fifth contains 7 (supposedly equal) semitones out of an octave of 12 semitones, so that 12 fifths should ideally just fit into 7 octaves. If they did, a fifth (3/2) multiplied by itself 12 times should exactly equal an octave (2/1) multiplied by itself 7 times. But the twelfth power of 3/2 comes out slightly more than the seventh power of 2 approximately by the ratio 519:512 — a sad and shocking little "blot" upon the presumed perfection of the spheres. Shades of Pythagoras — and Aristoxenos' *leimma!*

So to escape the absurdity of a 72-note scale and create a uniformly flexible one in which a piece could be played in any key without noticeable distortion or difference from any other key, compromises had to be made through which the tonal frequencies could be adjusted slightly to make all semitones really equal. In this way, our modern practical scale of "equal temperament" was created, in which twelve proportionately identical semitones exactly span each octave, and A♯ and B♭ (which differed by a *Pythagorean comma* in the diatonic scale) are one single note, as are all other contiguous sharps and flats. Not just one single note in each scale, either, but one single note all over the world — since A above middle C has been standardized by international agreement at exactly 440 vibrations a second.

Such an even scale had been dreamed about and suggested many times in history, but it was not until the first quarter of the eighteenth century that Bach, while living in the little town of Arnstadt in central Germany, adopted it for its practical advantages and brought it into general acceptance. Not only did he tune his domestic clavichords and harpsichords in this tempered way, making every black key both a sharp and a flat, but he wrote a series of preludes and fugues to demonstrate the virtues of his chosen scale. Called *The Well-tempered Clavier* or

"Equable Keyboard," this famous work of 1722 included pieces in all twelve major and twelve minor keys and pointed to the great advantages of unhindered transpositions. While admittedly a step down from absolute diatonic mathematical perfection (in which an ideal fifth is 7.019550008654 times the size of an ideal semitone, which is one twelfth of an octave or $\sqrt[12]{2}$ or 1.05946), the tempered scale smooths out all inequalities into a uniform imperfection (its fifth exactly seven times its semitone) so slight that the musician's ear and most piano tuners accept it with scarce a quaver. Even the different keys, harmonically standardized though they be, seem to imaginative minds to retain some of the flavor of the original diatonic individuality and to strike different emotional chords — D♭ major suggesting dignity and majesty, F a lighter mood and B♭ minor (to some) a spirit of licentious abandon.

A usable scale, however, can hardly be more than a tool of music. What then makes a melody? Is there any reason back of the rhyme that turns tones into tunes?

This complicated subject was once regarded as so elusive it was beyond all logic, but modern science, by considering music in its essence as a form of communication, has learned to apply mathematics even to the mysterious art of composing. Through a certain "technique of engineering," in fact, a musically inclined mathematics teacher at Columbia University named Joseph Schillinger early in this century evolved a numerical system that George Gershwin is said to have used while composing *Porgy and Bess.* The basic principle is the application of the entropy concept to melody and beauty. For it is becoming more and more evident that musical notes have many of the characteristics of elementary particles and that their patterns of relationship, mysteriously parallel to the structure of the universe, follow the same laws of thermodynamics as crystals, gases or any other measurable substance. A composer, then, with an eye to physical theory need only make the entropy (disorder) of his music low enough to give it some recognizable pattern yet at the same time high enough for an element of suspense and individuality, and he may be well on his way toward a judicious compromise between the Scylla of wanton discord and the Charybdis of dull monotony!

The finer points of entropy-oriented composition, of course, have not yet been worked out by the mathematicians, although a series of electronic randomized-pattern "composers" has started grinding out a weird evolution of tunes which, some claim, have already improved to the point of sounding almost as stirring as the dreariest of medieval chants.

The matter of volume of sound has turned out to be easier to understand — its general law: that the energy of a vibration is proportional to the square of the amplitude. And both volume and pitch interrelate with timbre or tone quality, which is a kind of sum-product of all the details through which vibratory energy distributes itself — as, for instance, the stick-slip-stick-slip-stick friction waves of the violin bow upon a string that blends its

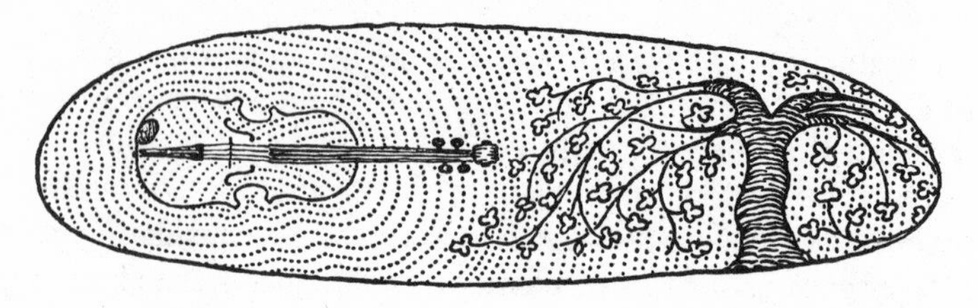

fundamental tone (voiced by resined horsehair on sheepgut) with its harmonics, flowing into the maple bridge and headlong down the spruce soundpost, spreading thence impetuously across the chamber of the body and through the cells of the surrounding planewood, beech or pine, shaped by interrelated contours, graceful *f*-holes, invisible nodes, and on outward into the air in all directions at better than 1,000 feet a second, to reverberate anew to the shape of the hall or the natural resonance of surrounding trees and buildings. Thus the strains you hear are literally the living compend of a hundred factors from the hand of the musician to the pitch of the balcony — even to the tensions in youı own ear and brain. A trumpet call takes but a tiny fraction of the total noise of pressure from the blower's breath — just those components that can be amplified by the natural vibration of the instrument — and molds them progressively outward with the help of all the molecules they meet, expressing thereby a sort of tuned aspect of the entire local world, including, of course, every other concurrent sound. In the confluence of waves, there may even be tones of inaudibly low or high pitch that react upon each other through resonance or beats to produce clearly audible frequencies. And the relatively pure mid-passage parts of notes from, say, cellos, oboes and organs may be virtually indistinguishable, leaving it to the irregular timbre of their beginnings and endings to tell the listener whether he is in the presence of bowing or blowing.

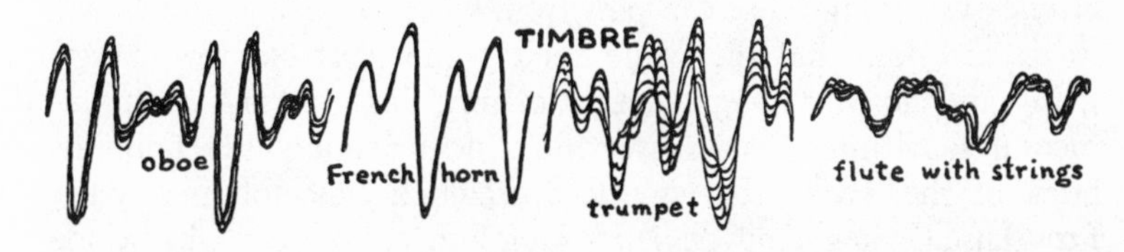

If such fleeting details look meager in print, their sound is actually quite ample in expressing individual character, for there is no known noise or music that cannot be created by the correct superposition of simple harmonic waves — from the cosmic thunder of an H-bomb to a sensitive set of coordinated organ pipes which have been recently taught to say "papa" and "mama."

This general discussion of waves and music may have seemed digressive at times, but I think it was really needed as an introduction to our next delve into the essence of matter. Obviously, most of the macrocosmos tends to form rhythms and cadences which we see everywhere as rows of mountain chains, constellations of stars, banks and ripples of clouds, grids of city streets, processions of trees in the forest, pebbles on the beach, waves at sea, schools of fish under the sea, flights of birds, hair on heads, grain in wood . . . and from here on downward, less and less obviously, more and more mysteriously, into the microcosm, to cells in the body, to fibers of muscle, serried molecules, crystal lattices, networks of atoms, "orbits" of electrons, even to the nebulous maneuverings inside atomic nuclei. Indeed, is there reason to think these redundant waves cease anywhere or anywhen? And if not, why might they not be made of the same potent "substance" with which Lao Tzu "tied" his philosophic knots "without rope"?

We have already noted that a wave is a shape in motion — not normally built of matter (though there are cases, as in surf riding, where matter may be carried by a wave) but rather built of energy — energy that can pass from wave to wave as individual waves are born and die while their group goes on, the energy living and flowing independently of any single form — a fundamental something that is ever more demonstrably the building brick of the world, that mystic abstraction that Job may have first sensed when God demanded of him out of the whirlwind, "Hast thou walked in the search of the depth?" or "Knowest thou . . . who hath put wisdom in the inward parts?"

It would be a help, of course, if man could handle an electron like a billiard ball or view it like the moon, taking its measure in a leisurely way. Even if it were as remote as Jupiter or as strange as Saturn, its music would be a lot easier to comprehend on the macrocosmic scale — as we calculate Jupiter's overtaking of Saturn every twenty years (each time in a different place) as only the overtone to a deep fundamental throb when the passing of these giant worlds works around to a nearly exact repetition once in something over nine hundred years.

But the electron is much more elusive than any planet or palpable object in the macroworld, if in fact it is an "object" at all. As we saw in the last chapter, it is definitely not free to follow any orbit it might happen into, like a celestial body or a sputnik, but must "choose" one of exactly seven concentric shells or energy states that surround its nucleus (see pages 338–339). Nor can two electrons collide like two cannonballs, for they keep a mysteriously inviolate interval between themselves in obedience to Pauli's exclusion principle (see page 324), which has never been known to harbor an exception. Besides, an electron's energy (whatever it really is) has proved to radiate itself outward in the strictly quantized vibrations or waves called photons whose energy has a wonderfully constant relationship to their frequency or pitch. This curious abstract relationship between the "smallest possible" parcels of energy and their frequency (comparable to the magnitude-frequency ratio of pulsing Cepheid stars) was discovered by Max Planck in 1900 and is the key to his famous quantum theory which the world could not believe for several years after he published it, but which has since revolutionized it inside and out.

To use again our analogy of Babylon, it is as if the Lord High Planner of Royal Works had somehow discovered and proved to King Nebuchadnezzar that the size of the Babel brick was absolutely dependent on its color, so literally so that if the king ever wanted to change the tint of his grand palace, he must also (to a very difficult and expensive, if slight, degree) alter the palace's dimensions. One would hardly blame Nebuchadnezzar for contemplating beheading his Lord High Planner before swallowing such a fantastic revelation. Yet Planck handed the scientists of 1900 virtually the same ultimatum, which, under the strong influence of Einstein, they eventually accepted. For by this time, it was common knowledge that color is a kind of pitch or wave frequency, and the energy quanta that form the "bricks" of our whole material world have turned out to be cut most exactly to their own pitch. Planck, appropriately, was an excellent amateur pianist and understood musical theory, which knowledge undoubtedly helped him in working out a precise mathematical figure for the constant relation between frequency and his minim of energy. This increasingly important constant he designated by the letter h — now one of the commonest symbols in physics — a symbol, you will admit, a lot easier to write or read than its fully established equivalent of .000000000000-

0000000000000006547 . . . erg seconds. One full erg second, by the way, amounts approximately to the action required to blink your eye once.

If h seems an awfully small quantity of something, it is reasonable that it should, for it is the very smallest physical quantity known, namely, one quantum. One quantum of what? Not of energy (since it is a relation between energy and frequency) but of *action*. Physicists use the term action for *energy* (in ergs) × *time* (in seconds) in measuring simple oscillation, or for *momentum* × *distance* in most other uses. Thus a 2-gram ball rolling at a speed of 3 centimeters a second has a momentum of 6 centimeter-gram-seconds. And, in moving a distance of 4 centimeters with that momentum, its action is 24 centimeter-gram-seconds, which would be many septillions of h. That is the working meaning of quantized action, and hundreds of the most exacting microcosmic tests have affirmed and reaffirmed the amazing fact that there are no fractions of h in the material world — while every year turns up fresh evidence of h's reality as a profound abstraction that can reveal important truth in almost any branch of science.

When I said that h signified action rather than energy, I did not mean that energy and other aspects of matter are not also quantized. For this seems to be a thoroughly quantized world in every size range from galaxies to stars to apples to electrons. Furthermore, h is a very versatile relation and, multiplied by frequency (usually designated ν) in the form $h\nu$, it stands for a *quantum of energy* — the smallest possible mote of the basic stuff all matter is presumably made of. In other words $h\nu$ may be the unsplittable ultimate of the material universe — a kind of abstract primeval atom of energy that embodies, much more truly than our now-familiar chemical atom, the "part" that is "partless" dreamed of by Leucippos and Democritos while strolling the provocative strands of Greece.

It was with thoughts somewhat along these lines that a young French prince got to musing about Einstein's equation $E = mc^2$ in 1922, wondering whether such a declaration of the fixed ratio between mass and energy meant that light has mass as well as energy. He was Louis de Broglie, Prince of Piedmont, a learned historian and part-time physicist. Though not quite the kind of man most people would have expected to inaugurate a new revolution in fundamental science, which was already staggering under the impact of Planck and Einstein, to say nothing of Rutherford, Bohr and others, he nevertheless realized the serious discrepancies in Bohr's theory of the atom with its patched-up correspondence principle and mystic quantum jumps of an electron from one classical orbit to another without really passing through the intervening space. How could even an electron-sized Jupiter become a corresponding Saturn without following some sort of comet's path from one planetary precinct to the other? Must one not choose between Newton's clockwork universe and the strange abstract world of Planck and Einstein without trying to blend them into one theory?

De Broglie thought so. And, even though his idea of a massive photon of light (based on $E = mc^2$) turned out to be pretty wide of the mark, it led him to a discovery of the first magnitude. He was familiar with the evidence that light has wave motion (which we will look into in the next chapter) and from there he reasoned: if light also has mass, which would make it some sort of aspect of matter, why doesn't all matter have wave motion? Why indeed may not matter consist entirely of waves?

This may seem a logical question today, but more than a quarter-century ago the concept of waving matter was utterly fantastic and de Broglie knew he could never get anywhere with it unless he somehow brought it down to solid earth with the aid of respectable mathematics. So he began to rummage around in mathematical notation, which, appropriately, is related to musical notation. He noted that Einstein's $E = mc^2$ and Planck's $E = h\nu$ could be put together as $mc^2 = h\nu$, which is easily transposed to $\nu = \frac{mc^2}{h}$, which is just an abbreviated way of saying that frequency (ν) is the equal of mass (m) times the square of the speed of light (c^2) divided by h. If the ν for frequency naturally reminded de Broglie of Pythagoras' vibrating string, it also gave him a sense of being on the right track — for is not the humming harp a prime example of waving matter?

But what exactly did ν refer to mathematically? It meant frequency, yes — but the frequency of what? De Broglie was reasonably sure it must be the frequency of some aspect of matter, since both Planck's and Einstein's equations had had to do with matter and matter's mysterious counterpart, energy. And, as he visualized the vibrant electron looping or whooping around the nucleus, quite manifestly the most active working part of the atom, he naturally picked the electron as the likeliest object of ν rhythm.

Burrowing deeper into his mathematical reserves, he then worked out something of the form and functioning of the electron's frequency. As it must be a wave frequency rather than a particle frequency, it was induced to emerge not as an orbiting monad's rhythmic revolutions but as a kind of centralized throbbing. And the actual mathematical equations turned out to be interpretable in three ways: either as (1) a concentrated heartbeat; (2) an explosive spherical pulsation; or (3) both. De Broglie assumed the last alternative: that an atom not only had a kind of localized heart of stable matter but also broadcast an expanding pulsation "forever in step with it and extending all over the universe" something like the now-known rise and fall of the whole ocean under each advancing wave. And since his atom itself was materially representable as a localized group wave that potentially could move up to the speed of light, he concluded that the individual, constituent phase waves (of which the group

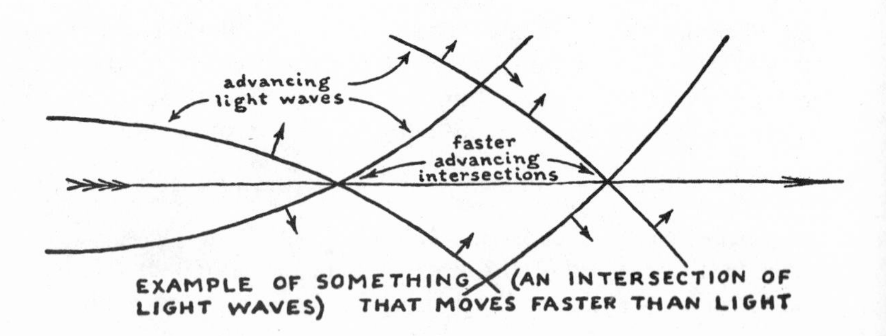

EXAMPLE OF SOMETHING X (AN INTERSECTION OF LIGHT WAVES) THAT MOVES FASTER THAN LIGHT

wave must be composed) might move still faster (inevitably in some nonmaterial way) at velocities ranging from the speed of light upward to infinity!

Naturally, all this was too much for any but the most untrammeled minds in the scientific world, yet, by what seemed almost a divine fluke, within a year and a half of publication of his theory, it was amazingly verified by experimental proof. Neither de Broglie himself nor anyone else seems to have known how to conduct such an experiment, but a researcher in the Bell Telephone Laboratories in New York named C. J. Davisson, with L. H. Germer, his assistant, had an accident in April 1925 while bouncing a stream of electrons off a piece of nickel in a high vacuum. A flask of liquid air exploded near by, wrecking the apparatus and spoiling the nickel surface so that it had to be heated a long time before the vacuum test could be resumed. Not realizing that this particular heat treatment would fuse the myriad microscopic nickel lattices into a few much larger crystals, Davisson and Germer were amazed at the diffraction patterns made by their next blast of electrons — patterns which immediately suggested, and soon proved to have been produced by, the very matter waves deduced by de Broglie.

The Davisson-Germer experiment became a classic and led to a Nobel prize for Davisson as well as de Broglie, for it not only fully confirmed de Broglie's theory of the wave nature of electrons but even verified the extremely short electron-wave length he predicted, which is so much shorter than that of light that, after a bare decade of development, it could provide man with a two-hundredfold further enlargement of microscopic vision through the marvelous new electron microscope.

 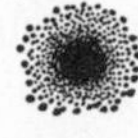 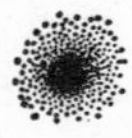

Such a radical turn of events seemed to many scientists only to perpetuate the growing paradox of particle and wave, because the wavelike photon had so recently developed into a particle while now the most solidly established of particles, the electron, had turned out to be a wave. It was a profoundly baffling paradox too, with waves and particles so opposite in nature (at least to the macrocosmic view) that an attempt to label them collectively as "wavicles" hardly seemed to get to the root of the question.

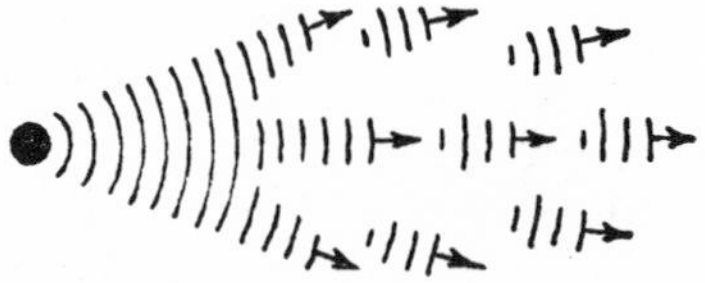

Yet a Viennese physicist named Erwin Schroedinger at Zurich University saw a wonderful interrelation between a wave's (lateral) front and its "rays" (paths) of outward advance that seemed to clarify the whole enigma. Considering these two dimensions (approximately at right angles to each other) as the warp and woof of matter, he showed how the wave front would reflect or bend around obstacles in typical wave fashion while an advancing segment of the same wave, like a foaming whitecap at sea, could be regarded as the trajectory of a moving particle. In his own words, "if you cut a small piece out of a wave, approximately 10 or 20 wavelengths along the direction of propagation and about as much across, such a 'wave packet' would actually move along a ray with exactly the same velocity and change of velocity as we might expect from a particle of this particular kind at this particular place, taking into account any force fields acting on the particle."

Schroedinger naturally realized that he must somehow tie the pattern of his wave-particles into the shape of the atom, expressing the relation mathematically. For this purpose he chose the model of a small tub of water with its complex of reflecting and interlacing waves, which he described as an "analogue" of electron waves in an atom-sized basin. "The normal frequencies of the wave group washing around the atomic nucleus," he explained, "are universally found to be exactly equal to Bohr's atomic 'energy levels' divided by Planck's constant h. Thus the ingenious yet somewhat artificial assumptions of Bohr's model of the atom, as well as of the older quantum theory in general, are superseded by the far more natural idea of de Broglie's wave phenomenon. The wave phenomenon forms the 'body' proper of the atom. It takes the place of the individual pointlike electrons which in Bohr's model are supposed to swarm around the nucleus."

The Schroedinger equation, in fact, endowed matter with a natural and simple beauty by revealing quanta essentially as waves of resonance, which, being basically Pythagorean in their relationships, had to have integral dimensions in order to exist. Like the moon tides that ring the earth in exactly two complete waves, the crests being on the sides toward and away from the moon, the troughs at the quarters, the electron's quantized crests and troughs always come out even around the atom. This gives the atom the harmonic resonance and integrity of a plucked string with fixed ends. For although the electron's course does not seem to have an end, one could think of it as having two ends fused together in a circle, like the rim of a bell, which thereby achieves its own resonance as it vibrates in a strictly integral number of waves. This is vital in holding the atom (or bell) together. If a bell did not thus oscillate in a cardinal multiple of whole sections, it would have to be broken or at least cracked — which would destroy its ring, removing its very reason of being.

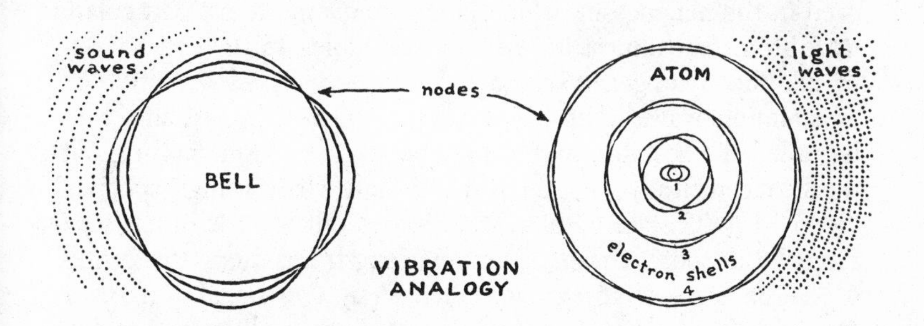

Another way of explaining quanta by Schroedinger's wave mechanics would be to say that each quantum is a discrete segment of vibration bounded by nodes. Such nodes need not be as simple as point nodes in a harp string, which are points of nearly

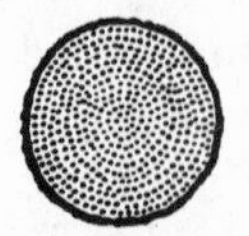
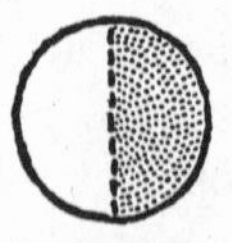

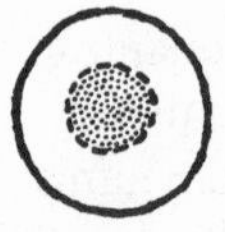
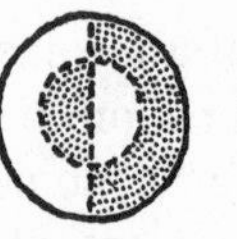
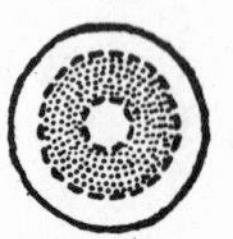

MODES OF VIBRATION SHOWING NODAL LINES
of a drum membrane (above), a brass plate (below) – the nodes varying according to where the surface is touched

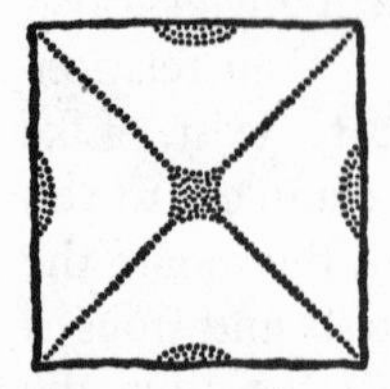
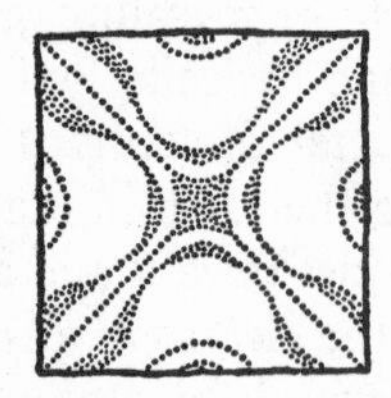
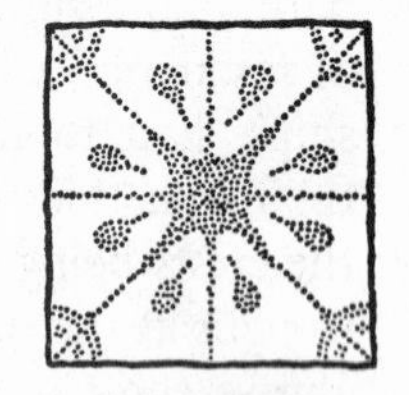

zero motion, for more often, they are line nodes, like the lines of "zero motion" in a vibrating drum membrane or metal plate, which may come in any of several modes or patterns (see illustration), each with its characteristic quantized frequency depending on exactly where and how the drum was struck or the plate held. Even more often, nodes must be surface nodes, amounting to two-dimensional areas of relative motionlessness among vibrating solids—and doubtless volume nodes of three and more dimensions...

It is hardly possible to visualize all these strange functions even in the macrocosm, while their patterns in atoms, particularly large atoms, are almost hopelessly inscrutable. In the easiest atom of all, the hydrogen atom, Schroedinger analyzed several types of standing waves that could develop. The simplest are what he called the *S* states, with purely radial wave forms pulsing outward concentrically—either as the 1*S* state with its single spherical "loop," the 2*S* state with two loops, one outside the other, separated by a nodal surface about two angstroms from the center, or the 3*S* state of three loops divided by two nodal surfaces at two and seven angstroms out, and so forth—all these with an additional outer nodal surface at infinity. If you wonder exactly what is waving when such waves come into being—what medium they are in—you probably feel much as Schroedinger did when he midwifed them, for it seems he could only imagine their unknown medium as a mysterious "essence" of mathematical space, which he simply designated by the appropriate Greek letter psi, ψ.

Next in order come his *P* states, in which the "northern" and "southern" hemispheres of the atom are in opposite phase bounded by a nodal surface intersecting the equator. The 2*P* state has two wave loops, one in each hemisphere, while the 3*P* state has an added spherical node dividing each loop into two segments, the 4*P* has two such spherical nodes, the 5*P* three such nodes, and so on.

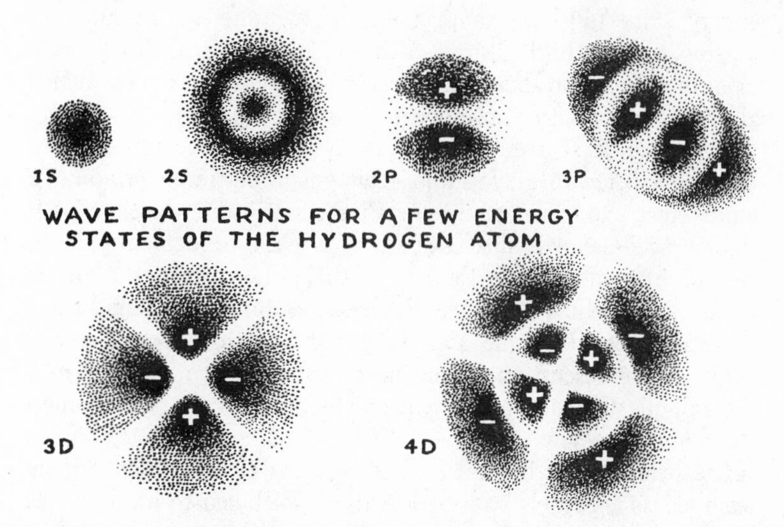

Then there are the *D* states, with two nodal surfaces sectoring the atom like a quartered apple in any of various ways, including a state in which it is sliced nodally through the tropics of Cancer and Capricorn with waves phased alternately + − + and − + − in the resulting polar and equatorial zones, each state combinable with still other increasingly complex states featuring various numbers of added spherical nodes, 4*D*, 5*D*, 6*D*, etc.

These waves of matter that can thus form the atom are of a kind that stand fixed in one spot like the waves around a big rock in a river or a lenticular cloud in the lee of a mountain. Their various numbers of nodes 1, 2, 3 . . . represent the possible energy states of the atom with corresponding quantum numbers 1, 2, 3 . . . , so that if you want to know the energy (quantum) state of an atom all you need do is discover what type of standing waves it has. This can now be done quite easily in a laboratory. The quantum conditions come naturally from the fact that, as with Saturn's rings, only certain wave lengths and frequencies are possible — that the electron, which pulses some 124,000,000,-000,000,000,000 times a second, steps up its frequency by about 2,470,000,000,000,000 pulses in going from the first "shell" or energy state to the second, then up by different amounts in passing to the third, fourth, fifth shells, and so on, each state or orbital position being a measurable quantum of frequency above the next with "nothing" in between.

Since the days of de Broglie, you notice, the electron has not been considered to move in an orbit but rather in an *orbital* — a word more expressive of a generalized, statistical, wavelike path than a well-defined line, indeed, of a path fundamentally uncertain in some details though mathematically true for all its fuzziness, and fully consistent with observation. The paradoxical profundity of the vague and wavy orbital in its ψ essence will come up for discussion in the next chapter, when we look into the famous uncertainty principle, which seems to be a cornerstone of natural law.

Meantime, let it be said that the waving electron may still be regarded as a particle so long as this is not taken to imply it has an exactly knowable "orbit" — for its particle aspect is in fact sometimes both useful and important. In the light of Schroedinger's wave equation, matter in the electron can be most reasonably visualized as occupying the known wave forms and following the orbitals that cut perpendicularly across the nodal surfaces. At least the mathematical "correspondence" between wave and particle works out that way, and its statistical conclusion indicates that the body of an electron is "most often" to be found where a wave's intensity is greatest, giving matter a probability texture that should ultimately explain it more completely than ever before.

Putting off the presentation of probability until the coming chapter, however, I can only say here that the whole modern atomic theory has been built up, to a great extent, upon Schroedinger's wave mechanics yet without rejecting any of the proven *workable* parts of Bohr's model, parts that are now collectively relegated to the atom's aforesaid and useful "particle aspect." The evidence is indeed overwhelming that every atom is somehow made of durable waves that vibrate continuously in hierarchies of energy — neat terraces of binding resonance between ψ frequencies. Although we have described some of the simpler dimensions of the simpler waves of the very simplest (hydrogen) atom, all atoms have comparable wave patterns, and the bigger, heavier ones could be likened to complex musical instruments or even whole orchestras on which many notes are being played simultaneously as chords from the contrabass levels to the outermost shell's altissimo. And of course, atoms vibrate as a whole also, and complete molecules generate their own cohesive wave systems, as do entire crystal lattices in beautiful interweaving integrated regularities, and all larger objects and organisms — including men, stars and, for all we know, the universe.

It is in these and comparable ways, as we are becoming increasingly and redundantly aware, that all matter tends toward its natural rhythms, ranging from the simple mechanical oscillation of pendulums and springs and falling drops of spray to the farther-fetched ups and downs of weather, where sunshine increases evaporation which thickens clouds, bringing rain and coolness which retards evaporation, clearing the sky for more sunshine . . . from electric oscillation of capacitance–inductance –capacitance–inductance to the chemical rhythms of the heart regulated by enzyme feedback — to such ecological interaction as the shark-sole frequency of the sea in which the abundance of soles provides such ample food for sharks that the shark population increases rapidly on its sole diet until the resulting scarcity of soles starves off enough sharks to enable the sole population to increase again to supply the next generation of sharks . . . even to our impalpable waves of radio and television that the senses cannot directly detect but which we mentally accept as real because *something* must convey the forms we finally hear and see, just as a "ψ essence" can be said to wave wherever matter is.

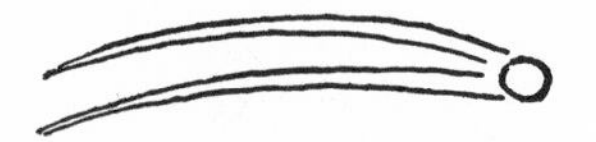

If it still seems incredible that an electron cannot be a simple "billiard ball" of ultimate smallness, one may reflect that such a jot of simplitude could hardly have meaningful continuity in time or place, for it could no more be identified twice than you could take in the same breath twice (since some of each breath becomes part of your body). Does the moving spot of light from a flashlight continue to be the *same* spot? An elementary *particle* obviously cannot be tagged or painted red or recognized as an identity from moment to moment. Yet a wave, being in its nature vastly more complex than the undifferentiated chunk of particle, can easily be imprinted with a recognizable, individual label. What is the coded signal from a lighthouse but a name-tab? Three flashes two seconds apart followed by a single flash separated by six-second intervals — what could it be but Malaga, Spain? Or when you recognize your mother's voice by telephone, do not the electromagnetic and sound waves carry a positive identification?

The "billiard-ball" concept of matter thus turns out to be actually more unearthly than an assemblage of waves, and it does not satisfy the newer, truer equations of physics. Very careful experiment has repeatedly shown that an electron's orbital is not deformed by perturbations in the way Uranus' orbit is deformed by Neptune — that the waving radiations coming out in expanding spheres from such an electron basically *are* the very stuff the electron is made of.

The wonder of such a discovery that radiation really is a material cannot possibly be fully grasped by a human mind, nor is mystery dispelled solely by compounding it upon more mystery. So it is hard to know whether the relation between quanta and continuity could in any sense resemble the relation between, say, apples and applesauce. Or if there is a "surface" between matter and nonmatter, exactly where is it? And assuming we have begun to know what matter is made of, what is nonmatter made of?

These deep questions seem to go on and on. But we will try to tackle them as best we may. Mathematics is of enormous help, but it is often hard to decide whether it represents reality in the way de Broglie's mathematics turned out to do. One wonders, for instance, whether abstract terms themselves may have a structure in objective nature? We have learned that energy grows according to its pitch, a principle of music, and that momentum literally has a wave length, but are these beautiful thoughts part of the material universe? How do they relate to structure? We know that a man can eat meat, wheat, salt or *any* of a thousand things out of the animal, vegetable and mineral kingdoms, all of them serving as structure of his body cells. But body cells contain millions of atoms. What about the structure or substructure of a single atom? Can something as small as an electron be made out of *any* material available? And does an electron's body include de Broglie's phase waves whose velocities may have *any* value so long as it is not slower than that of light?

It will take a lot of doing to find answers to such problems, which are as much philosophical as scientific — and which have been voiced down the ages since God challenged Job about the "wisdom in the inward parts?" — a challenge that echoes to this day in the laboratories as, "How does the electron choose its path?" or, "Are effects really caused by 'causes'?"

If answers there are, they must be found at last, I suppose, in the ubiquitous patterns of energy and thought, somewhere in the interflow of waves that crown the flood rivers of abstraction — somewhen among the consonances and equations that pulse the sleepless symphony of the worlds.

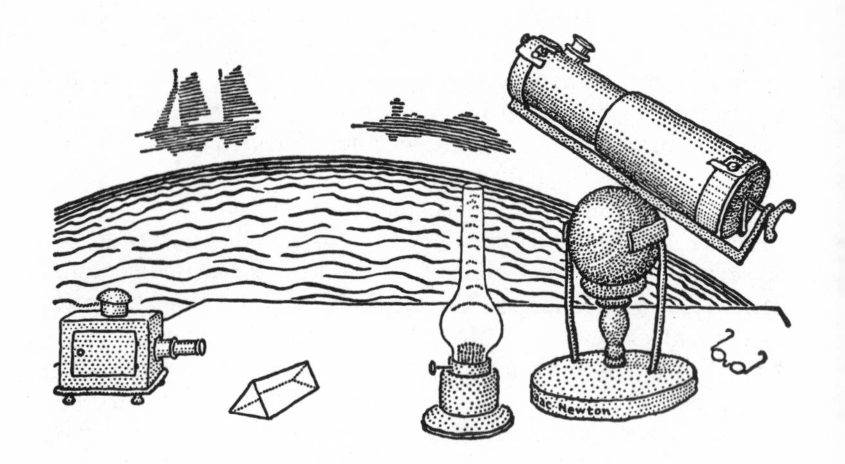

12. Of Light and Color

MY INSTRUMENT PANEL AND ALL THE WALLS of this space station are illumined by powerful built-in fluorescent lighting. This is necessary because out here in the void there is no twilight. I mean everything not in direct sunshine or artificially lit is shrouded in jet-black shadow because of the lack of what you'd call normal atmospheric diffusion. And earthly eyes obviously have not been prepared by evolution for such harsh contrasts.

As a matter of fact, until I got away from Earth I did not really know what total light could be. For of all the familiar phenomena of Earth, light now seems to me more nearly absolute than anything else. Not only does it travel mysteriously in the vacuous dark, its velocity the almost unbelievable constant that has been made a cornerstone of modern physics, but its immateriality is probably the most measurable connecting link between the physical and mental worlds.

Light is also the common stuff of vision, the messenger of form and, along with the rest of radiation, the swiftest-known medium of energy transport in the universe — its speed the limit of the propagation of material influence anywhere, its accomplishments far beyond human comprehension. Eight minutes ago the light you now see reflected from the butterfly's wing was actually inside the sun, where each photon of it is calculated to have spent something like ten thousand years wildly milling about on its way to the solar surface. The strange potency of even a little burst of light is shown by the fact that a single flash from a photographic flash tube can kill a sick rat, while a slightly longer exposure from an ordinary light bulb can stop a cocklebur from flowering or trigger a bird into singing a song before setting forth on a 10,000-mile flight! Moreover, as we shall see in the next chapter, any stream of light from a candle ray to a star beam is sufficiently substantial that, if unsupported, it will literally fall of its own weight or bend under stress. And it may be twisted like a cable or pumped (through valves) like a gas. It can even make itself felt in familiar economic units as was suggested by the engineer who figured out that light on Earth now costs approximately $400,000,000 a pound delivered.

In ancient days, men naturally wondered what light is and why it behaves so capriciously. Such an authoritative explanation as that God had said, "Let there be light" and there was light, was not sufficient for all. Nor could the real sun, promoted to godhood on six continents, satisfy every man's curiosity merely with his warmth and light, which often came to be worshiped as a kind of divine bestowal or manifestation. And so it happened that the imaginative Greeks became the first people to consider light as a subject suited to the scientific treatment.

Empedocles it was who proposed in the fifth century B.C. the modern-seeming theory that the radiating particles of light must have a finite velocity! This idea could hardly have been based on observation but, more likely, on reasoning and intuition, the most respectable sources of philosophical doctrine in those prodigious days. And Aristotle in his turn added that, if light takes time to move, "any given time is divisible into parts, so that we should assume a time when the sun's ray was not as yet seen but was still traveling in the middle space . . . before it reaches the earth."

This semimystical hypothesis was a great forward step toward the understanding of light — so far forward, in fact, that it could not be confirmed or disproved for nearly two thousand years, not even by the first scientific attempt to measure the velocity of light in 1667 with a method suggested by Galileo in which men flashed lantern beams vainly back and forth from distant hilltops. Yet only eight years later, in 1675, the Danish astronomer Olaf Roemer got the first definite evidence that Empedocles and Aristotle were right. For some ten years Roemer had been keeping very precise records of the eclipses of Io, the innermost of the four then-known moons of Jupiter, who disappears routinely behind the giant planet every 42 hours, 28 minutes. Roemer noticed that Io had the peculiar habit of being a little late in her rendezvous during each half-year in which the faster earth pulled away from Jupiter yet always perked up and became correspondingly early whenever the earth began to overtake him again. On the basis of this evidence, timed by his new pendulum clocks, the thoughtful Dane concluded that the light from Io to the earth must take just about "twenty-two minutes" to traverse the diameter of the earth's orbit!

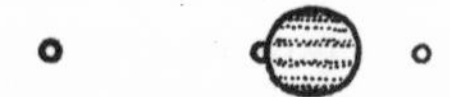

Such an unexpected gift of a precise scientific measurement of light's velocity, albeit of a fantastic magnitude, was naturally very exciting to all the scientists of the day. And to no one more than Christian Huygens of Holland, who had long been experimenting with lenses and lately had taken to searching for an explanation of why light moves in straight lines, and of how its rays can cross completely through one another without apparent hindrance.

Roemer's discovery, if it meant anything, reasoned Huygens, must mean that light literally travels. And travel, of course, implies the "transport" of something tangible from place to place. Yet in view of light's unaccountable ability to cross its beams without mixing them, how could light possibly be a transport of actual matter?

If Hugyens could not fathom light as a material, however, neither could he conceive of its being completely immaterial. So he let the paradox simmer in his mind for a while until, following a suggestion of Robert Hooke's, he saw a way to resolve it by postulating light as a "successive movement, impressed on the intervening matter." In this way light "spreads," he wrote in his *Treatise on Light*, "as sound does, by spherical surfaces and waves: for I call them waves from their resemblance to those . . . formed in water when a stone is thrown into it and which present a successive spreading as circles."

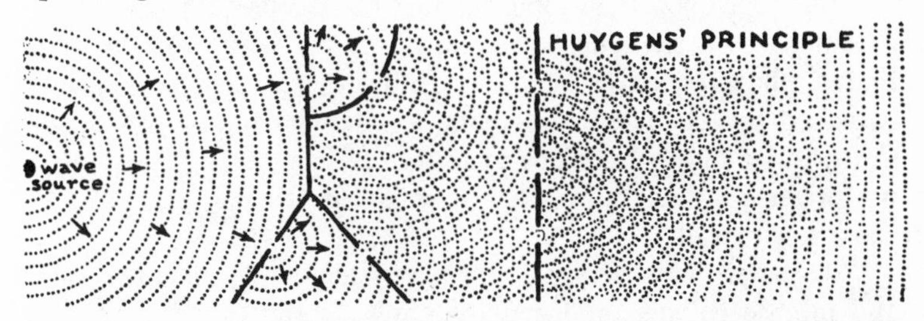

On this premise Huygens brilliantly developed the wave theory of light into an important concept still known as Huygens' principle, which recognizes that each point of any advancing wave front is in effect the source of a fresh wave, while all such fresh waves together continue onward as the advancing front. It explains why a loud sound originating, say, a mile outside and beyond the open door of a room produces something like a fresh sound source (as far as persons in the room are concerned) at the doorway.

But Huygens did not find light to be so amenable as sound to his principle, since light does not disperse evenly through a room and, unlike sound, moves in straight lines, creating sharp shadows. To explain shadows, he had to presume that light waves are extremely small and, to account for their capacity to carry across interstellar space, he considered also that an "infinitude" of them at each instant "unite together in such a way that they sensibly compose one single wave only, which, consequently, ought to have enough force to produce an impression on our eyes. Moreover," he added, "from each luminous point there may come many thousands of [successive] waves in the smallest imaginable time, by the frequent percussion of the corpuscles which strike the ether at these points: which further contributes to rendering their action more sensible."

It was a bold but reasonable attempt to explain one of the most baffling phenomena in nature. Young Isaac Newton, meantime, in Woolsthorpe had been engaged in some significant research on the same general subject, using both lenses and prisms. "In the year 1666," he reported in his first scientific paper, "I procured me a triangular glass prism, to try therewith the celebrated phaenomena of colours. And in order thereto, having darkened my chamber, and made a small hole in my window-shuts, to let in a convenient quantity of the sun's light, I placed my prism at its entrance, that it might be thereby refracted to the opposite wall.

" It was at first a very pleasing divertisement to view the vivid and intense colours produced thereby," he touchingly admitted, but a few minutes later began to apply himself more "circumspectly" to the question of what made the light separate into these colors, a mystery that naturally involved the deeper question of what sort of stuff or granules or atoms composes the ultimate inmost parts of light.

" Then I began to suspect," wrote Newton, " whether the rays, after their trajection through the prism, did not move in curve lines, and according to their more or lesser curvity tend to divers parts of the wall. . . . I remembered that I had often seen a tennis ball struck with an oblique racket, describe such a curve line. For, a circular as well as a progressive motion being communicated to it by that stroke, its parts on that side where the motions conspire, must press and beat the contiguous air more violently than on the other, and there excite a reluctancy and reaction of the air proportionably greater. And for the same reason, if the rays of light should possibly be globular bodies, and by their oblique passage out of one medium into another, acquire a circulating motion, they ought to feel the greater resistance from the ambient ether, on that side, where the motions conspire, and thence be continually bowed to the other."

Although he was unable to detect the "curvity" he suspected of his "globular" particles of light, Newton did soon realize, by placing prisms and lenses in many relationships, that at least the separation of colors in refraction is really a sorting of light into its fundamentally different "rays, some of which are more refrangible [bendable] than others," the "least refrangible" producing "a red colour" while the "most refrangible" show "violet." Thus he concluded that "colours are not qualifications of light derived from . . . reflections of natural bodies, as 'tis generally believed, but original and connate properties, which in divers rays are divers."

And "the most surprising and wonderful composition of all," he found, "was that of whiteness" because it could be made by remixing "all the colours of the prism . . . in a due proportion," thus confirming ordinary "white" sunlight as literally compounded of many parts. Even "the odd phaenomena of . . . leaf-gold, fragments of coloured glass, and some other transparently coloured bodies, appearing in one position of one colour, and of another in another," Newton reasoned, "are on these grounds no longer riddles. For those are substances apt to reflect one sort of light and transmit another; as may be seen in a dark room by illuminating them with . . . uncompounded light. For then they appear of that colour only, with which they are illumined, but yet in one position more vivid and luminous than in another, accordingly as they are disposed more or less to reflect or transmit the incident colour."

Thus, described in their expositors' own words, arose the two principal and opposing theories of light — Huygens' wave theory and Newton's corpuscle theory — which were to compete, sometimes bitterly, for dominance in optical science for two hundred years before they could be happily combined at last into a comprehensive whole by Einstein and others in the twentieth century.

At first Newton's easy-to-visualize explanation of light's straight rays and sharp shadows as due to the very fast courses of corpuscles of light was generally accepted, partly on the strength of his growing reputation as the inventor of a successful reflecting telescope and later as the author of the universal laws of motion and gravity. But as the eighteenth century came and went, more and more evidence appeared to support Huygens' wave idea — particularly new aspects of refraction and diffraction.

Refraction means the change of direction any radiation takes when passing obliquely from one medium into another in which its speed is different. Newton, of course, saw that oars appear bent where they enter water and knew that light can refract also on entering glass, but he could think of no way to account for it other than to guess that light (like sound) travels faster in liquids and solids than in air, and that was hardly a convincing explanation.

Diffraction, on the other hand, is the wave effect produced at the edge of shadows made by light coming from a single point. Although much less obvious than refraction, it too was known to science as early as 1665 in the form of a mysterious series of tiny fringe bands of alternating bright and dim light, a phenomenon particularly annoying to lens makers because it limited the resolving power (thus the practical magnification) of their optical instruments. However, neither Huygens nor Newton paid much attention to such an apparently minor detail, and it remained for thinkers of succeeding generations to begin to wonder if diffraction could possibly be caused by waves of light bending around the edges of whatever cast a shadow — even as sound waves behind a door or sea waves around the end of a breakwater.

The culmination of this classic issue between corpuscles and waves came with the famous "interference" experiment of Thomas Young, the English scientist, in 1800, in which he passed light from a single point source through two tiny parallel slits on its way to a screen. The fact that the striped diffraction pattern, produced when both slits were open, darkened certain parts of the screen that had been lit up when only one slit was open showed that there must be some kind of interference between the two light paths — which no one could logically explain except by the wave theory of light. The phenomenon may be compared to a silencer that dampens sound by matching the crests of one set of waves with the troughs of another, for some light from one slit obviously must be combining with light from the other at such a distance that the two wave frequencies were 180° out of phase, inevitably canceling each other and producing a band of darkness!

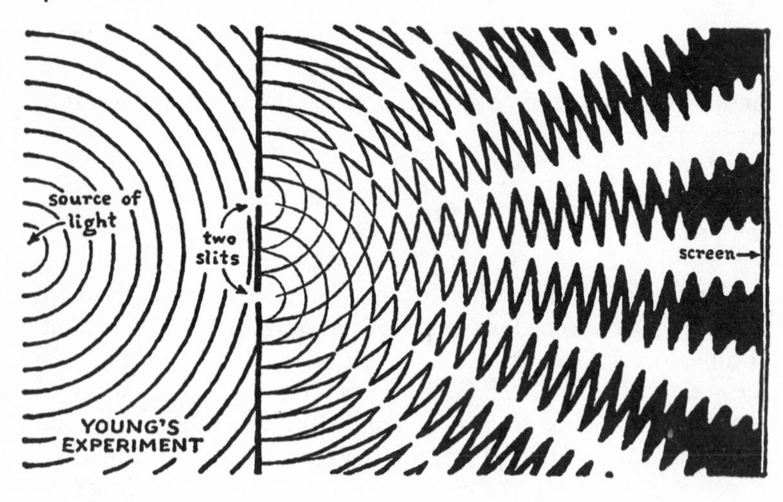

After several similar experiments by Young and some later ones by Augustin Fresnel, the wave theory of light became so firmly established that the corpuscle concept was pretty well discredited. Yet, clinging desperately to Newton's coattails, corpuscular light managed to get a "final" hearing when the French scientist Armand Hippolyte Louis Fizeau devised a way of mechanically measuring the speed of light by reflecting it back and forth between the teeth of a rotating wheel, a method which another Frenchman, Leon Foucault, improved the following year by substituting a revolving mirror for the toothed wheel and providing test tanks for comparing light's speed in water with that in air or other materials. By such means, in 1850 a "crucial experi-

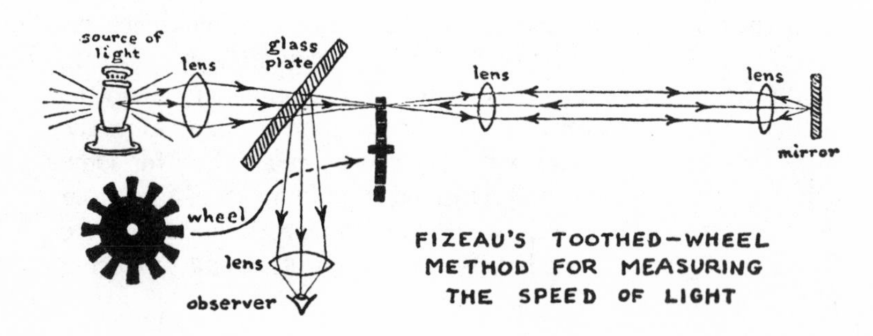

FIZEAU'S TOOTHED-WHEEL METHOD FOR MEASURING THE SPEED OF LIGHT

ment" was performed by Foucault to determine "once and forever" whether light is "a material body" or "a wave of disturbance," the issue to be decided solely by whether light's velocity proved to be greater in water, as Newton predicted, or in air as Huygens' calculations required. It happened at about the earliest date in the zooming evolution of technology that could muster the precision to cope with such a sophisticated comparison and the scientific world was agog for the verdict — which turned out to show quite definitely that light is swifter in air than in water, thereby effectively ruling in favor of Huygens and the wave theory.

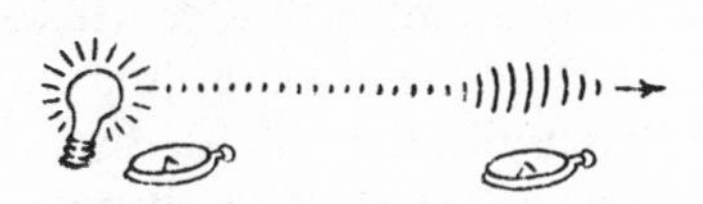

From then on, so far as the velocity of light was concerned, only technical refinements were needed to narrow it down to its present established value of 186,282 miles a second, which has been repeatedly checked to within a fraction of a mile a second by many methods, including the latest (suggested by a Russian in 1958), which requires nothing but an electronic flash tube, mirror and photoelectric cell. In this system, the mirror serves only to reflect the tube's flash back to the photocell, which is rigged to fire the tube automatically again, repeating the flash over and over at a frequency that by simple calculation, gives the speed of light.

One might think it futile to hope for any more direct observation of the motion of light than that, but astronomers occasionally are privileged to view a traveling parcel of light at such vast range that it appears to be standing still and can be clocked as easily as a moon or a comet. The first time such snail-like light was noticed was in August 1901, close to an ebullient "new" star known as Nova Persei. This exploding sun had been discovered the previous February 21st by a Scottish parson just after it had increased 60,000-fold in brightness (from 14th to 0.0 magnitude) in two days. And by August it had sprouted a "faint nebula" that, on close examination, showed a continuous, growing texture of concentric rings, each of which kept expanding at an angular rate between two and three seconds of arc per day. As no proper motion among stars even one fiftieth as fast as this had ever been known, it was quickly deduced that the rings must be periodic outbursts of light that were illuminating successively more and more distant parts of the nebulous surroundings of the star—in other words, the rings were actual rhythmic flashes of light moving outward through space in a stately procession at 186,282 miles a second!

Even on Earth, light has recently been observed moving in comparable parcels which, in this case, had to be deftly snipped off by a pair of electrically controlled polarized filters and photographed by a camera with a similar shutter system that could expose its film for only one hundred-millionth of a second. But by such means, Dr. A. M. Zarem of Stanford Research Institute caught a beam of light scarcely ten feet long spang in mid-passage, its shining presence "frozen" by the camera like a veritable fleeting flick of time itself, while both its ends (front and back) could be seen fading mysteriously away into the surrounding darkness!

Such a vivid demonstration of the moving reality of light hardly needs further confirmation, but I must mention the strange phenomenon of "aberration" discovered by James Bradley, England's Astronomer Royal, in 1726 when he was searching for clues as to the distance of stars. Finding astonishing and unmistakable evidence that stars all over the sky shift their positions with annual rhythm up to a maximum of 20″.47 near the ecliptic poles, he realized this difference could not possibly be the parallax of distance he was looking for, as it was in exactly the wrong direction, not to mention its being clearly independent of any randomness in the stars' distances. For a time, he suspected it might be the effect of nutation, a nodding of the earth's axis when the attracting forces of sun and moon periodically try to pull earth's equatorial bulge into the plane of the ecliptic, but it turned out to be due solely to the speed of light in relation to the earth. In

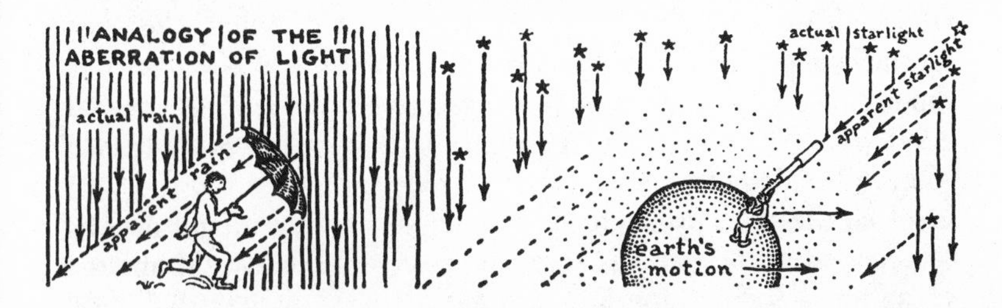

brief, just as an airplane pilot flying through a shower of vertical rain sees the rain slanting almost horizontally against his windshield, so an astronomer on the earth revolving through a continuous shower of light from the stars sees the starlight slanting at an angle that depends on the relation between the perpendicular component of his orbital velocity (perhaps 18 miles a second) and that of light (186,282 miles a second), thus making all telescopes aim slightly ahead of the true direction of the stars they see, in effect just like a gunner "leading" a duck with his gun.

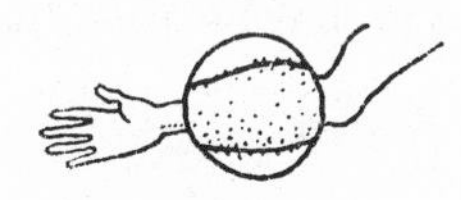

An entirely different aspect of light that has long occupied scientists is the magnification of visual images. The ancient Greeks observed that glass spheres filled with water could enlarge anything seen through them, and Ptolemy elucidated the optical properties of various materials in his famous *Optics*, which offered, among other things, a complete table of refractions for different angles of incidence. Although these surprisingly advanced discoveries in ancient science could hardly have been generally appreciated at the time, they do suggest how lenses with a focal length of nine millimeters came to have been in use in Pompeii in A.D. 79, probably for what Seneca described as the "magnification of writing."

After a little further development by Muslim and European scholars (including Roger Bacon in England), the practical manufacture of optical instruments was at last permanently established in northern Italy about A.D. 1286 when the first spectacle maker opened shop. One might think that the telescope should have quickly followed the eye glass, or at least have been developed by such an inventor as Leonardo, but it does not seem to have been so obvious that making very near things look bigger involved the same principles as making very far things look nearer — not even in the eyes of those who accepted the heavens as part of nature — for nothing approaching a telescope appeared until about 1590.

Then out of Italy (probably Naples) telescope making evidently spread to Holland from where, in the year 1609, in Galileo's own words, "a report reached my ears that a Dutchman had constructed a telescope . . . and some proofs of its most wonderful performances were described." Eagerly garnering material on "the telescopic principle" from this source, the only one he knew, Galileo spent a few months in "deep study of the theory of refraction" before preparing a "tube" of lead 9½ feet long by 1⅝ inches in diameter, "in the ends of which I fitted two glass lenses, both plane on one side, but on the other side one spherically convex and the other concave. Then bringing my eye to the concave lens I saw objects satisfactorily large and near, for they appeared one third the distance off and nine times larger than . . . natural."

Within a few more months, having constructed a second telescope "with more nicety, which magnified objects above sixty times," and a third that "magnified nearly a thousand times . . . I betook myself to observations of the heavenly bodies . . . with incredible delight." Almost immediately he discovered that the moon's face is "full of hollows and protuberances," that "the planets present their disks round as so many little moons," that there are "four sidereal bodies performing their revolutions about Jupiter — the four Medicean satellites, never seen from the very beginning of the world up to our own times," that "two huge protuberances" extend out from the two sides of Saturn, that in every direction there are "stars so numerous as to be almost beyond belief," including at least 80 new ones close to Orion's belt and 40 unsuspected Pleiades, while the Milky Way is revealed as "nothing else but a mass of innumerable stars . . . the number of small ones quite beyond determination."

The following year, adapting his telescopes to very short working distances, Galileo effectively founded the history of the compound microscope, while the 1,600-year-old simple microscope or magnifying glass reached something of a culmination of its long development at the hands of Antony van Leeuwenhoek later in the same century. A Dutch genius and great pioneer of the microcosmic world of life, Leeuwenhoek ground and polished his own exquisite single lenses, with focal lengths ranging down to a twentieth of an inch, through which he discovered in 1674 the strange "little animals" we now call protozoa and bacteria.

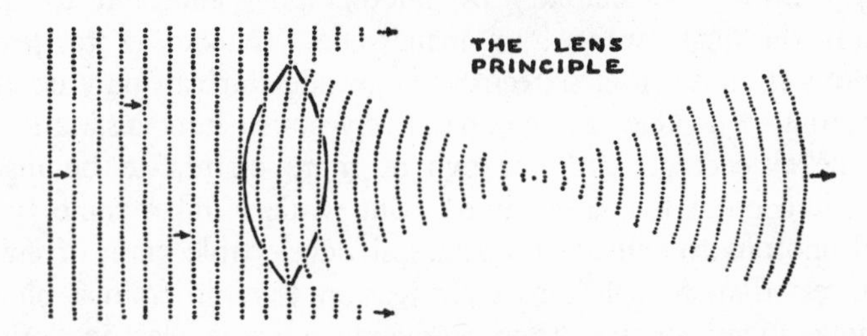

Thus the seventeenth century turned out to be tremendous in the history of light, not only in measuring light's velocity and explaining its nature but in the practical matters of using refraction and reflection, with the aid of lenses and mirrors, to magnify things out of other worlds, both far and near, into the consciousness of this one. And telescopes and microscopes have been improving ever since as their designers learned that the sharpness of an image depends not only on bringing light to where there ought to be light but on bringing darkness to where there ought to be darkness. With light waves tending to spread in all directions, the latter task is the more difficult. But it has been accomplished in ingenious ways, one of the latest developments being the electronic amplification of light through the principle of resonance or sympathetic vibration (see pages 386–87).

When the successive accomplishments of the famous 40-inch, 60-inch, 100-inch and 200-inch reflectors demonstrated that a practical limit was being reached in the size of visual telescopes on Earth, not only did radio waves come to augment light waves in our new radio telescopes (see page 191), but electronics offered a way to convert light into electrons (forming a "picture" by fixing static electric charges on a nonconducting surface) that is far more efficient than chemical photography. This was important because the twinkling of stars (due to refraction among the shifting densities of the air) inevitably blurs their images on time-exposed photographic plates, and "night sky glow" (from chemical phosphorescence in the upper atmosphere that sheds twice as much light on earth as all the stars combined) greatly reduces the contrast between a dim star and its background glim, effectively limiting detectability by photographic emulsion to stars within the first twenty-four magnitudes. According to Jeans, photons from even "a sixteenth magnitude star can only enter a terrestrial telescope at comparatively rare intervals, and it will be exceedingly rare for two or more quanta to be inside the telescope at the same time." But while it takes some thousand photons to activate the smallest developable grain of emulsion, less than ten photons will eject an electron from a photocathode metal surface, thus producing a static electric "print" with sensitivity improved a hundredfold. From here, the contrast between image and background may be vastly increased by electronic means, whereupon a television-type scanning beam can read off the amplified picture and transmit it to a screen for viewing or photographing.

By similar methods, it is calculated that a 200-inch telescope should be able to see as far into space as an unaided telescope with a 2,000-inch reflector, which would mean being able to photograph the creases in a newspaper held at least fifty miles away. Of course a 2,000-inch (167-foot) reflector of solid glass is out of the question on Earth at present, but here in space, free from the stresses of gravity, no doubt very large weightless optical reflectors (perhaps of some sort of lacquered sheeting) will be constructed for future observatories in which they can float unsupported in continuous focus upon unwinking stars set deep and fast in the glowless pitch of "eternity."

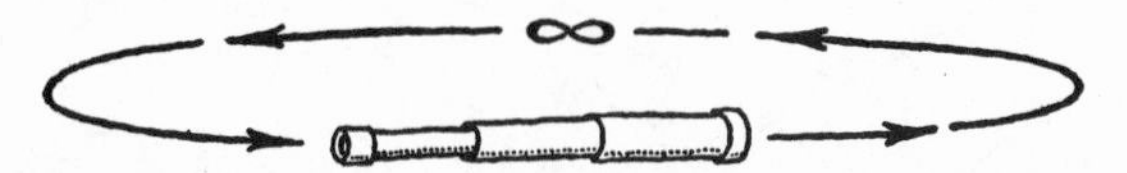

Eternity, admittedly, is a questionable word to apply to the travels of starlight in outer space, and perhaps I should not have used it. For if starlight really has anything like an eternity in which to travel, there logically arises the question: why is our night sky so dark? And why is the sky always dark as seen from up here? Why isn't it dazzling with the light of the infinitude of stars?

This rather insidious puzzle was first posed seriously in 1826 by the German astronomer Heinrich Olbers. When he was answered, "Why shouldn't the sky be dark with all the stars so far away that only a few of the nearest ones are even visible?" he agreed that the light of the average star is very feeble but pointed out that, if stars in general can be presumed to be about evenly distributed through boundless space, the number of them at any radius or distance must be proportional to the square of that distance (for the same reason that the surface area of any sphere is proportional to the square of its radius), which just counteracts the known fact that the intensity of light from any star (or other source) is *inversely* proportional to the square of its distance away. Thus the remoteness of stars should exactly balance their plenitude and, if you think of them in the ancient fashion as fixed in concentric spheres, something like the skins of an onion, each additional sphere or star layer (say, a thousand parsecs thick) should shed just as much light as the nearer layers of equal thickness inside or outside it, these spheres of equal light coming to Earth in endless succession from greater and greater depths of space until the stars in back are completely blocked from earthview by the stars in front, leaving the whole sky as bright as the "disk" of the sun.

Seeing that the sky, fortunately, is not that bright — which would incidentally burn us up as surely as if we were inside the

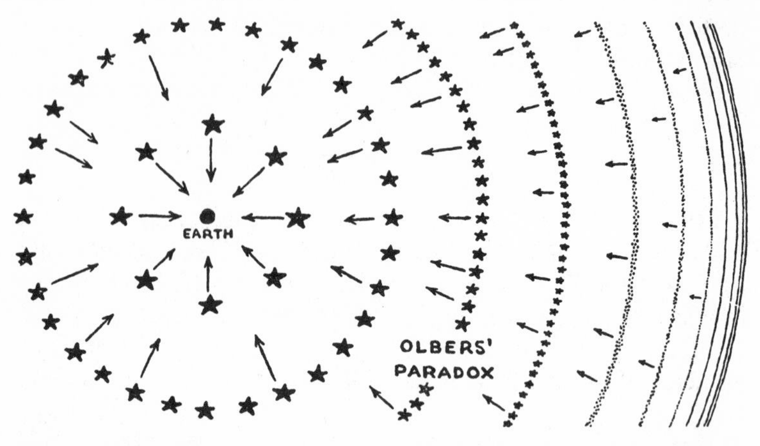

sun — one cannot suppress the suspicion that something must be rather wrong with this line of reasoning. In Olbers' day, as a result, astronomers concluded that the stars could not be distributed at random through space but must be concentrated largely in the Milky Way, and this presumption "solved" the question for the time being. However, when the twentieth century opened up the vastness of transgalactic and supergalactic space, it inadvertently opened up Olbers' paradox all over again.

Naturally new theories arose once more to deal with it: that the universe may be still so young that the majority of its starlight has not yet had time to reach us — and, in the 1920's, that it is expanding on the large scale at a rate that may be called explosive. The latter hypothesis, based on the most careful spectroscopic observation, appears, to a pretty conclusive degree, to have finally resolved the mystery. At least there is now a consensus among astronomers that starlight fades out into infrared invisibility at a range of a few billion parsecs (say 10,000-000,000 light-years), where the average galaxies are presumably retiring at about the speed of light (in relation to us) which, as we shall understand in the next chapter, effectively cuts off any prospect of their light ever reaching us here.

Along with such cosmic revelations, if the investigator into light should also condition himself to expect an intermittent assortment of odd earthly and unearthly phenomena to keep popping up, the explaining of which will add to his store of knowledge on this difficult subject, he is not likely to be disappointed. He might look, for example, at the commonplace specks of sunlight on the ground under a large tree on a summer day. I don't mean big patches of light but rather the small single light spots that come and go as tiny gaps among the leaves open and close to let darts of sunshine through to Earth. One might think these isolated flecks of light would be of all sorts of random shapes as the haphazard apertures chance to appear in the quivering foliage, yet surprisingly they all turn out to be the same: all smoothly rounded — in fact, they are exactly similar ellipses. How come?

It is a question known to have been asked by Aristotle in the form: why does a square hole admit a round image of the sun? And the explanation, first hinted at by Francesco Maurolico in the sixteenth century, is that every tiny opening between leaves, even though square or irregular, acts like the aperture of a pinhole camera, which will take remarkably sharp pictures without a lens. In fact, the sun, being no mere point but a round disk half a degree across, as it appears from Earth, is being "photographed" by the tree so that its round image appears wherever the shafts of its light get through to the "film" of Earth. In confirmation of this, during a partial solar eclipse all the little sun pictures are observed to be crescent-shaped and completely inverted (as required by optical theory) in respect to the real sun.

Something comparable happens also in the case of shadows of small objects. You have heard of the *umbra* of total eclipse when the moon passes before the sun, this being the very dark, cone-shaped, full shadow, which is surrounded by the much larger and fainter semishadow or *penumbra* of partial eclipse. Such a distinction among shadows is likewise attributable to the sun's not being a mere point, and the penumbra, which spreads out as it reaches farther from its source, obediently forms an increasingly realistic sunshadow picture — which explains why a butterfly's shadow becomes round like the sun as he flies higher above the

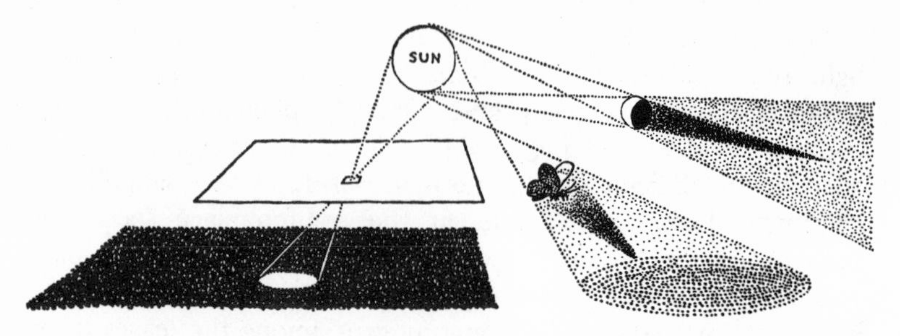

ground — and why the shadows of your outstretched fingers tend toward the shapes of claws during a semieclipse.

Such phenomena are pretty easy to reason out because the main light rays involved go in straight lines without reflection or refraction. But when you investigate what happens in mirrors, light begins to get more mysterious — which may explain why women and magicians regard the mirror as one of their important tools.

Did you ever wonder why a mirror makes your reflected right hand into a left hand, in effect reversing things sideways right and left, yet without reversing things up and down? Why should a mirror have a preference for swapping hands but refuse to swap heads for feet? The puzzle is not made easier by the fact that there is a kind of mirror that does not reverse right and left, and which both Plato and Lucretius described as a rectangle of polished metal bent into the concave form (shown on the right, below.) In such a mirror you see yourself as others see you and a printed page reads perfectly normally.

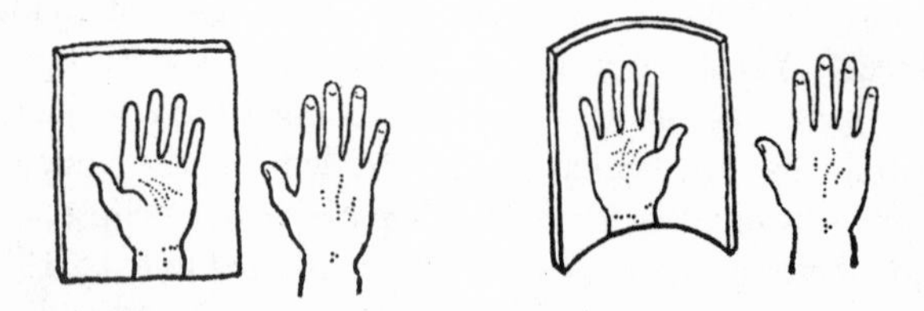

The explanation involves the laws both of reflection (which is really a special case of refraction) and of symmetry, which we will deal with near the end of this chapter. Reflection's familiar law — that the angle of incidence must equal the angle of reflection — solves the simpler mirror problems quite readily, for re-

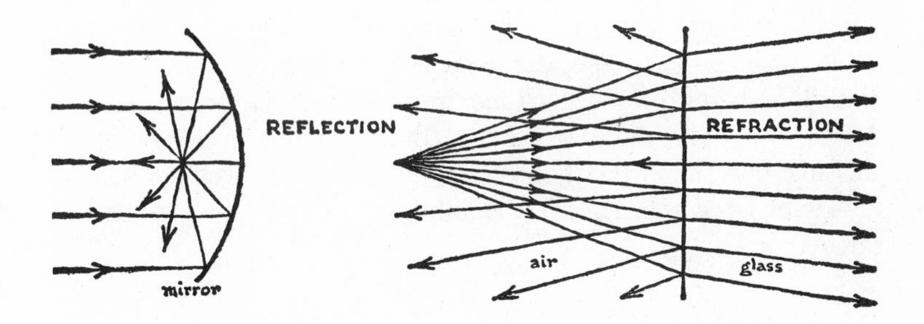

flection is essentially a bouncing process and a rubber ball truly reflects off a floor. Reflection's relationship to refraction, moreover, may perhaps best be illustrated by the shooting competitions that used to be held on Lake Koenigsee in the Alps where I once spent a summer. Surrounded by high mountains, the lake is so calm that a special feature of the sport was to have the marksmen aim not directly at the target but rather at its reflection in the water, counting on the bullets to rebound from the surface to hit it. This, amazingly enough, the bullets apparently did — and fully as accurately as if they had been shot straight through the air all the way.

But the oddest thing about the phenomenon is that it was not a case of simple reflection of bullets off the top of the water. For a scientist who hung a series of screens in the lake proved that the bullets penetrated well below the surface where, obeying hydrodynamical laws, their pressure of friction upon the liquid depths forced them into smooth curves upward until they emerged into the air again at exactly the angle they had entered! One may well ponder whether photons reflecting off a mirror move in some such manner while in contact with the vibrating electrons of its polished surface.

In any case, the angle law is enough to account for every kind of reflection, from the so-called "universe" mirrored in a garden globe to the glass strands that conduct light into the stomach for photographing ulcers. By simply tracing the incidence-reflection angles upon the mirror globe's surface, you can see why it acts as an optical instrument with an ideally large "aperture," presenting to your eye the whole earth and sky confined within one visual face of the sphere, the parts of the universe farther behind the globe being progressively compressed into narrower and narrower distortion at its apparent edges, leaving nothing totally missing but the little spot of background lying exactly behind it — which becomes relatively smaller as you withdraw farther and farther away.

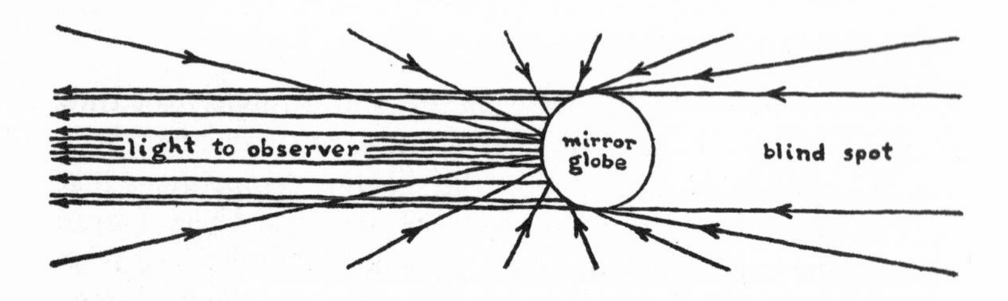

The glass strands, on the other hand, carry light around sharp curves by letting it reflect back and forth internally from surface to surface dozens or hundreds of times in the manner of shortwave radio impulses that bounce from ionosphere to ground to ionosphere to sea as they follow the curvature of the planet. And

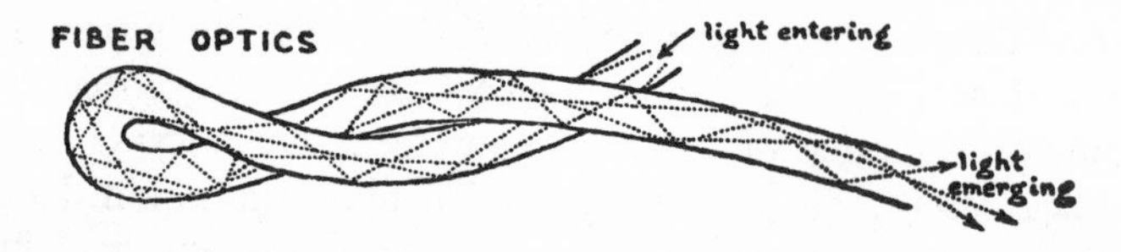

the gastroscope that illumines the stomach is only one of many applications of the new "fiber optics" using hair-thin strands of optical glass as light carriers. Great bunches of glass fibers are assembled to bear precise images through difficult routes, a method that works as long as the fibers keep in the same relative positions even if they have been tied in knots. And fiber messages can also be coded by scrambling — disconnecting and twisting the bundle a certain number of degrees — and later decoded again at the other end by countertwisting it the same number of degrees.

Still another aspect of light reflection is the phenomenon of polarity or a predominance of direction in light's vibration. This kind of polarity does not occur in the case of sound waves, which are longitudinal, but it is often a major factor among transverse or crosswise waves such as those of light. Just as a whale can swim either by flipping his tail up and down or wagging it sideways, light waves have been found to oscillate in different directions, sometimes more in one than another — which affects the quality of the light somewhat as the timbre of a violin depends on whether you pluck or bow it. When light vibration has an excess of 3 percent in one direction, we say the light is 3 percent polarized. A fair proportion of starlight is polarized 10 percent and most light from the blue sky a good deal more, because of reflections off innumerable dust particles in line with the sun.

The easiest way to see light's polarity without using special polarized material is to look at the reflection of some blue sky in a flat piece of glass with a dark backing. Where light strikes this glass at an angle of about 30°, the reflected light will be almost completely polarized. That is to say, all of it that is in the common plane of incidence and reflection will leave the glass at an angle exactly equal and symmetrical to the one at which it arrived, thereby denying itself any lateral vibration in that plane — which would mean it could vibrate only outside, and at right angles to, the plane. Thus the various natural (transverse) oscillations of the light waves will be sorted (polarized) all into the one direction.

If you hold the glass so that it reflects the blue sky from straight above you at about a 30° angle, meanwhile turning successively to each point of the compass, you will see that the reflected sky is brightest when you face toward or directly away from the sun (letting the naturally polarized vibrations reflect intact) but darkest when you stand sideways (cutting them off with counterpolarization). This shows that most of the zenith light is polarized to oscillate perpendicularly to the plane that includes the sun, the zenith and you.

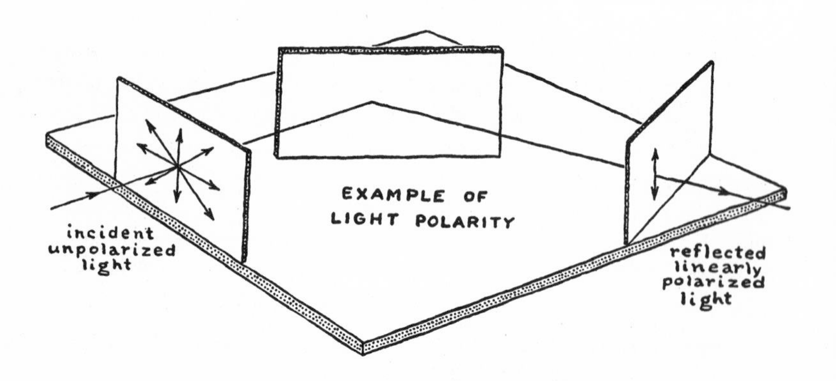

With a little practice, you can even see sky polarity with your unaided eye, particularly by staring at the zenith for several minutes in clear twilight when the whole sky may pervade your consciousness as a polarized network. Then you should become aware of a remarkable retinal vision known to science as Haidinger's brush, an hourglass-shaped figure of faint yellowish light squeezed at the waist between a pair of blue "clouds" and always aligned with the sun. It can be seen much more clearly through green or blue glass, and there are various optical instruments which can accurately measure its angle, such as the new sky compass for Arctic navigation that shoots the azimuth of the sun even when it is far below the horizon.

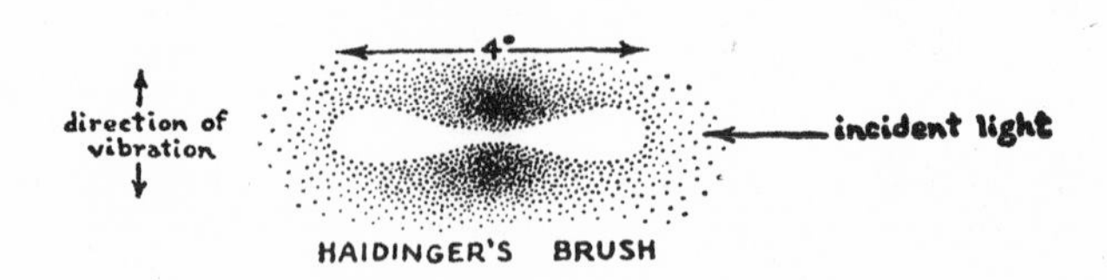

A still more striking, and more unearthly, example of light polarity is the monstrous "jet galaxy" called M 87 in the supergalactic womb of Virgo. Its central "blue jet" is 2,500 light-years long, evidently bursting with the most powerful cosmic rays known, and the light of it arrives on Earth about 35 percent polarized in a unique bluish continuum, something like that of the Crab Nebula (blasted forth by the supernova of A.D. 1054), only ten million times bigger. Some astronomers have speculated that the extraordinary polarity of this potent starlight may derive from its passage through space clouds of needle- or disk-shaped dust that has been uniformly oriented by the force lines of a vast galactic magnetic field.

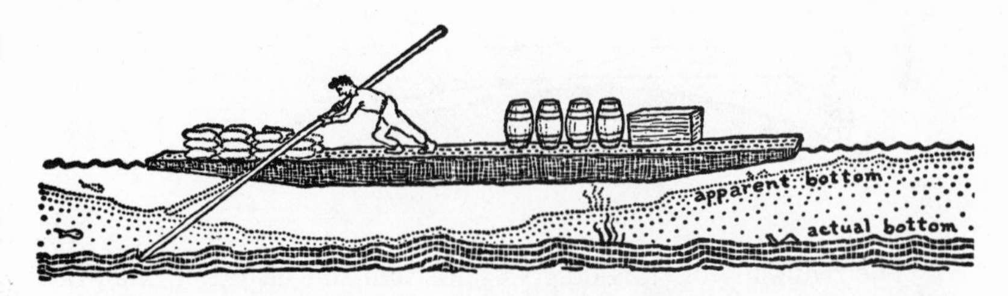

After touching on such varied aspects of reflected light, it may be as well now to mention the more general subject of light's refraction: the familiar bending of its rays that appears to "lift" the bottom of the canal before and behind your barge, that widens the setting sun to your earthly vision, that twinkles the stars, that spreads mirage lakes upon the desert or floats enchanted castles in the sky.

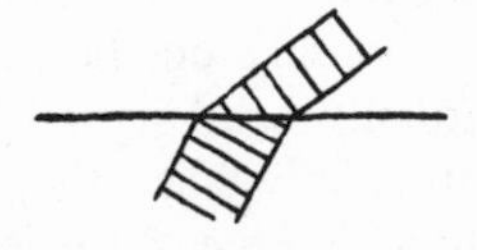

All these are variations of the aforementioned curving of light, when one side of a beam of photons is slowed down relative to the other side by encountering a denser medium, just as a column of soldiers tends to veer left when the left side of the column is slowed by marshy ground, waving girls or any other impediment.

Thus the slanting, convexed rays from the canal bottom seem to lift it everywhere but directly below your eyes where there is no slant to refract it.

The case of the setting sun is similar except that the medium is air gradually thinning into space instead of air and water joined at an abrupt surface. As the light from the "lower" part of the sun passes a longer distance through the earth's curving atmosphere than the "upper" sunlight, inevitably the "lower" sun appears raised by the refraction curves to a greater degree than does the "upper" sun. Thus the top and bottom of the sinking sun are optically squeezed together, giving it its characteristic look of lateral distortion.

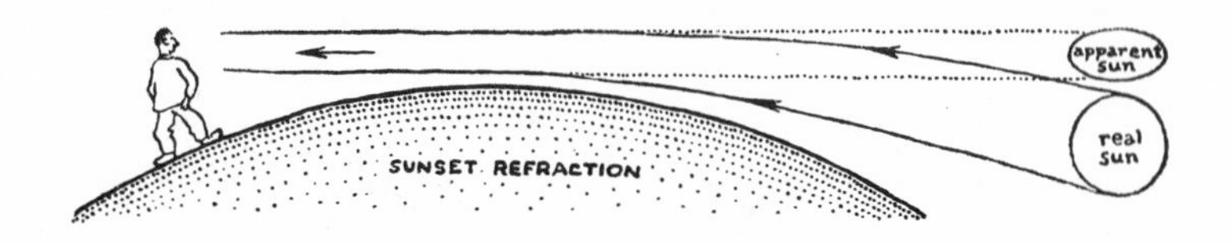

The twinkling of stars differs again from sun distortion mainly in that the cone of light to your eye from these distant suns usually subtends an angle of not *half* a degree but something less than a *millionth* of a degree. This is to say that the apparent width of the average single naked-eye star about equals the width of a dime viewed from a thousand miles away. Such a tenuous cone of light is bound to be seriously distorted, indeed intermittently cut off by the slightest atmospheric turbulence, by the innumerable flowing "bubbles" or strands of relatively warmer air called striae that rise wavelike on the wind. Such local differences in air density naturally bend light by refraction, and that is why stars twinkle, particularly when they are low in the sky so that their light passes through more of the lenslike striae. You can see the same effect when looking from near the ground at distant railroad tracks in the sun, for the lively striae just above the hot surface make the tracks seem to twist and shimmer and sometimes actually wink like stars. You can see it negatively in the shadows of high towers too, which often squirm and tremble from the irregular refraction. You can even measure the striae (usually between four and sixteen inches thick) when a searchlight many miles away shines its beam against a white wall, revealing the undulating shadow bands of demarcation — or by timing the successive blinkings of the Pleiades, like valves of an engine, as each wave of distortion drifts by.

The kind of refraction that creates the common inferior mirage of an illusory "lake" upon sun-baked ground is produced by the great contrast in temperature between the hot surface and the air that may be 25° F. cooler only half an inch up. Since the density of the air varies with the temperature, light rays bend fairly sharply in such disparate layers, letting you see the down-dipping rays of the sky after they have skimmed low over the hot ground, then curved up to your eye, striking you exactly as if they had been reflected off calm water. Thus it is the sky that you really see refracted upon the distant highway or sunny plain, not water, just as it is the sky that you really see reflected upon the surface of a far-away real lake — which explains why the two phenomena look so alike.

But in contrast to the down-dipping light rays of the common inferior mirage over a warm surface, there are the opposite, up-arching rays of the rarer superior mirage over a cool surface such as the sea in spring or a frozen plateau. The reason the superior mirage is less common is that it requires not only a cool surface but increasingly warmer layers of air as you go higher above it: a condition meteorologists call an *inversion* of the usual decrease in temperature with altitude. When horizontal light is slowed by

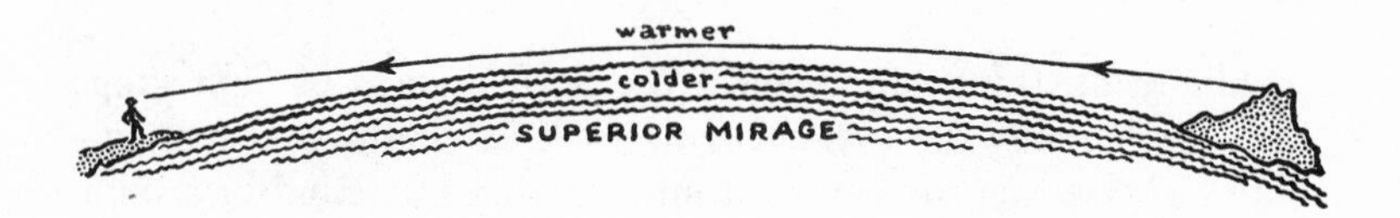

lower, cooler, denser air, its rays naturally arch or bow themselves convexly across the sky, sometimes leaping clear over intervening obstructions like the horizon — which produces the startling optical effect of islands or mountains rising spectacularly into view from positions previously hidden behind the earth.

When there is a cool layer of air *between* warm ones, many light rays may reach your eye from the same part of a distant knoll but by traveling different routes, some of the rays curving under the cool lenslike air, some going through it, others arching over it, all these visual lines, inferior and superior, meeting at your eye from their vertically various directions to give you the impression that the knoll is greatly elongated vertically — a kind of visional tower. Or the opposite may happen: rays from the different altitudes on the knoll may curve and merge toward one another through a warm air stratum between cooler layers until they reach your eye almost as one ray, giving the knoll a very condensed, stubby look. Or the rays from top and bottom of a ship may actually cross each other, making the ship look upside down in a natural mirror image of reality upon an absurdly inverted horizon.

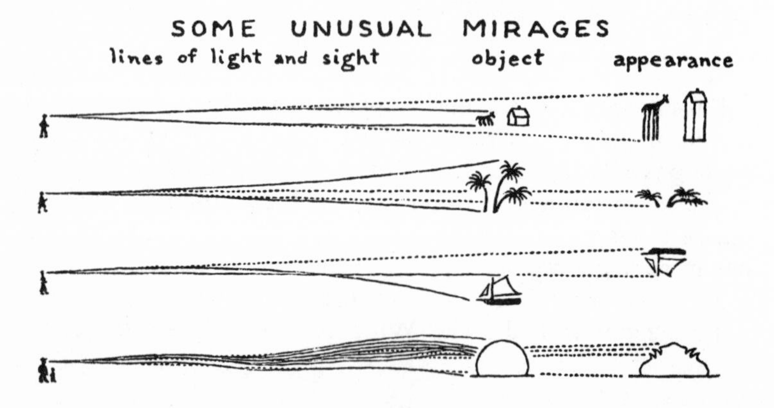

Almost anything can occur as the flexible paths of light grope their way among the ever-changing folds of density and temperature. The setting sun can turn into a stepped pyramid or a crimson pineapple, often sliced into layers with nothing between — and a few cliffs or crags perched above the far shore of a still sea may be glorified by vertical elongation into majestic castles in the air interspersed with soaring towers, minarets and battlements — bewitching fairylike scenes that spasmodically crumble and evaporate, then form anew like dreams, producing a deep sense of nostalgia, a profound and trancelike wistfulness — the fata morgana!

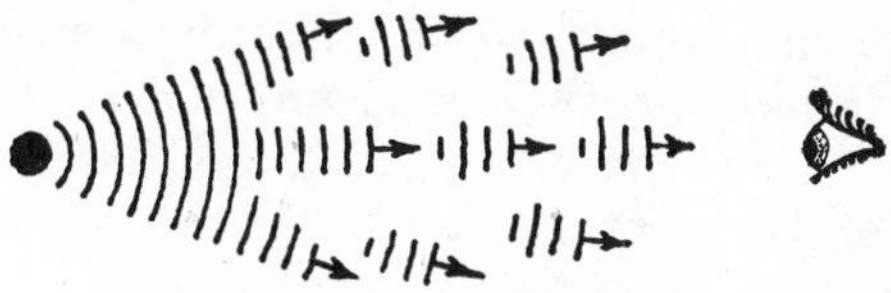

Leonardo once wrote in his notebook that "as soon as there is light, the air is filled with inumerable images to which the eye serves as a magnet." It was a rashly imaginative attempt to explain vision, yet, fantastic as it seemed, it was nowhere nearly fantastic enough to be literally true. For Leonardo had no way of observing the still secret world of the microcosm or of knowing how to deal with things beyond the generalizations of common experience. It is true he had an approximate idea of how a stone curves when thrown through the air. And a cannonball's flatter trajectory was not completely strange to him, since the cannonball could usually be seen in flight from the cannon. Also one might discover where it landed and, on occasion, where it passed through a tree, flag or windmill blade on the way — thus, by interpolation, its whole passage. But should one assume that smaller, subtler or more elusive forces must necessarily follow the laws of stones and cannonballs?

Galileo did not think so, as we learned from his principle of similitude (see page 233), and, even though high-speed photography and powerful microscopes have since made it possible successively to observe Brownian movement, molecular and atomic gyrations, even (by magnetic spectrometer) the motions of protons and neutrons inside the atomic nucleus, science cannot yet presume it safe to generalize any basic physical laws over such a range of sizes.

Faraday, you will remember, found he could not generalize Newton's macrocosmic laws of motion to explain the microcosmic behavior of electromagnetism but had to postulate unheard-of "tubes of force" extending invisibly outward from a terminal or a magnet like tentacles that reached with diminishing though unending strength to the uttermost limits of the universe. That was the original concept of "field," a mysterious immaterial force that could influence matter and which Faraday, not having been trained in mathematics, envisioned or invented as the most logical way to give form and continuity to an evident cause and effect of nature.

Maxwell was too young then and too imaginative to be one of the many scientists who scoffed at Faraday's "amateurish" unorthodoxy. Instead, he grew up to develop Faraday's field idea into its modern mathematical structure that defines electrical and magnetic flux in terms of components of force and direction through the now famous Maxwellian equations $\nabla \cdot E = \rho$; $\nabla \cdot H = O$; $\nabla \times E = \frac{-\dot{H}}{c}$; $\nabla \times H = i + \frac{\dot{E}}{c}$. And, strange to relate, among the ensuing mathematical implications was the odd requirement that whenever an electric field is put into motion or retarded, it must not only set up a magnetic field at right angles to it but must send out (perpendicuar to both) an impalpable electromagnetic wave that travels through space with the speed of light and has other important properties derived from Huygens' wave theory of light.

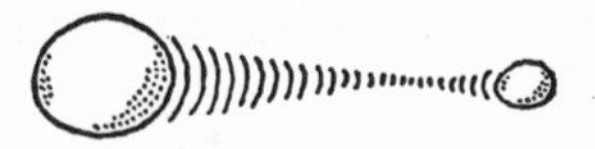

This unprecedented development struck Maxwell as exciting in its potentialities. Could such electromagnetic waves — if they turned out to exist — really be a kind of light? Was light, in fact, a form of vibrating electromagnetic energy which might have other forms at other frequencies? Maxwell definitely concluded so, but his health was failing and, try as he would, he could not succeed in detecting any kind of space waves in his few remaining years and died before he had vindicated his theory. Yet only seven years later, in 1886, a young German physicist named Heinrich Hertz, who had thoroughly studied Maxwell's equations under the great Hermann von Helmholtz, proved the close relationship between light and electricity in a now famous experiment in the physics lecture hall in Karlsruhe. Using a Leyden jar (condenser) whose electrical discharge was well known to be not a one-way current but a rapidly alternating oscillation, Hertz rigged it up with a spark coil (model-T Ford type) so that the powerful sparks jumping back and forth between the two brass knobs of his foot-long broken-rod transmitter would, he reasoned, send out the mysterious waves predicted by Maxwell. He had

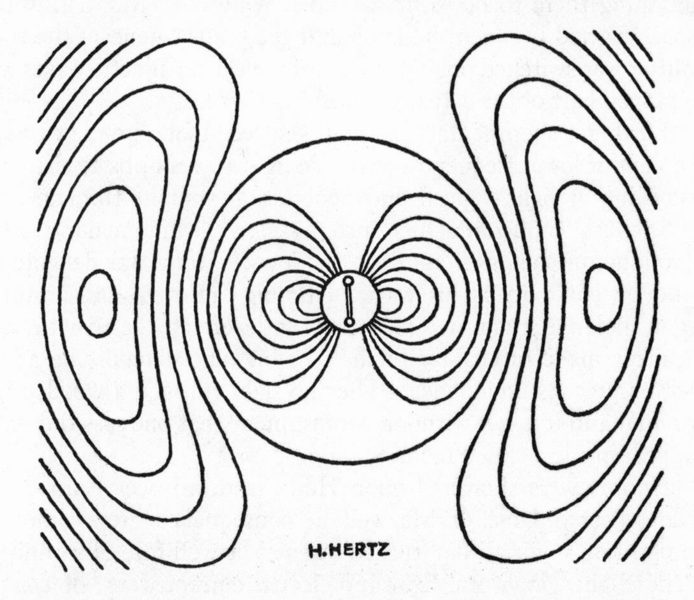

brilliantly calculated the exact shape these theoretical waves should have as the current, oscillating at a frequency of slightly more than 500 megacycles, broadcast its subtle "influence" outward in all directions, and he saw no reason why his receiver, a similar divided brass rod with a delicately adjustable spark gap between a knob and a sharp point, set up at the other end of the hall, should not show some kind of electrical response — assuming the waves indeed possessed any sort of material reality.

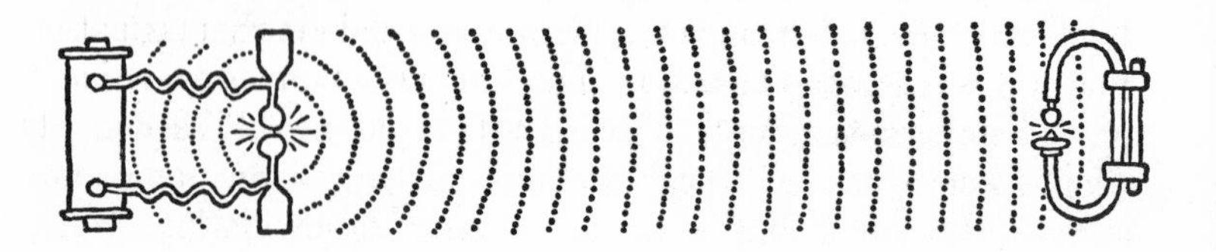

It was a tense moment, therefore, when Hertz put out the lights in the great room and, letting his eyes adjust themselves to the dark, turned to peer intently at the receiver's spark gap. Was there anything there to be seen? Could it really be — yes, a tiny blue spark flashed between the knob and the point whenever the transmitter was switched on. So it was true: electromagnetic waves were a proven part of the actual physical world at last!

Further experimentation soon showed that the new waves, though of lower frequency, were like light waves in basic character, traveling at light's speed and penetrating straight through some substances (nonconductors such as glass) while making others (conductors such as iron) cast "shadows." Hertz was delighted to find he could focus his waves with big "mirrors" of galvanized iron, polarize them by reflection and refract them with large "prisms made of coal-tar pitch." In his official report he added, with scarce-restrained glee, "They go right through a wooden wall or door and it is not without astonishment that one sees the sparks appear inside a closed room."

Honors were showered upon Hertz from all over Europe, and the reflected fame of Maxwell in consequence grew almost in proportion — for all had turned out as Maxwell foretold: sunlight, candlelight, glowworm light and electric current were "of a kind."

Hertz was soon experimenting with the photoelectric effect (see page 455) and might well have discovered the electron before J. J. Thomson if he had not developed bone cancer and died on New Year's Day, 1894, at the age of thirty-seven, evidently without having ever taken seriously the growing talk of long-distance communication by "Hertzian waves" — a world-revolutionizing development that Guglielmo Marconi would bring into exciting reality with his first wireless messages the following year!

The advent of practical "instantaneous" communication across trackless space in fact proved so stirring to the human imagination that it was accompanied almost simultaneously by a whole series of related discoveries; x-rays in 1895, radioactivity in 1896 with its constituent alpha, beta and gamma rays, and at the same time experiments with cathode (negative) and positive "rays" revealing the first subatomic particles, followed by the quantum theory in 1900, which threw an entirely new light on all these developments. Thus was to end the comfortable, 300-year-old mechanistic era so inspiringly launched by Galileo and Kepler and so solidly established under Newton with his neatly geared universe of orderly understandable motion. And thus, into the vacancy, was to dawn the new age of restless unsubstantiality and profound abstraction.

Several of the key steps in this great change-over seemed to come entirely by accident, as when Wilhelm Roentgen, a physics professor at the University of Würzburg, experimenting with a platinum plate fixed inside a large vacuum tube in which a raylike electric current of electrons was shot against it, noticed that a screen covered with fluorescent salt (that just happened to be near) started to glow every time he switched on the tube. Since it was known that the stream of electrons (then called a "cathode ray") could not penetrate the glass walls of the tube, it was evident that some other kind of radiation must be being produced, probably by the electrons from the negative (cathode) terminal as they hit the metal plate. Roentgen tried blocking off the screen with a fat "bound book of about one thousand pages" but it still "lit up brightly" and did not diminish much until shielded with plates of copper, silver, gold, platinum — or when virtually blacked out by lead more than 1.5 millimeters thick. Later, venturing to hold his bare hand in front of it, he was thrilled to see the screen clearly reveal "the darker shadow of the bones within the lighter image of the hand itself."

Finding he could not reflect, refract or polarize his new rays by any of the usual methods, Roentgen guessed that they might be ascribable to "longitudinal vibrations in the ether," but so much mystery attended them that he rather wistfully named them "x-rays." Even after the medical profession eagerly adopted x-rays as a prime tool for diagnosis of internal injuries, seventeen years were to pass before their true nature as high-frequency electromagnetic waves was to be established in 1912.

By that time, many scientists had tested x-rays with diffraction gratings (surfaces on which thousands of very fine parallel slits have been scored per inch), looking in vain for the successions of spectra that visible light of all wave lengths diffracts by these means. Then unexpectedly, Max von Laue, another German physicist, came to realize that if x-rays were waves at all, they must be a lot shorter than had been suspected, and he suggested that man-made gratings were too coarse for the job but that "a grating provided by nature in the form of crystals might catch waves less than a thousandth as long."

Sure enough, as soon as von Laue had worked out the mathematics of his theory, other physicists confirmed it in a sequence of astounding experiments in which precisely controlled beams of x-rays passing through the almost unknown three-dimensional lattices of zinc sulfide, calcite, phosphorus, uranium, mica and other substances diffracted the beautiful and characteristic patterns of each crystal upon photographic plates as a kind of unheard music, triumphantly proving thereby that the wave length of x-rays is scarcely one ten-thousandth the wave length of light. Or, putting it another way, not only had the deduced regular spacings of the crystals "taken the pulse" of the x-rays, but the x-rays simultaneously returned the favor by revealing the exact chemical interrelationships of each crystal.

And as the culminating act in this drama, a brilliant young English physicist named Henry Gwyn-Jeffreys Moseley succeeded in measuring the x-ray diffraction patterns of practically the whole

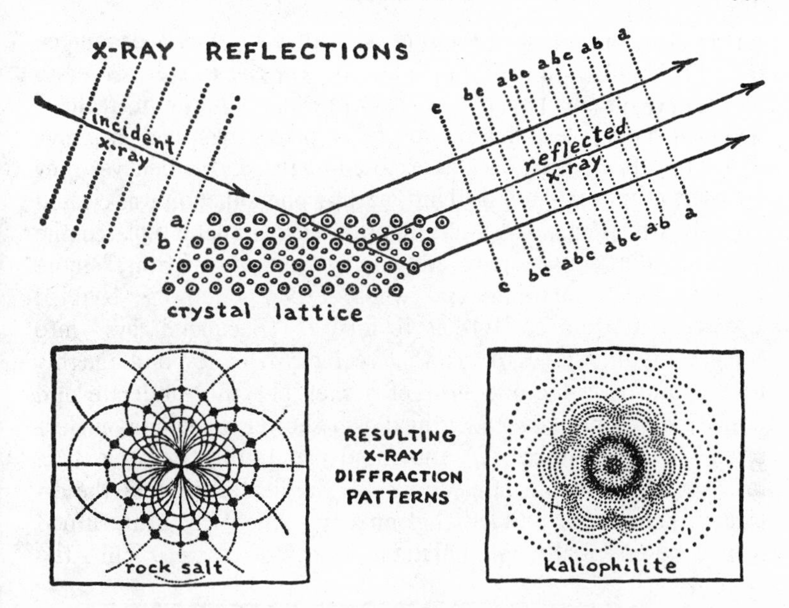

table of elements, discovering a hitherto unknown order in the square roots of their vibration frequencies that precisely conformed to the periodicity of the series of atomic numbers from 1 for hydrogen to 79 for gold — an exquisite and sweeping confirmation of the ancient, but no longer legendary, music of the spheres. The only flaw history seems to have allowed in connection with this noble accomplishment is the heart-rending irony that Moseley was prevented from testing the elements beyond gold only by the British army's recruiting office, which perfunctorily called him into His Majesty's service in 1914, just in time for some hasty training in the use of the rifle and bayonet before shipping him to Gallipoli to be impersonally cut down by a Turkish bullet at the age of twenty-eight — a sickening recapitulation of the senseless martyrdoms of Archimedes in the fall of Syracuse and Lavoisier under the Revolutionary Tribunal of France.

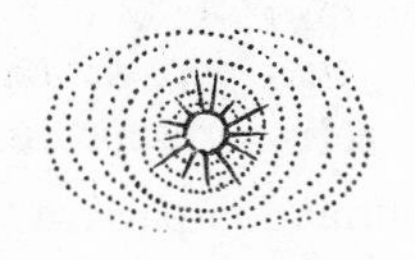

Another "accidental" discovery was that of Henri Becquerel, the French physicist, who, in trying to solve the mystery of x-rays by looking for them in phosphorescent salts of uranium, realized one day in 1896 that his photographic plates, wrapped in heavy light-protective paper, became exposed *in the dark* whenever they were left near the uranium. This puzzling phenomenon was cleared up only after Pierre and Marie Curie two years later isolated the element radium that "radiated" energy "millions of times" more vigorously than uranium and whose experimental use enabled Ernest Rutherford and others to analyze "Becquerel rays" into three components (α, β, γ) that seemed to correspond roughly to the smoke, flame and heat of a fire: (1) the moderate and positive alpha ray (later identified as a shower of helium nuclei) serving as the smoke; (2) the more penetrating negative beta ray (actually a stream of electrons) as the flame; and (3) the extremely penetrating and lethal gamma ray, which, like heat, turned out to be essentially a nonmaterial aspect of matter. Thus the

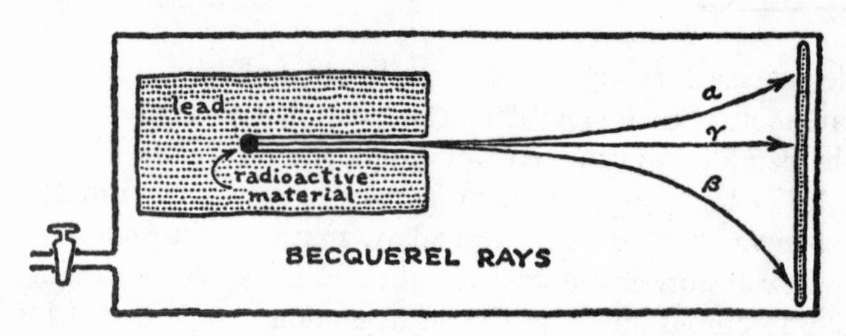

last is the only one of the three that is, strictly speaking, a ray, being in fact a very short electromagnetic wave of radiation, specifically only about a hundredth as long as an x-ray wave and emanating straight out of the atom's nucleus. This means that radioactivity is a dispersal of elementary particles along with high-energy radiation, and it turns out that, at any given moment, half the remaining dispersable energy of any radioactive element will be released within a constant period called its half-life. Which may explain, for example, why a waning supernova loses half its luminosity every 55 days, that period being the half-life of radiating beryllium 7, a radioactive isotope that disintegrates to form lithium.

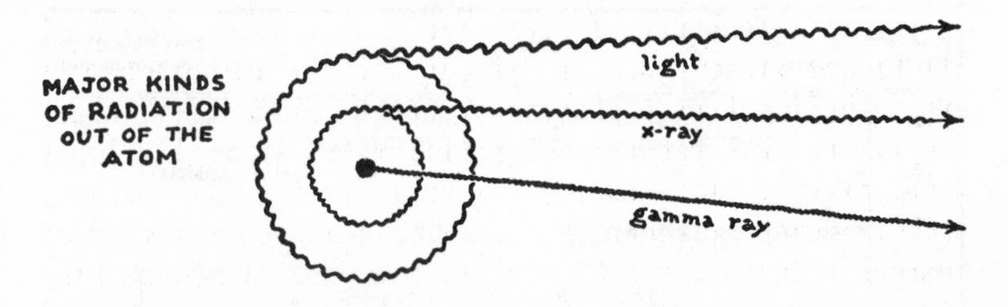

By the time most such phenomena had been sifted out, the whole spectrum of electromagnetic radiation, including visible light, was becoming pretty well established as a continuous scale of wave frequencies — a thought-provoking analogue of the musical and atomic scales, only much vaster, being not just seven or eight octaves in extent but stretching over more than 70 octaves plus several overlappings — all the way from the very long "double-bass" radio waves that vibrate only a few beats a second through the "tenor" of infra-red rays, visible colors, "alto" ultra-violet waves and "soprano" x-rays up to the shortest and highest pitched of "altissimo" gamma waves at frequencies in the quadrillions of megacycles per second.

It is, of course, a tremendous, not to say fantastic, spectrum and one which would surely have delighted Newton as well as Pythagoras, Aristotle, Galileo, Huygens and many another great pioneer of science. Its mere existence speaks great truths about the interrelationships of the world and the nonmaterial nature of its inmost texture. And our new realization that man's unaided eye cannot see even one full octave of the seventy known to exist is as heathfully humbling to the human ego in its way as was Copernicus' demotion of the earth from the center of all creation to the third satellite of an average star.

Viewed in the perspective of the whole sweep of radiation, then, invisible ultra-violet waves 3,700 angstroms long are harmonically close to and in the same "key" as invisible infra-red waves 7,400 angstroms long, being exactly one octave higher with all humanly visible light neatly bracketed between. Maybe that is why the opposite red and violet ends of the visible spectrum seem to join into a natural color circle through magenta and crimson-purple. Certainly this single obvious octave of frequen-

THE RADIATION SPECTRUM

Frequency in cycles per second	Types of waves or rays	Wave length	Wave length in angstrom units
10^{-2}	UNKNOWN WAVES OF PULSING STARS, GYRATING GALAXIES AND THE UNIVERSE (?)		10^{20}
10^{-1}		one million kilometers	10^{19}
1 one cycle per second			10^{18}
10	ALTERNATING CURRENT WAVES		10^{17}
10^{2}		one thousand kilometers	10^{16}
10^{3} one kilocycle			10^{15}
10^{4}	LONG RADIO WAVES		10^{14}
10^{5}		one kilometer	10^{13}
10^{6} one megacycle	AIRCRAFT CONTROL NORMAL BROADCASTING BAND		10^{12}
10^{7}	FM WAVES SHORT RADIO WAVES		10^{11}
10^{8}	TELEVISION ULTRA HIGH FREQUENCY WAVES	one meter	10^{10}
10^{9} one thousand megacycles	HYDROGEN IN SPACE		10^{9}
10^{10}	RADAR	one centimeter	10^{8}
10^{11}			10^{7}
10^{12} one million megacycles			10^{6}
10^{13}	INFRA-RED		10^{5}
10^{14}		one micron	10^{4}
RED / VIOLET	VISIBLE LIGHT	RED / VIOLET	
10^{15} one billion megacycles			10^{3}
10^{16}	ULTRA-VIOLET		10^{2}
10^{17}		one millimicron	10
10^{18} one trillion megacycles		one angstrom	1
10^{19}	X-RAYS		10^{-1}
10^{20}			10^{-2}
10^{21} one quadrillion megacycles			10^{-3}
10^{22}	GAMMA RAYS		10^{-4}
10^{23}		one fermi	10^{-5}
	UNKNOWN WAVES INSIDE THE ATOM'S NUCLEUS (?)		

cies contains all sorts of striking color intervals, beautifully analogous to music, like the harmonic fifth of red and violet, the fourths of red-blue and orange-violet, the complementary thirds of red-green and orange-turquoise and any number of chords beginning with the simple major triad of red-green-violet. Color chords, however, as you must have noticed, are not seen by the eye in a musiclike multiple form but rather as single blended tones: most often dull brown or gray or some other averaged pitch of light.

If you ever thought the key of a plucked harp string takes too much effort to visualize in abstract numbers of vibrations per second, you can be certain the color or pitch of light is many millions of times further removed from easy comprehensibility. Should you glance for just one second, for example, upon an ordinary yellow dress, the electrons in the retinas of your eyes must vibrate about 500,000,000,000,000 times during the interval, registering more oscillations in that second than all the waves that have beat upon all the shores of all the earthly oceans in ten million years. And if the dress were green or blue or purple, the frequency would be that much higher. Or if x-rays were used instead of light, it would be increased a full thousand times. If gamma rays, a million times!

This quantitative explanation of radiation that portrays color as the look of frequency is probably close to what Newton was seeking when he experimented with his prism in 1666, for it explains every kind of iridescence from rainbows to the changing hues of oil upon dirty harbor water. By now, of course, the rainbow has long since become well known as a series of refractions through all raindrops in sight that are aligned at certain angles (particularly near 138° and 129°) from the sun, each color being formed very precisely as an individual frequency cone of directions for each eye that sees it.

But oil iridescence is quite different in origin. It is the result of interference between the light rays reflected from the upper and lower surfaces of the oil film, these two sets of waves creating a phase disparity that amounts to a particular frequency and varies directly with the oil's thickness. Perhaps you have noticed that the delicate rainbowlike colors of oil freshly dumped on water always disperse as the oil spreads out its irregularities and thins down into a layer of uniform thickness and color. Then, if the oil is not prevented from continuing its spreading and thinning, its color gradually rises in frequency from red to yellow to green to blue to violet in conformance with the decreasing wave length of the interval between the oil surfaces, finally disappearing into ultra-violet when the film is too thin to be picked up by any wave of visible light. After this, it can be seen only while its slipperiness still reduces the friction of the wind enough to keep it a smooth, unrippled slick amid surrounding relative roughness.

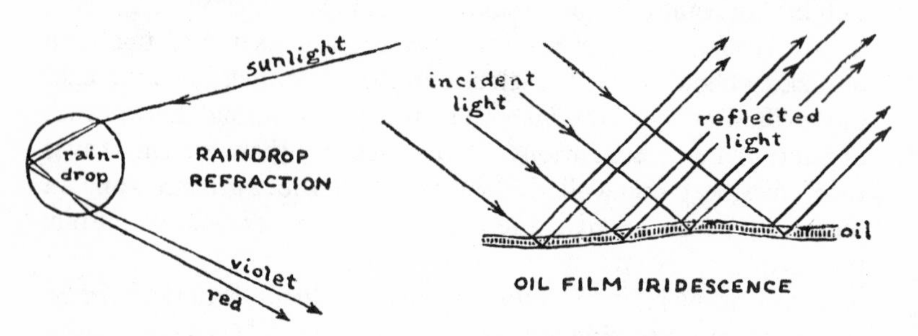

This matter of the oil film's becoming too thin to be seeable is related to the fact that most crystal lattice patterns were too fine to be discovered before von Laue thought of looking at them with x-rays — and to the inability of any sort of waves to accept modulation or gather information while the waves themselves are as long as the things they encounter. This is why a pipe between rooms can be used as a speaking tube to convey intelligence so long as the waves are much shorter than the tube, but waves long enough to tune into it immediately turn the pipe into a whistle.

On the other hand, if you are walking at night along a beach with heavy sea waves rolling in, at any place where the waves suddenly diminish you can be reasonably sure there must be a reef or something big out at sea that absorbs or reflects the waves' energy, thereby sheltering you in its "shadow." But if, instead of a reef, there should be only a thin pole out there sticking rigidly from the bottom up through the waves, you would never notice its effect at the beach. For quite obviously its thickness would be much too small in relation to the big waves to impress any very lasting influence upon them. Even if there were thousands of poles, placed close enough together to add up to a kind of reef, the resulting shadow could not hint of any individual poles. Only if the sea were nearly a flat calm but laced with tiny ripples from some whispering wind could single poles cast shadows that would appreciably persist to leeward. And it would happen then only because the length of the ripples had become smaller than the diameter of the poles.

The principle here, you see, is that of the artist who wants to paint a Persian miniature. He does not choose a barn brush but looks for a fine camel hair suited to the delicate texture of his work. And just so do the white clouds, made of water droplets whose diameters are greater than the wave lengths of light, disappear from sight as soon as their droplets evaporate (break up) into separate H_2O molecules too small to modulate the light.

X-rays, however, are fine enough to "see" water vapor in "empty" air — for instance, in samples of air taken fresh from a cloudless sky. On the other hand, although clouds can be easily seen with ordinary light, their microscopic droplets are too small to affect most radar waves, which are proportionately so long they are blind to them, while the much larger drops of falling rain can be "seen" collectively by radar yet do not provoke any reaction from the still longer waves of broadcast radio.

I mention such relationships because they give some idea of the meaning of transparency, which occurs when light (or other waves) pass through a substance without being appreciably influenced by it — which is to say, without being harmonically related to it or tuned in on it. And conversely, they show the meaning of opacity, which is the stopping of the waves to an extent that leaves a shadow whose sharpness, significantly, is strictly proportional to the shortness of the waves.

Opacity also means either reflection or absorption of waves, and commonly both, since something must happen to waves that are stopped — which something is a prime key to the appearances of things. Light waves are absorbed, for instance, when their frequency is in resonance (sympathetic vibration) with the atoms or electrons of anything they meet. As the atoms of most solids and liquids vibrate at frequencies of somewhat less than three hundred million megacycles (300,000,000,000,000 vibrations) a second, which is in the infra-red frequency range, the infra-red radiation of ordinary heat is absorbed by tangible objects in general. And as electrons and other *sub*atomic particles have much higher frequencies, in fact, considerably higher than those of any visible light, short ultra-violet as well as most of the longer x-ray radiation is also readily absorbed by virtually all materials.

Thus there are two main absorption bands of opacity in the electromagnetic spectrum — in the longer infra-red range and in the shorter ultra-violet and adjoining x-ray ranges. But this leaves prevailing transparency in three other zones: in the low-frequency bands of radio waves (which pass fairly easily through nearly anything but metals), in the very high-frequency regions of x-rays and gamma rays, which penetrate almost everything and, more obviously if less completely, in the intermediate (between infra- and ultra-) frequency range of visible light where many substances like glass, quartz, water, alcohol, glycerin, gasoline, salt, fluorite, diamond and various plastics are manifestly transparent. This middle band of transparency, moreover, is in effect extended a little way into its bordering frequencies by such devices as rock-salt lenses for the shorter infra-red waves and quartz prisms for the longer ultra-violet ones.

If you wonder now why metals are so opaque, not only to light and radio waves but to any radiation of lower frequency than short x-rays, it is easily explained in that they have loose electrons in their outer "shells" capable of vibrating in response to all but the very highest frequencies. And it is these same loose electrons, you may remember, that earlier explained such phenomena as the magnetism of iron, the conductivity of copper and, perhaps even more significantly, made it possible for Hertz to jot down a certain brief note during his famous experiments with electromagnetic waves. In this note, I am told, Hertz recorded that whenever he let the light from the flashing sparks of his transmitter shine directly upon his receiver, the tiny answering sparks in its gap became slightly brighter.

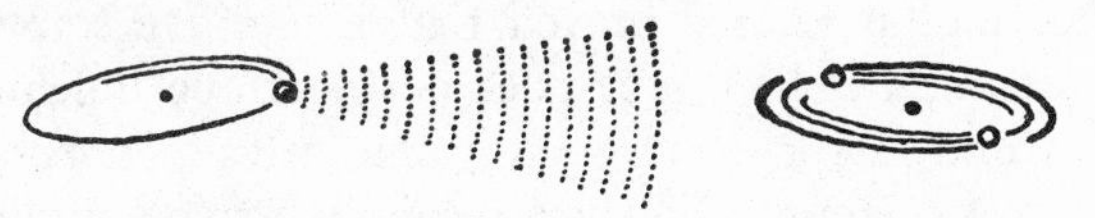

Even though Hertz's mind was evidently too occupied with other things at the time to sense much significance in this curious effect, the phenomenon could have been none other than the very photoelectric emission of electrons by light incidence that Einstein would make famous nearly two decades later when he published his concept of photons in 1905. This brilliant intuition of Einstein's was actually an extension of Planck's five-year-old quantum theory into the shattering postulation that the energy in a light beam does not spread out evenly into the universe along with the fields of its electromagnetic waves (as leading scientists then supposed) but is concentrated instead into the almost infinitesimal packets called photons.

Millikan's contemporaneous photoelectric experiments, indeed, had shown that if a piece of metal, insulated from the ground, is negatively charged with electricity, the charge (excess electrons) will leak off through the air when light shines on it— and the more so if the light is of a short-wave color such as violet or ultra-violet. What could this mean but that pieces of matter called electrons were being knocked away by pieces of light, and in proportion to the number of them? And what were pieces of light but the units Einstein called photons?

Einstein's photoelectric law, it is true, does not specifically banish or deny the existence of light waves, since a photon is considered to have a frequency, and its energy (directly proportional to its frequency) is transferable in a mysterious wavelike way to and from electrons, but every step of Millikan's painstaking investigation consistently and triumphantly supported the existence of photons and quanta, most of all through the ejected electrons' speed and kinetic energy which were completely independent of the light's intensity yet always in exact agreement with the formula of Einstein.

Likewise, as Einstein pointed out, a little-known but "crucial experiment" had confirmed the quantization of light as early as 1902, when Phillip Lenard proved that a sensitive screen being withdrawn from a light source does not continue indefinitely to absorb less and less radiant energy but instead, when its rate of absorption has stepped down to a certain extreme minimum, it refuses to go any lower until it suddenly drops to zero! Was such granulated energy consistent with the diffraction wave patterns of light? Einstein evidently accepted both without really explaining or reconciling them.

But how could light be both particle and wave? If some atom in space, for example, radiates just one quantum of light, the wave theory says that after one second this energy will be spread very thinly over the surface of a sphere 186,282 miles in radius with the original atom at its center. And then the quantum of light may happen to be absorbed. But we have found that a quantum cannot be divided. By definition and by all tests it must be absorbed as a unit by a single localized atom. So arises the awkward question, how can all the light energy spread over a sphere thousands of times bigger than the earth be gathered up instantly into a single atom? Could any mortal mind really believe such a miracle is part of normal everyday nature?

The puzzle amounted to a big enough dilemma to prevent the general acceptance of space-localized quanta until 1923, by when Arthur H. Compton had discovered that reflection can lower the frequency (or increase the wave length) of radiation by absorbing some of its energy. This he observed while watching x-rays deflect off carbon and other light elements, a process in which the reduction in frequency ranges from zero on upward as the angle of deflection correspondingly increases. And Compton demonstrated mathematically that this is exactly what one should expect from Einstein's photon theory if photons are really material objects that have momentum enabling them to "bounce" off electrons or other particles according to the mechanical laws of billiard balls. In a very glancing blow, he noticed his x-rays were only slightly deflected in direction and gave up little energy to the electrons they hit. But in a nearly head-on collision, they were knocked away at a sharp angle after relinquishing much more energy — a fact made obvious by their considerable drop in frequency.

The compton effect, as it came to be known, practically clinched the photon as a proven material particle — incidentally explaining fluorescence as a slightly lagging "reflection" at lowered pitch and phosphorescence as a longer delayed "reflection" (really an absorption followed by a re-emission) that also reduces the radiation to a lower-frequency color.

But there were lots of other aspects of light that still needed clarification, including some that had been under intensive scientific development for decades. Did you ever read about the singularly unfortunate position philosopher Auguste Comte put himself into last century, when he rashly announced that man must reconcile himself to "eternal ignorance" of the composition of the stars?

His words had hardly been spoken before the astronomical spectroscope eloquently refuted him in what has been called the language of the atom. It was a language in refracted colors that explained, among other things, exactly what the stars are made of, how hot they are, whether they are coming or going and, rather conclusively, that they are chemical cousins of the sun and earth.

The letters of this atomic alphabet had been first remarked by William Hyde Wollaston in 1802, followed by a poor glass-polisher's apprentice named Joseph von Fraunhofer in 1814, in the form of mysterious dark lines among the separated colors of sunlight refracted through a prism. After Fraunhofer, starting at red, had designated these lines A, B, C, etc., it was discovered that each one originated in the absorption of light of its own frequency by a particular kind of gas in the atmosphere of the sun or the earth and, within a few years, every chemical element was shown to possess its own indelible, spectral signature: a fixed set of absorption and emission lines and bands that are as insolubly keyed to it as the lines of your fingerprints are keyed to you. Thus Fraunhofer's red lines A and B (at wave lengths of 7,594 and 6,867 angstroms respectively) proved to come from the earth's most plentiful element, oxygen. His red line C (6,563A) is made by hydrogen, his yellow lines D are sodium, green E is iron, blue F hydrogen again, violet G both iron and calcium . . .

Probably the easiest way to visualize the workings of such absorption lines is through the analogy of music. If you have three tuning forks in a room, each keyed to a different note in the major triad C – E – G, and this same chord is played in the next room with a large peephole opening between the rooms, your three forks will start to hum the chord by sympathetic vibration through the air. But if another E fork is placed exactly in the peephole and the experiment repeated, the original E note will be found missing from the resultant dyad, which will now consist only of C and G, the middle fork being silent because the new E fork in the peephole hums in its stead, having absorbed most of the energy of E frequency as it tried to pass through the opening. The three original tuning forks in this analogy correspond, of course, to the spectroscopic screen and the missing E note to the dark line of a frequency that did not arrive because it was absorbed by its resonance with some substance it passed through on the way.

Analyzing such spectral fingerprints geometrically, and concentrating on those of the simplest element, hydrogen, an obscure Swiss mathematics teacher named Johann Jakob Balmer became so fascinated with the riddle of their musiclike intervals that by 1885 he had succeeded in working out a "ladder" or climbing sequence of the frequencies of all the known hydrogen lines, each frequency being a definite fraction of the ladder's limiting frequency (3,287,870,000 megacycles a second) beyond which the sequence does not go. And these fractions in turn formed a beautiful harmonic series, each denominator a cardinal square in natural succession, each numerator four less than its denominator — $\frac{5}{9}$, $\frac{12}{16}$, $\frac{21}{25}$, $\frac{32}{36}$, $\frac{45}{49}$, $\frac{60}{64}$, $\frac{77}{81}$ . . . — a kind of Bode's Law for hydrogen, smallest of the inner spheres, since, if one recalls the mysterious number 4 that Bode added to each doubled planetary distance (see page 85), one may well wonder, could Bode's 4 be in any sense related to the 4 that Balmer subtracted from his denominator?

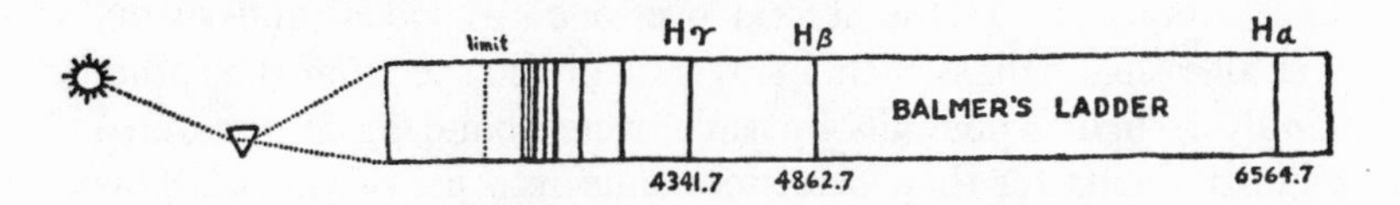

Whatever the truth might be, few spectroscopists had ever heard of Bode, and Balmer's Ladder appeared to be nothing but a mathematical curiosity even to the leading physicists of the day. Yet as more and more spectral frequencies of more and heavier elements were measured in succeeding decades, it became increasingly obvious that every element has its ladder or ladders and most of them are much more complicated, though no less beautiful, than Balmer's Ladder for hydrogen. Furthermore, an unexpected Rosetta stone of the atom's language with a complete key to the meaning of the spacings of the rungs in each ladder suddenly turned up when Niels Bohr proved just before World War I that the relationship between these colored lines corresponds exactly to the quantum intervals of the electron shells within the atom.

Thus the descent of an energy-radiating electron downward and inward, shell by shell, orbital after orbital, from a higher to a lower energy state as Bohr explained it (see pages 338–341), had its perfect visual expression in these dainty footprints of light down Balmer's Ladder! It was a breath-taking revelation of actual quantum divisions in action — one might say a glimpse into the electron's secret garden — a vision significantly comparable to, and no less lovely than, the sight of Saturn's giant rings resting palely, placidly upon the black counterpane of space.

As the dynamics of the atom further unfolded, it became evident that the series of spectral lines Balmer studied is not only quantized but comes specifically from the light emitted by the hydrogen electron in dropping from the third, fourth, fifth and higher energy levels to its *second* shell near the atom's nucleus. And this led to the realization that the wave lengths of all radiation producing visible light are functions of the distances between the atom's shells down to shell 2, the drop from shell 3 to shell 2 radiating as red light, from shell 4 to 2 (the first overtone) as blue, from shell 5 (the second overtone) as violet, and so on. Yet all visible light is such a tiny part of the spectrum that other kinds of "light," normally invisible, were bound to be discovered as instruments for their detection came into use — which is how it happened with the so-called Lyman series of ultra-violet light emanating from the narrow limits of shell 1, with the Paschen series of short infra-red light from the not-so-narrow shell 3, the Brackett series of longer infra-red from wider shell 4, the Pfund series of still longer-waved radiation from still wider shell 5 . . .

Each of these invisible rainbows of spectral music, you see,

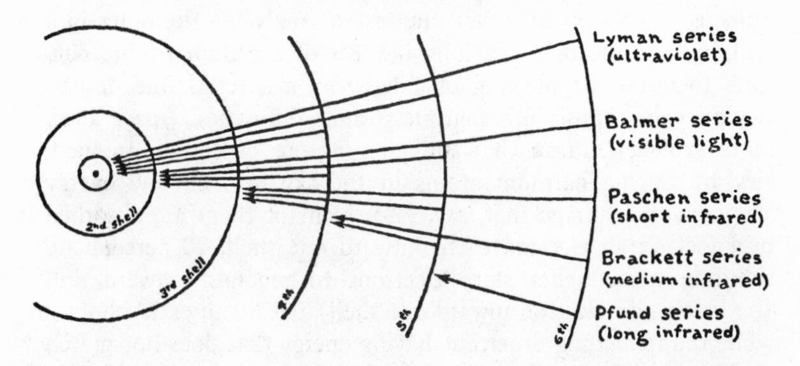

is a world unto itself that took legions of scientists on several continents many years to develop. Special towers and observatories were built for spectroscopy of the sun and stars, modern counterparts of Stonehenge, Cuzco and the great pyramids of Egypt and Mexico, and their powerful series of prisms, lenses and reflectors magnified and widened the spectrum into literally a hundred thousand lines and bands that now reach a total linear spread of more than a mile. Even the very simple Balmer series on visible hydrogen has become an extension ladder of thirty-one rungs, some of them traceable to hydrogen's extra shells called "virtual" orbits because the hydrogen electron never attains them under earthly atmospheric conditions but only when the atom is vastly expanded in vacuous space. And some of them originate not with the single H atom but with the H_2 molecule and its added frequency of rotation about an axis perpendicular to the line connecting its two H nuclei, not to mention the interference beats from their combined vibrations.

If the visual light of simple hydrogen can produce so many lines, is it any wonder that a hundred elements, with their much more complicated and multiple systems of electrons create several thousand times more? The problem of deciphering such units, which further research is ceaselessly subdividing into still smaller units, is patently greater than the task of analyzing the notes in a symphony. Indeed, if the long low B♭ of a certain timbre conjures to mind the blowing of a bassoon, a spectral line of particular color, position, sharpness and diffuseness may, even more explicitly, mean that sunlight photons of certain frequencies are hitting hafnium atoms in the sky in such low energy states on the average that, say, 72 percent of them are absorbed by knocking an electron each outward one shell, 20 percent are colliding with higher state electrons to rebound onward and (by batting an electron inward one shell) release an extra photon, while the remaining 8 percent having energy that does not match any energy difference between two electron levels are simply passing through the hafnium without reaction.

Winding up this subject with a few sample rules of spectral analysis, I will add that when the characteristic absorption lines of iron are found in starlight, the source star must be at least as hot as molten iron. And the star's color or frequency will show directly whether it is red hot, yellow hot, blue hot, ultra-violet hot or to what degree it may be putting out x-rays or gamma radiation. Broad bands, whether from Earth or sky, indicate a solid or liquid source since overlapping and blended (i.e., broad) radiation is an unavoidable output of the denser states of matter, while thin separated lines tell of the lonely freedom of gas molecules or isolated atoms that rarely interfere with one another. And the qualities termed *sharp, diffuse, principal* and *fundamental* are clues to the subtle subshell differences described in Chapter 10 (see page 340), just as the Stark and Zeeman effects (particular splittings of lines) indicate electric and magnetic fields respectively, while the doppler effect (a change in frequency) shows whether the source is approaching or receding. The last phenomenon, known as the red-shift in the case of spectral lines from remote galaxies, is our main evidence of the expansion of the universe (page 215). It is the spreading of waves or lowering of frequency that is familiar, in the realm of sound, as the descending pitch of a passing train's whistle, and its appearance now likewise in the electromagnetic realm is virtual proof that all moving stars and galaxies compress their light toward the short violet waves ahead of them while trailing it out in long low-pitched red waves behind.

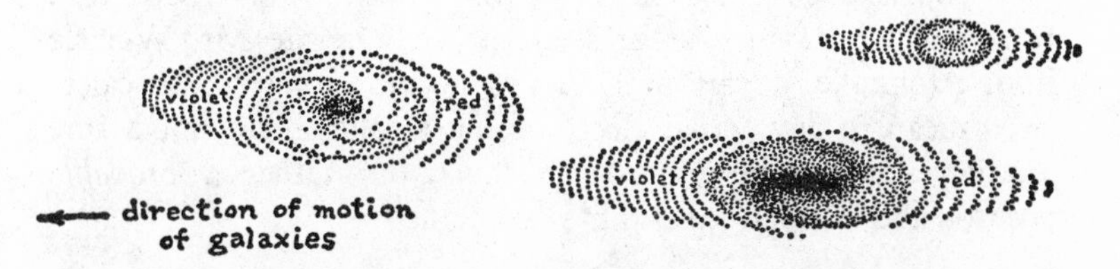

Here is a good place, I think, to glance back upon the full scope of the radiation spectrum, for by now, we should have perspective enough to see beyond it at both ends. If in a long view of the ocean, say, one could regard the tides as nothing but extremely long, low ocean waves of a super order and very slow period, so might the end waves of radiation be conceptually extended into other orders of space and time: down into the deep infra-bass frequencies of slow-flashing variable stars and whole gyrating galaxies, and — why not? — up into the still less understood ultra-altissimo frequencies that quite possibly vibrate somewhere far inside the inmost hearts of atomic nuclei.

This spectrum, as you surely realize, is not only much more than light. It is much more than radiation in the usual sense. Indeed, it is fundamental to all matter, to all energy and to all the interactions between things and things in the world. Somehow one cannot help but feel it is at the heart of the mystery of the particle-wave duality of matter — which remains to this day almost as mysterious as ever. There is no doubt that a German professor named Max Born had thoughts along this line when he was teaching physics at the University of Göttingen in the 1920's and trying out ways of expressing all the abstract and unvisualizable aspects of the atom. Balancing one attempt against another in his efforts to define the photon or check the elusive electron's position and velocity for every instant of time, he took to the tactic of cornering his facts by statistical methods. As it seemed to be impossible, for example, to pin down an electron, why not try pinning down a billion electrons, he thought. A man might be able to get a real grip on a billion electrons. And then afterward, in the hope of coming somewhere near putting his finger on a single electron, he could divide his results by a billion.

Of course, such a statistical electron would not be strictly real. It would be just an *average* electron, a composite Mr. Wavicle from Atomville, a sort of actuary's norm. Applying its calculated mean to any actual electron, one would obtain not a sure position nor a true term of existence but rather a *probable* position and maybe a kind of life expectancy.

This turn of thinking inevitably brought the classical theories of probability into modern focus and raised not a few philosophical eyebrows at the newly discovered pores of uncertainty inside matter. Just what was this demi-dawn in the world's interior that offered mankind yet another fundamental lesson in humility, this abstract kingdom of chance that loomed through the gloaming of the germs? Is chance, they asked, really the antithesis of certainty or law? Is Lady Luck but another name for ignorance?

One might conclude so on reflecting how many people annually pray for rain, whose atmospheric sources are still shrouded in confusion, yet how few think of praying for an eclipse or a sunrise, which are comparatively much simpler to understand. The immeasurable thus we naturally attribute to chance, or God, and only the measurable, the clearly calculable, to science or human reason. As we cannot correctly measure the impulse we give the roulette wheel, we consider its spinning energy a matter of luck. Raindrops fall "at random" because we cannot accurately measure their formations. Even the postman's methodical rounds through a crowded city street are apt to appear random to the eye of a pilot flying half a mile overhead. And are not the coursing stars but studs upon a kind of great roulette wheel named the zodiac?

While the ancients are not known to have considered stars to have been sown at random, there is some reason to think that stars are actually about as random as molecules. Yet if their randomness seems the greater for their large numbers, it may be their numbers again that ultimately overcome their randomness. For laws of chance activity on a large scale have a remarkable degree of certainty about them, as is now well established. We call it pure chance, for instance, whether a tossed coin falls "heads" or "tails" because we do not know how to predict such alternatives. Yet we are very sure, in the long run, that the number of heads must nearly equal the number of tails. This is about the simplest possible expression of the law of probability — a kind of paradox in which a few large simplicities are literally composed of a host of small complexities — in which, as numbers increase, uncertainty steadily approaches certainty.

On a slightly more sophisticated scale, if we flip six coins, we may get anything from six heads to six tails and, by doing it hundreds of times, we discover the results form what is known as a probability curve. This is based on the fact that, out of the 64 possible ways in which the six coins can fall, one way produces all tails, six ways show 5 tails and 1 head, fifteen ways 4 tails and 2 heads, twenty ways 3 of each, fifteen ways 2 tails and 4 heads, six ways 1 tail and 5 heads, and the last way shows 6 heads. It can be expressed in numerical probability terms as:

Number of heads	0	1	2	3	4	5	6
Probability	1	6	15	20	15	6	1

or graphically as in the curve on the next page, which is strikingly similar to that produced by a "probability board" made to display the random distribution of falling shot.

A century ago, Maxwell startled the scientific world by using a similar distribution curve to confirm the famous law discovered experimentally by Robert Boyle in the seventeenth century that "the pressure of a gas is inversely proportional to its volume." Maxwell did it by analyzing the presumably haphazard velocities of molecules for the first time, plotting them on a graph according to Gauss' "law of error" derived mathematically from probability theory.

From here, Ludwig Boltzmann took an important step further with his "law of probable population densities for molecules,"

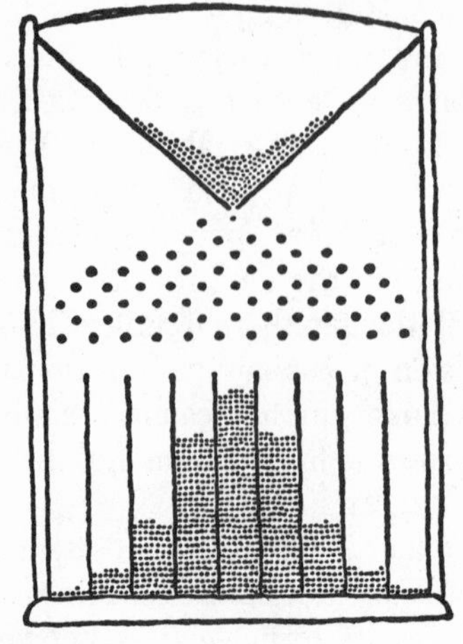

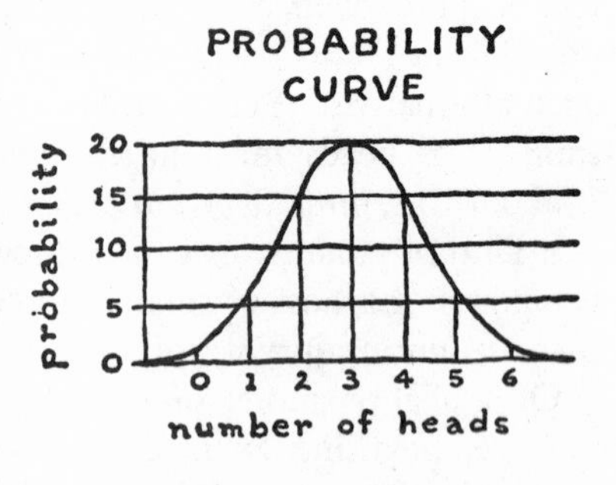

which has since become a kind of common law of the atmosphere. Perhaps of more general application in chemistry than any other principle, Boltzmann's law is the basis for calculating the evaporation rate, the energies of molecules — for practically our whole modern conception of air in which every molecule hits multitudes of other molecules a total of some three billion times a second while traveling an average of 1/160,000 of an inch between collisions. And although each oxygen and nitrogen molecule, each H_2O, each CO_2, each argon, neon, helium and krypton atom is much freer than a rolling die to go its own way — any way "luck" will take it — the net effect of them all in vast quantities is as dead certain as death itself. If this were not so, any air might explode or collapse any time, and no man could be sure of his next breath.

The workings of probability in our familiar macrocosmic realm, however, are in some ways quite fantastic — certainly not in accord with horse sense or any sort of common intuition. In a simple game of coin tossing that lasts for weeks or years, for example, you might think that, if the coin is perfectly impartial and the tossing truly random, first one player will be ahead in the score, then the other, the lead seesawing frequently between them to give each a winning edge about half the time. But this expectation is entirely wrong! The most careful probability research proves that in 20,000 tossings it is about 88 times more probable that one contestant will lead in all 20,000 cases than that each player will lead in 10,000. And "no matter how long the series of tossings, the most probable number of changes of lead is zero."

In the microcosm, on the other hand, the functioning of probability turns out to be not only different from in the macrocosm but, significantly, even farther removed from common sense. When, for example, in macrolife two golf balls are tossed at random into a bin whose bottom is partitioned into two equal compartments, there is one chance that both balls will fall into the first compartment, one chance that both will fall into the second and two chances that a ball will fall in each compartment. In the microcosm, on the other hand, substituting electrons for balls and quantum states for compartments, the chances turn out not 1:1:2 but 1:1:1.

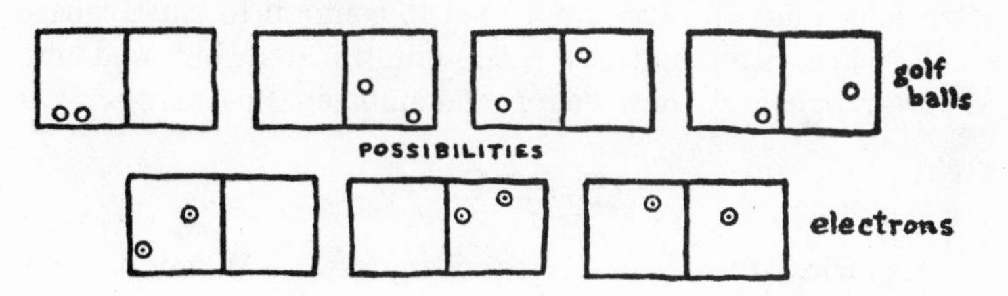

A curious difference, you'll agree! And it struck one of Professor Born's bright young graduate student assistants as a very important clue to the true nature of matter. His name was Werner Heisenberg and he not only played the piano in his off hours but had been "playing" with both the wave and particle concepts of matter at the university, "going from one picture to the other and back again," as he put it, in an attempt to find out if "nature could possibly be as absurd as it seemed to us in these atomic experiments."

The probability workings of atoms henceforth offered a way, in Heisenberg's fertile mind, of measuring the queer lack of individuality of any elementary substance — of expressing the difficulty (or was it really an impossibility?) of keeping track of which of identical molecules is which. Perhaps golf balls, he speculated, if they could be made smaller and smaller until they faded out of sight, would ultimately turn into something like marks or dollars in a bank account, which soon lose their tangible identity and become only "things to balance the books by." Thus the first ball in the first compartment and the second in the second compartment would be no different from the second ball in the first compartment and the first in the second. Each set of combinations would not only seem like the other but would literally *be* the other — having lost its identity materially as well as abstractly.

Of course, such an eerie hypothesis is not rigidly expressible in classical language where you are more or less stuck with words like "particle," "wave," "energy," "velocity," "time," "position," "frequency." These expressions represent concepts thunk up by us huge inhabitants of the macrocosmos, while atomic reality is not macrocosmic and seems not even to conform to our language which, when inappropriately used, unfortunately but undoubtedly tends to convey contradictory and ambiguous meanings.

Recognizing this, young Heisenberg saw a promising new light on Bohr's old correspondence principle, which had attempted, not very convincingly, to reconcile the classic and quantum worlds, and which Prince de Broglie had pretty well by-passed with his startling wave concept of matter — a concept that Schroedinger, in turn, unbeknownst to Heisenberg, was just now (1925) working into a full-fledged modern theory of the atom. What Heisenberg saw was that atomic events have a strange independence or integrity of their own — that they do not accept partitioning by space or time into parallel or successive events.

For a convincing example, he could review Thomas Young's famous experiment in 1800 (see page 421), which revealed the interference of light when it passed simultaneously through two slits in a screen. Even after he reperformed the experiment with modern apparatus, using the fewest possible photons of light, he observed that each photon, in effect, still must go through both slits — that the investigator who asks which slit a particular photon really goes through does not receive any answer. Quite obviously, the question in that form just does not make quantum sense. In Heisenberg's own words: "We have to remember that what we observe is not nature herself but nature exposed to our method of questioning." For, if you ask *inappropriate* questions, why should nature feel any compulsion to give you *appropriate* answers? No more does the plumber respond very efficiently to the lady who phones him for "a thingamajig to fix the tub in my bathroom" without any hint that there is a leak and that the job will need a hacksaw, two 8 by ¾-inch nipples and a ¾-inch union.

It is plain, then, that on entering the microcosm, we have sacrificed a good deal of our capacity to describe what we encounter in familiar classical language. But, on the other hand, we may have gained a fresh chance to describe in some more suitable (and still largely undiscovered) language wholly new features heretofore unknown to our macrocosm.

• ⁘ ⁘ ⁘ ⁘

Heisenberg naturally turned to mathematics as the most likely available source of this needed language. In particular he turned to certain abstract principles that were already nearly a century old and had been developed by the famous mathematicians William Rowan Hamilton of Ireland and Arthur Cayley of England. Hamilton was an especially appropriate choice from the standpoint of language, for he had been a prodigy reputed to have mastered thirteen languages by the age of thirteen, among them Latin, Greek, Hebrew, Sanskrit, Persian, Arabic, Chaldee, Syriac and sundry Indian dialects. At twenty-two he was appointed a professor and, in another of those mental projections that in retrospect seem to have a divine prescience, he borrowed a principle from Pierre de Fermat, who had observed of refracted light that "the ray pursues that path which requires least time," and perfected it mathematically as if in preparation for the undawned, undreamed atomic age and the even less dreamed advent of relativity. It was the ancient metaphysical *principle of least action*, developed also by Pierre de Maupertuis (who applied it to animal motion and plant growth) and by Leonhard Euler. But Hamilton formally adopted it for optics and dynamics, explaining that light navigates space as a sailor navigates the ocean, by just naturally seeking the great circle or shortest path — the easy, economical route of *least action.*

Hamilton used a simple mathematical notation of p for momentum (mass times velocity) and q for position (in terms of coordinates), by which he could express the dynamics of any particle in p's and q's. And with rigorous logic he forged and proved a link between the sciences of light and dynamics, incidentally using weird nonarithmetical quantities that multiplied by unheard-of rules which often made p times q come out different from q times p.

There is no need further to confuse you, or myself, with a lot of abstruse mathematical language we haven't learned the vocabulary or grammar of — so suffice it to say that Heisenberg took to Hamiltonian mathematics like a cat to fish. And as he tried to analyze an electron's behavior in p's and q's, separating these factors into their simple constituent sine waves by Fourier's method (see page 391) to get at the heart of the wave-particle paradox, he half-expected that the resulting list of sorted wave frequencies would match up in some normal, or at least reasonable, way with the frequencies of the radiation spectrum. When the two sets of frequencies apparently would not agree, he wondered why. If matter and light are essentially just two aspects of the same thing, should they not somehow correspond or have some kind of matchable relationship? It was, of course, Bohr's old correspondence problem in a subtly different context.

At this point, Heisenberg suddenly caught himself. "Perhaps," he reflected, "I am asking the wrong question again." And a slight shift of focus immediately came to mind: if Fourier's analysis (a classical relation) and Balmer's Ladder (a quantum relation) will not correspond in the obvious way, why not try an unobvious way?

So Heisenberg began casting about for fresh concepts. He tried to avoid thinking of electrons moving around orbits, since no one had ever seen an electron and, as de Broglie pointed out, any "orbits" one of them might or might not be presumed to move along could hardly be other than fictitious, to say nothing of being descended, more than likely, from ancient astronomical tradition. He mused instead on the well-observed ladders of light, which are about the solidest facts yet established in the atomic realm. And he got to thinking of them as profound spatial intervals, as irregular geometrical abstractions — sometimes simply as relative distances stretching this way or that way across the universe. When a man matches up distances in down-to-earth life, however, comparing the distances, say, between towns on a map, he has an interesting range of choices as to how he may go about it. "Might such an idea possibly give me my opening," wondered Heisenberg. At first the man may think only of a linear representation of distances, as, Berlin to Leipzig 90 miles, Leipzig to Nürnberg 145 miles, Nürnberg to München 95 miles. But eventually something will probably spawn a bigger or more sophisticated concept in his mind — perhaps a tabular form with its added dimension, its superior flexibility and wealth of information, not to mention a strangely fascinating symmetry about one diagonal axis. The tabular form may be expressed as:

A MILEAGE TABLE

Name of town	Berlin	Leipzig	Nürnberg	München	Name of town
Berlin	0	90	235	330	Berlin
Leipzig	90	0	145	240	Leipzig
Nürnberg	235	145	0	95	Nürnberg
München	330	240	95	0	München
Name of town	Berlin	Leipzig	Nürnberg	München	Name of town

Could this, by any chance, be what was needed for a meeting of the classical and quantum worlds? It was certainly neat and suggestive. It was a square table and just happened to suit Heisenberg's listing of his p's and q's, using frequencies and amplitudes in place of towns and miles. There didn't seem any point in worrying about just what it all might mean so long as it worked. After all, Heisenberg was barely twenty-four years old and would have plenty of time for checking up and philosophizing later.

Without any clear idea of where he was going, therefore, he plowed ahead into what turned out to be quite a complicated system of square tabulations of momentums and positions. And, sure enough, he found that, when he had laboriously devised a logical way to multiply momentum p tables by position q tables, $p \times q$ almost never equaled $q \times p$.

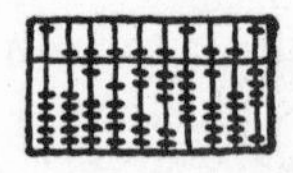

How old Hamilton (who died in 1865) would have exulted over this! And Professor Born eagerly pointed out to Heisenberg that Arthur Cayley, who had generalized Hamiltonian algebra and explored the frontiers of geometry, had also managed to invent a visionary "matrix calculus" in 1858 that was as if made expressly for Heisenberg—a curious theoretical technique for handling square number "matrices" such as the new p and q tables. In Cayley's day, this was about as esoteric and other-worldly a field of endeavor as could be imagined, with no real, practical, or even potential application in sight. And yet it was exactly the field Heisenberg would be reopening sixty-seven years later in the process of reconstructing the whole foundation of physics! By what slim odds could such a "coincidence" have occurred through random luck alone?

Professor Born and Pascual Jordan, a mathematical colleague, now became so intrigued by Heisenberg's matrix discoveries that they set to work to dig to the bottom of the mystery and, after "much extraneous assumption" while diligently minding their p's and q's, finally came up with the peculiar equation $p \times q - q \times p = \frac{h}{2\pi\sqrt{-1}}$ which says that the difference between the two ways of multiplying p and q is nothing but Planck's constant h divided by 2π times the imaginary number $\sqrt{-1}$. Wasn't this perhaps more than a little odd, to put it mildly? For 2π is simply what you multiply the radius of a circle by to get its circumference but, since there aren't any known numbers that can be multiplied by themselves to obtain a negative product, how can -1 have a square root?

The mystery went very much deeper than this, of course, as all mathematicians (including Heisenberg) well knew, and the word barely had time to spread abroad before many another theoretical physicist was drawn into the fray, notably the great Bohr and Pauli and that extraordinary but then still unknown Englishman, Paul Dirac. A year younger than Heisenberg, Dirac had just begun to wonder if he had enough aptitude to finish his engineering course when the news struck him like a bolt from Olympus. Instantly he forgot about becoming an engineer and threw his whole soul into the challenge of basic matter, quickly and independently developing an abstract generalization of Heisenberg's theory that more than equaled the combined work of Heisenberg, Born and Jordan. Like the Göttingen trio, he found the magic combination of h, 2π and $\sqrt{-1}$ in one lightninglike flash of revelation that was "the most exciting moment" of his life — but, more important, he saw beyond the "scaffolding" of Heisenberg's huge, square tabulations and deep into the essence of the p's and q's. He discerned that p's and q's were actually a new kind of number that followed quantum, instead of classical, rules and convinced physicists that henceforth they would have to deal with both kinds of number. Calling the new ones "q numbers," he demonstrated with rare elegance that what they stood for included not only Heisenberg's p's and q's but also such abstract dynamical quantities as time and energy that had been represented in the square tables.

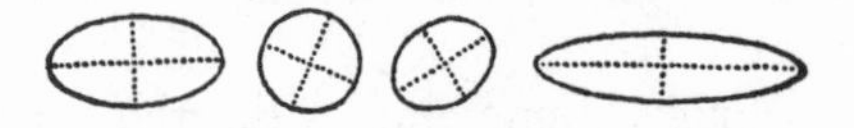

Thus the scientific world, which had been increasingly wondering what could ever replace Bohr's ailing correspondence principle, suddenly found itself in 1926 with not one but four new theories of matter, all profound, revolutionary and promising. It was an unprecedented turn of history, and the four physicists, deBroglie, Schroedinger, Heisenberg and Dirac, would all soon be rewarded with Nobel prizes — and Dirac with the chair of Newton as well.

Their four theories, moreover, were by no means wholly at odds with one another and came to be classified logically in two pairs: (1) the wave theories of de Broglie and Schroedinger, which descended from mechanics, and (2) the abstract particle theories of Heisenberg and Dirac, which developed more from optics. Besides this, there was the deeper binding influence of old Hamilton, who had boldly combined mechanics with optics in his mathematical theory, and whose "wave dynamics" had helped Schroedinger almost as much as his "particle dynamics" had served Heisenberg and Dirac.

Then one day Schroedinger had an inspiration about the abstract new quantum rules that govern the multiplication of p's and q's. He suddenly realized that these new numbers are not *quantities* at all but *commands* that mathematicians call "operators." For example, while 2 is a number, the command "multiply by 2" or the command "subtract 5" is an operator. This explains why the order of multiplication makes a difference in p's and q's for, if you apply these two operators (above) to a number, say, 8, the first order $(8 \times 2) - 5$ results in 11, while the second order $(8 - 5) \times 2$ results in 6.

$\sqrt{-1}$

Operators also put sense into the square root of -1, which had too long been a mystery. It is said that in the year 1572 a reckless Italian mathematician named Girolamo Cardano first put the apparently meaningless root of a negative number on paper, explaining that it provided an irrational and fictitious way of solving the otherwise impossible problem, "What two numbers add to make 10 and multiply to make 40?" The two numbers of his rather complex answer came out as $5 + \sqrt{-15}$ and $5 - \sqrt{-15}$, a "fictitious" solution that nevertheless proved itself mathematically.

During the next two centuries mathematicians were surprised to find $\sqrt{-1}$ and other roots of negatives increasingly useful, despite the great Leonhard Euler's appraisal of them as "neither nothing, nor greater than nothing, nor less than nothing, which necessarily constitutes them imaginary." But at the beginning of the Napoleonic wars a fancy-free Norwegian surveyor named C. Wessel tried $\sqrt{-1}$ out as an operator. Since the military command "about-face" results in a symmetrical 180° reversal of direction, evidently he thought the negative number -1 might be considered a kind of "about-face" of the positive number 1. It seemed a pretty far-fetched idea at first but it exactly fitted $\sqrt{-1}$ for, if you regard $\sqrt{-1}$ as the command for a 90° "right-face" or "left-face" turn, then $\sqrt{-1}$ times itself would be a 90° turn upon a 90° turn, which would make the product -1 or a 180° "about-face." Even the right-left ambiguity of direction is logical, as all square roots have an ambiguous sign. And of course the -1 of "about-face" fulfilled twice (or squared) results in the original 1 or forward direction.

Simple as this concept is — in fact *because* of its simplicity — it has had a profound influence on the evolution of mathematics. Today $\sqrt{-1}$ is such a seasoned tool it is commonly written only in its abbreviated form i. Most of this developed long before the day of Schroedinger, of course, but it was through Schroedinger and his new interpretation of the p's and q's that the i and the operator principle generalized wave mechanics enough to induce Heisenberg and Dirac separately to adopt Schroedinger's ψ essence of space. Whereupon, largely through Dirac's abstract insight, almost all the other important aspects of quantum mechanics were assimilated into one unified system. The mathematical ramifications of this combination — including such features as a "spinor calculus" devised to harness the spin of the electron's "waves of probability" — are rather technical to go into here, but I cannot resist mentioning that Dirac presided henceforth in Newton's professorship at Cambridge with such astringent calm and sphinx-like restraint that his students jocularly honored him with their own special unit of professorial reticence, the *dirac*. By definition: "one *dirac* = one word per light-year."

The mathematical language and structure that thus crystallize our present quantum mechanics, while obviously very important, can hardly give the average pedestrian much idea of what this stuff we call matter really boils down to. So let us try to see what can be visualized out of all these particles and waves and shells and spins and orbitals and probabilities. Is there more to matter's abstraction than the thought of square tables of p's and q's observed through "imaginary" i's?

Before trying to pin anything more definitely down, however, I think there is one additional check we should make. We should try to be sure we are facing our limitations squarely. That is just what Heisenberg did in his determination to resist being led up any scientific garden paths. And, quite significantly, it led him instead to his famous, and now solidly established, uncertainty principle.

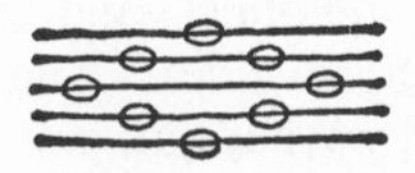

This principle is a mathematical expression of the limit of fineness of the atomic world, of the ultimate quanta grains of matter which cannot be further subdivided, of Leucippos' final "parts which are partless," of the unsharpenable bluntness of all material tools when they get down to exploring things as tiny as electrons and photons. Incidentally it casts profound light on why this world is quantized, why no action has ever been measured or computed to be less than one whole h.

Heisenberg could hardly have reached his great conclusion solely on the basis of any physical experiment, because, as you can well imagine, one cannot literally handle or measure the single electrons or photons that most clearly exhibit the uncertainty in question. Even to measure a giant molecule a million times bigger than an electron with the most delicate conceivable tool would be something like taking the temperature of a demitasse with a swimming-pool thermometer. The demitasse would be affected nearly as much by the thermometer as the thermometer by the demitasse — so the thermometer could not possibly register the original temperature of the demitasse. Much less could a single atom be measured by any material instrument. Even less again could an electron or a photon. In other words, in the microworld the observer and his instruments inevitably become an integral part of the phenomenon under investigation — even as the elephant who knew no way to test the health of his friend the beetle without squashing him to oblivion.

So, instead of a physical experiment, Heisenberg conducted what has been described as a "thought experiment." This technique, used earlier by Einstein, offers the advantage of an "ideal laboratory" in which any sort of gadget or instrument or condition is permissible so long as it does not violate accepted basic laws of physics. Heisenberg therefore equipped his mental set-up with an imaginary electron gun that would shoot a single electron horizontally across a vacuum chamber so perfectly empty it did not contain a single air molecule! And he "observed" his moving electron by the light of an "ideal" microlamp that could emit "photons of any desired wave length and in any desired number," watching it through a perfect mental microscope capable of transmitting any part of the spectrum "from the longest radio waves to the shortest gamma waves."

According to classical principles of mechanics, of course, any particle of matter must follow an elliptical (Keplerian) path, even as a planet or a pebble, and the fact that it is observed with the aid of light should make no difference. But in actuality, as we now know in this quantum age, light does make a lot of difference in the case of an individual electron — for as soon as a photon strikes it, the electron rebounds with changed velocity and, if subjected to many photons, it inevitably follows an unpredictable zigzag course.

In order to simplify the problem of keeping track of the electron's path, therefore, Heisenberg thought of reducing the frequency of his magic lamp so low that the photons striking his electron would have minimum energy, as in extremely long radio waves. Thus they would virtually cease deflecting it (energy being proportional to frequency), allowing it to go practically straight and at a fairly measurable velocity and momentum. But here arose a new difficulty. The longer the wave length of the "light," the less accurately it could "see" or locate any object — for the same reason that long sea waves are little influenced by a thin pole sticking out of the water, or that infra-red waves are blind to many things noticed by x-rays. Thus Heisenberg found himself faced with a baffling choice: either he could speed up the frequency of his photons to throw more energetic light on the electron, defining its precise location every time a photon knocked it off at some unknown speed and direction — or he could throttle down the photon's energy to stabilize and measure the electron's speed and direction at the expense of having less light to locate it by. In other words, he could accurately determine either position or momentum so long as he did not try to do both at once. He could have p or he could have q. He could not have $p + q$. And no combination of the two could amount to more than a peculiar compromise value for pq, the product of these two mutually exclusive parameters, a composite uncertainty that (it turns out) "can never be smaller than h/m or Planck's h divided by the mass of the particle."

Such is the real and mathematical essence of the principle of uncertainty. It is not just a problem of measurement. It is not a mere mental quandary, nor a plaintive ploy on two symbolic letters p and q that happen, most appropriately, to be mirror images of each other. Rather it is something like the dilemma of a photographer trying to shoot a fast-moving celebrity in a dark alley. Unless he has a flash outfit, he must make a hard choice between (1) using a fast enough shutter speed to stop the motion, which will not quite admit enough light to expose his film, and (2) using a slower shutter to gain sufficient light with the virtual certainty of blurring his subject.

It is a perverse paradox, a perpetual plight of the microcosm — the peculiar pq symmetry of mutual mathematical exclusion that knows many forms. If an electron is shot through an ideally tiny hole in a rigid screen, for example, its position in the hole at an exact time can be determined with great precision but not its momentum at the same instant. Yet by suspending the screen on an ideally sensitive spring scale, the electron's momentum can be precisely ascertained by the motion of the scale — only, however, at the price of uncertainty of the electron's position in a hole that is no longer rigid. Here again the total uncertainty can never *in principle* be less than h/m which, in effect, means that the electron's path cannot be regarded as a *line* but is rather a *band* of uncertainty nearly as wide as the radius of its "orbit."

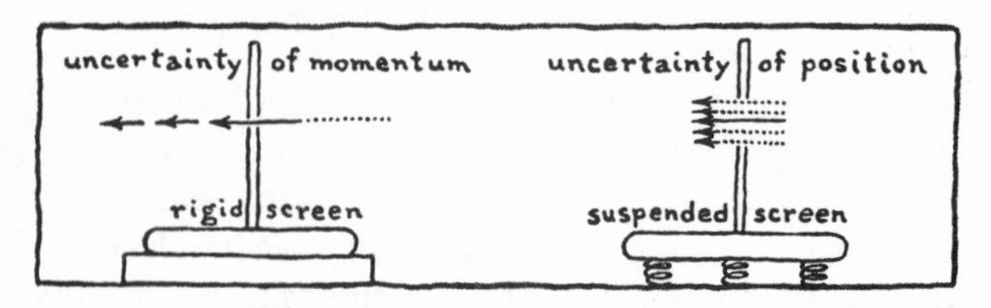

And from this it follows that a tiny grain of dust weighing one milligram should be measurable simultaneously to within a trillionth of a centimeter in position and to within a trillionth of a centimeter per second in velocity, yet no closer in either of these respects without suffering a compensative loss in the other respect. One might perhaps profitably liken this to the newly discovered macrocosmic symmetry of errors in space travel which, at high rocket speeds, makes errors in velocity more tolerable and errors in aim less tolerable whereas, at low speeds, the opposite is found true.

In any case, such symmetries of exclusion keep turning up as physicists continue exploring the abstractions of the quantum world, and probably most, or all, of them are aspects of one another. Have you ever considered, for instance, the curious symmetry between the incompatible qualities of precision and accuracy? Precision, technically, means "fineness of detail," while accuracy means "truthfulness," and an increase in either tends to be balanced by a decrease in the other. This relation is perhaps best exemplified by maps of various scales — scales whose range of usefulness turns out to be surprisingly narrow because of the practical need for both precision and accuracy in not too different proportions. If a map of the universe, let us suppose, should be drawn to show its features as looked at from infinity — taking an extreme case — the map could hardly help being absolutely accurate but also, unfortunately, it would have to be entirely blank because its precision could be no higher than zero at that unimaginable distance. If finite maps were later drawn to show smaller and smaller portions of the heavens on scales more and more magnified or — which is the same thing — with galaxies and stars seeming to come nearer and nearer, which (from the dynamic standpoint) would make their counterparts in real life appear to move faster and faster, accuracy would inevitably decrease as precision increased. And the ultimate map produced by this trend toward nearness and unlimited precision would presumably be one drawn with a perspective of infinitesimal closeness showing the minutest imaginable subatomic detail, virtually no part of which (because of the limitations of knowledge) could possibly be accurate.

From such a symmetry of the unknown universe and the uncertain electron, one can range onward into all manner of significant symmetries. Time-energy is an important one, for any experiment that measures the energy of any particle or system takes a certain length of time to perform and tends to alter the system's energy. Moreover, it turns out that the shorter the time period the greater the effect upon the energy, so that the more exactly the time of measurement is known the less certain can be the energy quantity measured. And again, the product of both uncertainties is never less than h.

Probably the simplest kind of symmetry to visualize is ordinary structural symmetry which has the quality of remaining unchanged when reflected in a flat mirror. A cube is symmetrical in this sense. So is an average table or chair. But structural symmetry can be nicely divided into three geometric classes corresponding to the *point*, the *line* and the *plane* and associated respectively with the three classic kingdoms of our world: *mineral, vegetable* and *animal* (see illustration, next page). Thus a basic mineral form such as a star, a rain droplet or (to a lesser extent) a tourmaline crystal tends to be symmetric about the point of its center, a gravity-resisting vegetable to be symmetric about the line of its vertical axis, a mobile animal symmetric on either side of the plane that separates the right and left halves of its body.

But the third of these evolving symmetries includes the curious asymmetry of handedness or, more exactly, the mirror twinship of

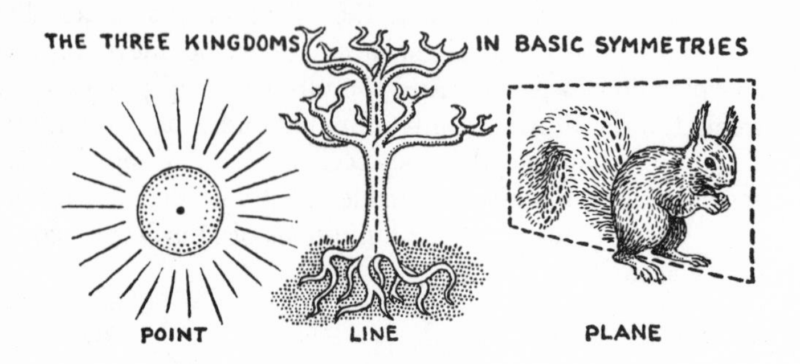

asymmetric parts in a symmetric whole. This abstruse distinction between rightness and leftness deeply perplexed Immanuel Kant, who asked, "What can more resemble my hand and be in all points more like it than its image in the looking glass? And yet I could never fit such a hand into the place of its original."

Handedness pervades the world — as much the microcosm as the macrocosm, not excluding (so far as we know) the galaxies, the supergalaxies nor, mayhap, the whole universe. As a geometric concept, handedness is found in abstract structures of four and more dimensions — in fact in structures of any number of dimensions. While a straight line, for example, is symmetric as a whole along its one dimension, a long segment of it followed by a short one makes an asymmetric (handed) pattern that may be mirrored by a point into a short segment followed by a long one. Three-dimensional crystals are commonly right- or left-handed as we saw in the case of quartz (page 288). And the vertebrate body, while bilaterally symmetric in general, has a leftward heart and a definitely handed spiral intestine. Even in the traditional realm of the soul, handedness has developed an asymmetric symbolism along the lines of mental and spiritual polarity, accounting for such familiar Biblical distinctions as those found in St. Matthew's account of the last judgment when "he shall set the sheep on his right hand, but the goats on the left" and the fortunate ones on the right "inherit the kingdom" while the outcasts on the left are condemned "into everlasting fire." Obviously, it is only this sort of handy symbolism handed down the ages that makes possible such a *hand*some pun as "Be right or you will surely be left."

Most words of ordinary speech are alphabetically asymmetric, since they must adapt their linear structure to the complex needs of semantics. Yet there are a number of palindromic words like *level, redder, noon* and familiar names like *Ava, Bob, Dad,* that read the same both ways. And there are symmetric sentences such as, "Draw pupil's lip upward" or Napoleon's famous retrospection, "Able was I ere I saw Elba." There are even complete conversations: such as Adam's first remark, "Madam, I'm Adam," to which Eve, sitting up and taking notice, appropriately replied, "Tis I, Eve. I sit."

Word symmetry can be vertical as well as horizontal, however, an example being the ambidextrous gambler's telegram "DOC HID DICE — CHECK BOX BED" which reads as easily in a mirror as outside one. This is likewise true of the girl and boy names AVA and BOB, but here appears a curious horizontal-vertical distinction, a polarity that permits the feminine gender to reflect only horizontally and the masculine only vertically. I hardly need point out that the name of the notorious hermaphrodite HOXOH inevitably reflects both ways.

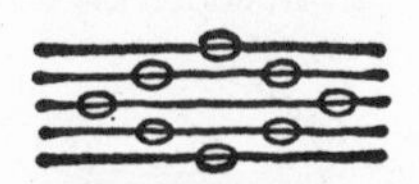

Music obviously lends itself to symmetry along the linear dimension of time, and Bach, Haydn and Beethoven all used temporal symmetry for contrapuntal effects, probably influenced by the somewhat harrowing fifteenth-century fashion for palindromic canons with complete dual melodies that imitated each other backward. And musical reflection in the vertical (nontemporal) direction can now be accomplished with startling ease by turning a player-piano roll upside down so that it plays forward but with treble and bass notes reversed. The ensuing unrecognizable melody, significantly enough, is just what would come out of a looking-glass piano if it could be played in the normal manner. And it often turns out not unpleasantly melodious despite the polar inversion which includes an unexpected transposition of major and minor keys. The classic example of such nether music is to be found in Bach's *The Art of the Fugue*, where the twelfth and thirteenth fugues are invertible. But Mozart wrote what probably is the tops of all turvy tunes with his doubly-dextrous canon featuring a second melody that is the first one both backward and upside down, thus handily (though unhandedly) enabling two players (one of whom may be Chinese) to read the same notes simultaneously from opposite sides of the sheet (with or without mirrors) in either direction!

The world is full of considerations related to symmetry, like the handedness of dice, there being two ways of placing the dots on each die's face so that its opposite sides always total seven — each way the mirror image of the other. And the fascinatingly knotty topological problem of proving that "a pair of mirror-image knots in a closed curve cannot be made to cancel each other by deforming the curve," no mathematician having yet succeeded with this obviously true but deceptively simple theorem.

If you feel you should shrug off all such matters as trivial, here may be a good place to recall that great thinkers from Pythagoras to Einstein have eagerly tackled equally humble questions — and not always victoriously. Certainly no man need be ashamed to look over Alice's shoulder into her mirror and to wonder, as she did, whether looking-glass milk is good to drink. Only half a century after her author's death, it was discovered that the milk we drink is literally endowed with handedness, being composed of counterclockwise or left-spiraling protein molecules which the enzymes of our bodies (with equivalent handedness) are able to digest. So our systems would undoubtedly react violently against right-handed looking-glass milk, if we ever could contrive to drink any. And, worse still, the recent discovery that many elementary particles have an even more fundamental handedness suggests that any physical contact between our world and a mirror one would inevitably trigger explosive annihilation wherever the irreconcilable right- and left-handed particles met!

What are we to make of this fantastic new idea of particle and antiparticle asymmetry? Of handed polarity?

If I said "new" idea, I must quickly qualify it. For in essence it is as old as history in both Eastern and Western civilizations.

The ancient Chinese monad of yang-yin ☯ symbolizes all basic dualities, including the light-darkness, positive-negative, male-female and right-left relationships of what Lao Tze called "Is and is-not coming together." And its simple and charmingly balanced asymmetry makes appropriate the fact that a pair of Chinese physicists (one of them named Yang) won the Nobel prize in 1957 for discovering (of all things) particle handedness!

Meantime the other side of the globe developed the Greek idea, expressed by Leucippos, that "*what is* is no more real than *what is not*" and that "both are alike causes of the things that come into being." He called his atoms "*what is*" and held that they move in the void "*what is not*," which is a *real* void.

Such a concept of nothingness is more basic and understandable to man, including primitive man, than you might think. Even the Micronesians have a legend of "Lowa, the uncreated" who said "Mmmmmmmm" and raised islands out of the unbroken sea to make the world.

And now the very front line of basic physical research directly faces the seeming ultimate in nadir fantasy: antimatter. Although this other-worldly stuff, composed of antiatoms made of negative (not positive) nuclei at their centers and positive (not negative) electrons moving around them, exists only theoretically so far, being extremely difficult to put together even in the best of laboratories in a materially opposite world such as ours, all its needed elementary antiparticles have been discovered separately, as you may remember from Chapter 10 (see page 327). And physicists now soberly admit that they see no reason why antimatter on a vast scale may not exist in other worlds somewhere in the universe.

Since particles and antiparticles have proved to destroy each other by explosive cancellation the instant they touch, antimatter could not possibly exist naturally on Earth, nor likely in the solar system if our sun and planets actually had the common origin now generally accepted. It is true a few physicists have speculated that the great Siberian meteor of 1908 and perhaps the older ones of Arizona and Quebec (see page 121) may have been made of antimatter, thus explaining the absence of tangible remains where they landed, but this idea has not been generally taken very seriously. Even the local Milky Way galaxy does not appear a logical place for antimatter. But distant galaxies and, more so, supergalaxies are so far apart they should not normally bother one another even if they were made of reciprocally opposite stuff.

The nearest to direct proofs of antigalaxies have been the violent radiation sources picked up by radio telescopes from a few points where galaxies are (or were) colliding or exploding (pages 212-13), the vast energies recorded being unexplainable by present knowledge except as radiation emitted in the wholesale annihilation of matter. But if there really are two such antipodean kinds of galaxies, theorists are hard put to explain how they got that way—how matter and antimatter were born and what force ever could have separated or kept them apart.

One theory holds that antimatter may exert antigravity, somehow enabling the rival structures to repel each other. But their double origin is a more difficult problem and apt to bring up the old question of Lemaître's primeval egg of the universe in the ylem days of the Big Squeeze (see page 216). At least, it will be hard to prove that antimatter was *not* created along with matter. Both stuffs are eminently stable when not in each other's company, and the laws of symmetry revealed by the paired creation of particle and antiparticle in the modern laboratory strongly indicate that both should exist in equal amounts.

Symmetry is far too big a subject to cover adequately here, as is also its application to particle life within the atom. But I might mention that since it is one of symmetry's strictest rules that heavy nucleons and their antinucleons are always born in pairs, no hermit antinucleon is likely to appear alone to explode a nucleon of ordinary matter, a destructive happenstance that, on the large scale, would amount to the cancerous infection of matter by antimatter and from which our material universe — for all we know — could completely vanish into radiation through progressive annihilation.

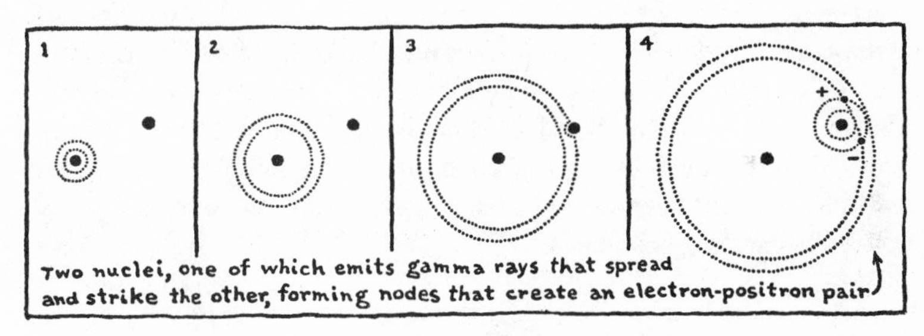

Two nuclei, one of which emits gamma rays that spread and strike the other, forming nodes that create an electron-positron pair

Electrons and positrons are also born in pairs, as well as mesons and other momentary particles. Likewise do they "decay" and die in pairs — and generally in mysterious symmetrical ways. It is as if empty space, the void, is itself somehow symmetric, not to mention literally teeming with potentiality — indeed, with a dual potentiality. It is as if the most perfect vacuum is still chuckfull of what physicists call zero-point fluctuations, or abstract subquantum uncertainty. This could mean that *in principle* there is no such thing as absolute emptiness — certainly not an emptiness in every sense. In other words, the electron may well be born out of nowhere, physically speaking — yet, because of the demands of symmetry, it leaves something behind it — something it came "from" — a sort of a hole in nothing that can move around like an invisible bubble. If it sounds crazy to say this, remember that there are no words yet in existence (in any known language, even mathematics) that can adequately express basic physical truth. So as near as we can say or think it, a hole in nothing is something. And at least some aspect of nowhere is somewhere. That seems to be one of the profound discoveries of modern science and philosophy. It is unquestionably abstract, but exactly what part of it may also have what material relation — who can say? And again and again out of the depths it keeps asserting itself. *A hole in nothing is something.* It is the root essence of symmetry in physics.

The known ramifications of this "fact" are already too many to grasp. Under their influence some of the strange fleeting particles have been found to have such strange yet orderly displacements of their expected group centers of electrical charge that physicists have adopted the new physical quantity of *strangeness* to describe this displacement. And the symmetry laws for the distribution of strangeness have worked out so consistently that a few years ago they had already given us our first definite clues to the existence of the then unknown lambda and sigma particles (of strangeness − 1), xi particles (strangeness − 2) and anti-xi particles (strangeness +2).

Another symmetry law, known as the CPT theorem, governs the three rather drastic, symmetrical operations of charge-conjugation (C), space-inversion (P), and time-reversal (T), which, in combination, turn any material process into its counterprocess in antimatter as observed in reverse motion through a mirror. The "mirror" aspect of such symmetries, however, has little to do with a looking glass, for the subatomic image, instead of bouncing off a surface, merely converges, then diverges straight through a single point as in the simpler pinhole-camera type of reflection. And other rules may deal with the mirrorlike symmetries of size, handedness, energy, momentum (both linear and angular), acceleration, oscillation, singularity (vs. plurality), gravity (vs. antigravity), emission (vs. absorption), heat (vs. cold) cause (vs. effect), numbers (real vs. imaginary), relativity, and so on.

Pythagoras would be fascinated, perhaps completely enthralled, by the modern reasoning that has grown logically out of the very principles of harmony and balance he once introduced. I mean such hyper-Pythagorean logic as: that since the symmetry of a sphere is the greatest that is possible, the symmetry of any *one* state may be greater, and cannot be less, than the symmetry of any of its corresponding *many* states. That therefore the symmetry of the *many* is a subgroup of the symmetry of the *one*. Or: as effects grow out of causes and show equal or higher symmetry, cause-symmetry is a subgroup of effect-symmetry, past-symmetry is a subgroup of future-symmetry, etc.

In numbers, as we have already hinted, one important kind of symmetry works to give each ordinary number, say, 9, its "imaginary" double, usually written as $9i$, which means $9\sqrt{-1}$. This is the same as saying that, since every ordinary number (plus or minus) is the square root of something, a number symmetrical to it can always be imagined which is the square root of *minus* the same something. Thus as $\sqrt{4} = 2$, so its symmetrical counterpart $\sqrt{-4} = \sqrt{4} \times \sqrt{-1} = 2\sqrt{-1} = 2i$, the "imaginary" double of 2. Therefore i is a kind of master key to the symmetry of all numbers: 17 and $17i$, $52/63$ and $52i/63$, 8.604 and $8.604i$ . . .

Such symmetry obviously depends partly on language, in this case the language of mathematics. But it is also deeper than language, for it involves two basic and alternative viewpoints, two profound aspects of the same thing — which just might be what Heracleitos of Ephesos sensed intuitively when he said that the knowledge of opposites is one. And if these opposites seem to be contradictory, who or what is to decide between them? I mean, when you open your door to a winter's night, are you letting *in* the *cold?* Or letting *out* the *heat?* Is a bubble rising in the sea a real thing-in-itself? Or is it mainly an absence of water? And how do you evaluate a sculptor's marble? Are the solid, positive chips he drops on the floor worth more, or less, than the negative chips of antimarble emptiness he so carefully chisels into the block to give it meaning and beauty?

At our present stage in history physicists are trying to unify the established fields of force (Gravitational, Electromagnetic and Nuclear) into one GENeral theory. This naturally encourages our budding periodic Table of Particles that is roughly comparable to Mendeleyev's periodic Table of Elements of last century, and other probings into the seemingly endless patterns of symmetry. One of the more significant recent efforts has been Dr. John Grebe's discovery that the mass ratio between muons and pions as well as between protons and sigma particles is equal to $\pi/4$. This produces a new constant (1.1288) based on $4/\sqrt{\pi}$, a tool that helps "explain" in the language of symmetry many of the mass relations of particles.

Heisenberg, for his part, is working toward adding a new constant of nature to the established c of light's velocity and the quantum h of action. "There must be a third such natural unit of measurement," he says, "which is conceived in present-day atomic physics as a length of the atomic order of magnitude — for example, the size of the diameter of simple atomic nuclei. The goal of atomic theory would be reached if one succeeded in stating a mathematical structure that does not contain any arbitrary constants besides these three natural units of measurement and from which the various known elementary particles with their proportions can be derived."

Most physical laws boil down to the conservation of something such as energy or momentum. And some, like the law of conservation of symmetry, are interrelated with other conservation laws applying to various associated symmetrical properties: spin, charge, nucleon number, handedness, strangeness, etc. Spin, for instance, is significant as one of the principal particle statistics, manifesting itself in magnetism and quantized behavior according to its own prim rules for each particle, pair or group of triplets, quintuplets, multiplets, etc. Since an elementary particle, whatever it may be, is certainly not a planet or a star but part of the microworld, it cannot spin in the easily understandable macromanner of Earth or sun. It cannot gradually slow down or speed up, and it knows no "seasons." Its poles sprout abstract appendages known as "axial cones of uncertainty." Yet, in its strange quantum way, it is affected by a "spin orbit coupling" analogous to the moon's tidal force, and it exhibits a curious interaction between its "day" and "year." Furthermore, if the spins of a lonely hydrogen atom's proton and electron, for example, happen to be aligned in the same direction (the two "north" ends parallel) there is a tendency for one of the poles to flip over until they are reoriented oppositely — and the resulting radiation

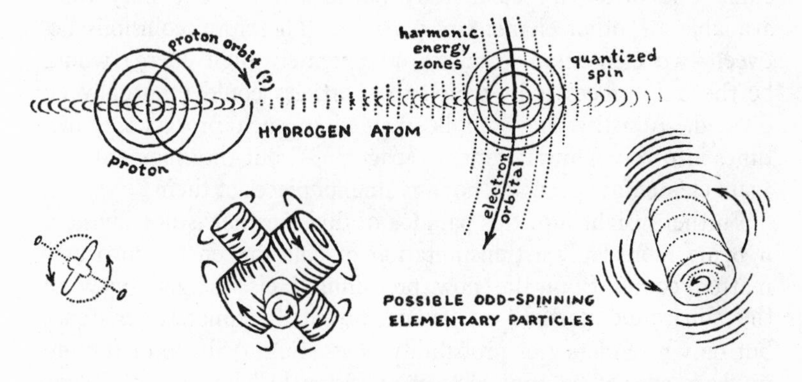

from space is the main thing "seen" by radio telescopes. The spinning of a "cylindrical" particle, on the other hand, may possibly be self-symmetrical: it may spin in both directions at once, its ends somehow rotating oppositely on the common axis which thereby achieves a continuously slipping twist. Other spins and counterspins, nodes and antinodes, turn groups and mirror-turn groups and all of the confusing ferment of translational and breathing or pumping vibrations, are balanced reciprocally as in an abstract mirror.

It may have been the many such demi- and semiunderstood factors, relentlessly cropping up like weeds in the atom's garden, that prompted Heisenberg to declare recently that we have not yet got any convincing proof that basic functions have to be simple. "The final equation of motion for matter," he ventured, perhaps a little imprudently, "will likely turn out to be some sort of quantized nonlinear wave equation for a wave field of operators that represents generalized matter, not any specified kind of waves or particles."

Asked if the so-called elementary particles, whatever they are, might someday be divided into smaller bits, he replied, "How could one divide an elementary particle? . . . The only tools available are other elementary particles. Therefore, collisions between two elementary particles of extremely high energy would be the only processes by which the particles could eventually be divided. Actually they *can* be divided in such processes, sometimes into very many fragments (page 333); but the fragments are again elementary particles, not any smaller pieces of them."

Further insight into this paradox of division that is not division, it being more truly a transmutation of collision energy into new matter, comes from the now prevailing Heisenbergian view of the subatomic particle as having no really concrete existence but only a tendency or probability of existing. "Shape or motion in space cannot be applied to it consistently," he says, and "certainly the neutron has no color, no smell, no taste. . . . Far more abstract than the atom of the Greeks, it is by this very property more consistent as a clue for explaining the behavior of matter."

Since the first photon that strikes an electron is "sufficient to knock it out of its atom," he explains, "one can never observe more than one point in the electron's orbit. Therefore there is no orbit in the ordinary sense. . . . It is of course tempting to say that the electron must have been somewhere between any two observations of it, and that therefore the electron must have described some kind of path or orbit even if it may be impossible to know which path." But this would amount to a complete disregard of the uncertainty principle and of the quantum nature of the microcosm. Uncertainty within the atom, we must not forget, is considered to be not just uncertainty of knowledge. Rather it is uncertainty of being — that elusive mutual exclusion of objective dimensions that seems to follow the lines of abstract symmetry. "Quantum theory does not contain genuine subjective features," continues Heisenberg. Although the mind must have its uncertainties, the uncertainty that is crucial to quantum physics is an uncertainty in the objective world. Indeed in a material experiment, the actual occurrences being observed *happen* at interactions between objects and measuring devices, not at registrations of these in the mind. "And since a device is connected with the rest of the world, it contains in fact the uncertainties of the microscopic structure of the whole world."

Even though Einstein had a somewhat different view, being loath to abandon his hope of achieving "a theory (of matter) that represents things themselves and not merely the probability of their occurrence," he eventually had to concede that "we must give up the idea of a complete localization of the particles," adding that "this seems to me the permanent upshot of Heisenberg's principle of uncertainty."

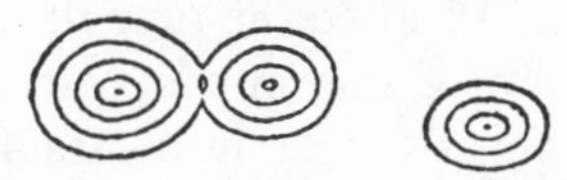

As for the real nature of ultimate matter, perhaps Einstein's best attempt described it as abstract fields "having modes of action in which there would be pulselike concentrations of fields, which would stick together stably, and would act almost like small moving bodies."

Obviously one does not need to ask what these fields or their "bodies" are made of. For what word could be used but "matter" or "energy" or some synonym equally wanting in final clarification? And if you would resort to such geometric idealism as "points in space," not only are "points" of very questionable reality but what could you actually mean by a point in space anyhow? You could hardly claim to pin down in space something which, by definition, occupies no space. Nor could a finger, needle or instrument hope to split a quantum distance, even in principle. Furthermore, how many points does a quantum hold?

Even if some instrument somehow could point to a single point in space at a single point in time, could anyone determine that that point remains exactly where the instrument located it, or exactly when, if ever, it had been there? Time too, we notice, is as indefinable as space and in the same basic way. A billionth of a second is about the most precise interval yet measured by man, and obviously there must always be some limit of measurement in this finite world.

So space and time become conscious human concepts amalgamated largely from measuring devices built of matter, not the reverse. Like a newspaper picture, which is an over-all concept formed of a white background containing a pattern of microscopic black dots, so are time and space the over-all relationship between events. And the nature of our minds synthesizes the structure of the world (dots, events . . .) into meaning (pictures, time, space . . .).

This is a way of saying that very likely fundamental entities of some sort exist independently of time or space or other illusions of limited perspective. Just as a thought can seem to be in two places at once because it is really no place, so may a single unit of matter exist in two places at once or a single event happen both before and after another event, as we shall see more clearly in the next chapter. Indeed, time and space now stand exposed by the success of relativity theory as probably not truly fundamental — and physicists no longer marvel when an electron (framed in the mirror of symmetry) is discovered "meeting itself coming back from a place it hasn't been to yet."

If we are here edging into a realm of thought that has more of metaphysics than physics in it, never mind — for physicists are coming to realize that, if they throw metaphysics into the fire, their own subject will soon smolder also — that not only philosophy but mathematics contains undecidable propositions and logic is far from unique.

It seems to be true, furthermore, that the abstract aspects of quantized matter that show up more and more prominently as we dig deeper into fundamental questions are more to be trusted than the old concrete easy-to-see illusions of classical theory. Certainly one cannot build a television station on sheer common sense. There is a real and growing difference here, you might say, between abstract and concrete things — a difference that is perhaps most easily exemplified in money transactions. If Amos owes Basil a dollar and Basil owes Amos a dollar, the abstract debts clearly cancel out to a result of zero. But if Amos has a dollar in his hand for Basil and Basil has a dollar in his hand for Amos, there are two concrete dollars which do not cancel out. Now if your imagination will allow you to think of light falling on an object as a shower of dollars descending upon a poor village, influencing it this way and that in a kind of wave motion, we can point out that the dollars will be more like real photons of light if those of opposite influence cancel each other abstractly rather than accumulating concretely.

The same holds for electrons or any other elementary particles of matter. The main difference between radiation and other matter is that radiation spreads spherically while other matter moves in a line, but each seems to boil down to the abstraction that Bertrand Russell calls "a string of events." Of course, our world contains some continuous series of events (say, a wave of water or a sunbeam) that do not all belong to the same piece of matter, in which case the change from one "event" to the next could not be considered simply as a particle in "motion." But in Russell's words, "strings of events exist which are connected with each other according to the laws of motion; [and] one such string is called one piece of matter."

Evidently "the unity of a piece of matter is causal," he goes on, "an observed law of succession from next to next. . . . If a tune takes five minutes to play, we do not conceive of it as a single thing which exists throughout that time, but as a series of notes, so related as to form a unity. In the case of the tune, the unity is esthetic. In the case of the atom it is causal."

There may be nothing at all to matter except just such causal series of events. And be the events "particles" or "ψ waves of probability" or "laws" or "symptoms" or just abstract relationships of unknown meaning, it makes little difference. You can call them anything. They are the "wisdom of the inward parts" in Job. When there are enough of them to create molecules and when the masses of molecules reach the dimensions of statistical illusion, the smooth flowiness of water will appear (if the molecules are called H_2O), or the transparency of air (if N_2, O_2, A, etc.) or something else.

But while statistical illusion increases with magnitude, there is an opposing increase in energy with minitude — which means a growing potency as we approach the nucleus of the atom. Energy being vital as the medium of passage between physical cause and effect, most phenomena become apparent only at a certain threshold of energy. That is why atoms that encounter one another at low energy behave like billiard balls, clicking each other's surfaces, their complicated interiors coming into play only at higher energies, the nucleus of the atom yielding to nothing less than 100,000 electron volts. Which explains how we can keep discovering deeper secrets of nature as we penetrate to higher and higher energy thresholds with no foreseeable end to the sequences of unexpected phenomena that ever become attainable from behind the endless veils of mystery.

Of all these shifting aspects of matter, however, undoubtedly the hardest to grasp is the factor of abstraction, which seems inversely proportional to size. As the macrocosm fades into the microcosm, indeed, the statistical certainty of masses always dwindles inexorably toward the uncertainty of single units. Macrofact devolves into microprobability. It is something like seeing a friend walk off into the distance. Exact knowledge of where he is and of what he is doing decreases. You cease knowing him so directly and you begin to know him more indirectly. Soon you know him only through telephones, letters, chance acquaintances, rumors . . . Sense has been replaced by symbolism.

Another way to express it is that there is no inherent "realness" to matter. There is instead an element of latency in things. Length and duration utterly depend on measurement, which utterly depends on an observer. These are aspects of things, just as mass and energy are aspects of the same thing. A piece of matter, therefore, no longer stands for something absolutely solid, nor something that keeps an identity, nor even "a hypothetical thing-in-itself known only through its effects." It is the "effects" themselves. We really do not know what makes electrons or protons or photons do whatever they do. We only know their effects. To us the effects *are* the electrons and protons and photons: the matter. In other words, as a particle or wave acts, so it *is*. Oscillation through space *is* light. Agitation of molecules *is* heat. Abstraction *is* reality. Matter *is* manifest behavior, palpable or deducible occurrence. And this may be the prime clue as to what the Apostle Paul meant when he said, "The world was created by the word of God so that what is seen is made out of things that do not appear."

Of course, he may have simply been referring to the established concept of atoms, which compose things without themselves appearing. On the other hand, he more likely intuitively had in mind the more mysterious spiritual aspects of composition: macrocosmic potentiality through the overcoming of quantum limitation, plural definition out of singular uncertainty, the beauty of unified structure, of binding energy, harmonic synthesis, love . . .

No one will deny that it takes more than four separate wheels, two axles, some springs and a body to make a cart. Or more than body cells to make a horse or a man. If you were given three or four individual notes on a harp, would it be possible to imagine the sound of a musical chord before you had ever heard one? If not, the whole of a piece of music must somehow be incalculably greater than its parts and, in a larger perspective, the whole of any composition perhaps infinitely more than the sum of what composes it. It hardly seems that this could fail to be the answer to why hydrogen and oxygen can convey no idea of water until the mysteriously intricate abstraction of combining them has been introduced. And, by the same reasoning, if a train runs one day along the single track from Irkutsk to Novosibirsk and if a train also runs one day back along the same track from Novosibirsk to Irkutsk, the combination of both trains running on the same day can surely amount to an event far different than the sum of their two separate runs, which, in the quantum world, would logically be a cancellation. It can, in fact, be a collision — with all that that implies of violence and woe and love and unfathomable consequence.

Thus the world that arises in a grain of dust, the sky that floats in a bubble of air, the waves of ocean and hill that can no more hesitate nor halt than a photon can stand still — and for the same reason — along with our thoughts and the mystery called God, what else is there? In the ceaseless struggle of matter and life and spirit that composes the universe, the clash and traffic of every parcel is a basic part — the eternal triangle of two pluses trying to cancel the same minus, the cross seas of probability that rise and meet and fade away, the symmetries that do not quite balance, the spins that know a bias, the polarity that yearns.

13. of space, of time

THIS SPACE STATION, I am coming to think, is just the place to contemplate space. And also time — and relativity. For our place up here seems nearly no place, my watch has turned quaintly arbitrary and the earthly calendar is about as pertinent as the old timetable of a railroad that once existed on some legendary continent.

Do you remember the ancient fable of the monk who wandered pensively a little way into a wood, only to hear a strange bird break into an ecstasy of song? And somehow the exquisite loveliness of the song carried him out of his world — for, when he returned to his monastery, the gateman knew him not. And he was amazed to find the place full of strangers who, as their old records proved, had lived there most of their lives and were equally astonished to discover their long-forgotten brother returning unchanged from an absence of fifty years!

Such tales, we are beginning to learn, are closer to the quick of life and modern physics than you might think. Before the advent of relativity earthly scientists would hardly have considered such a Rip-Van-Winklesome disparity in time fields as possible even in theory, but today they have accepted the principle that time is by no means the universal flow of sequence it always seemed. Rather must it depend on relative motion, and its different parts move at different rates according to different points of view.

Meditating this, one may well wonder what sort of time exists at the center of the earth? And how much of time extends inside the atom? What becomes of it outside the galaxy? Or, if the earth's motion should somehow ever be reversed, would that change the direction of terrestrial time in any sense?

And what of space? How intimately related is space to time? To gravity? If space should shrink to a point, would time endure without it?

The answers to such questions are not easily come by. And the dearth of words to express them may be less than the dearth of minds to comprehend them. Yet human reason need not be considered a rigid chest of logical drawers, for it is in many ways more like a waving tree that is just coming into blossom. In fact, man's collective mind has been shown to grow while learning, continuously generating in itself new potentialities for thought which had formerly been unimaginable.

I should probably mention space before time because, of the two, space is the more obvious, the simpler to visualize, and therefore it probably developed earlier in history as a conscious concept. Language definitely supports this assumption, for we do not talk about a "time of space" yet rather of a "space of time," and such spatial words as "long" and "short" now also refer to "expanses" in time, as the ancient mariner's spatial distinction between things located fore or "before the mast" and aft or "after the mast," used on a moving ship, may naturally have evolved the temporal distinction between events happening "before" and "after" other events. What could it be but our natural bias toward space that started us saying "hereafter" or "thereafter" instead of the more logical "thenafter"?

Although time was not recognized as the fundamental parameter of material change until late in the Middle Ages, the abstraction of space appeared at least as early as the fifth century B.C. After millenniums of nothing more abstract than field area units like the Sumerian "little acre" or *še* (named for the bagful of seed it took to sow it), there suddenly appeared the mysterious, ethereal void of pure numbers — the classic space of theoretical geometry. Aristotle credited this creation to Pythagoras, the first man known to have asked, "Are days and miles necessary?" Numbers can live without them, he argued. And so can circles, triangles, shapes of all sorts. And these things are real even though they exist in a peculiar ether of their own. Some of Pythagoras' followers may have visualized this unearthly space of geometric form as somehow related to matter, for Archytas, the Pythagorean, is reported (by Simplicius) to have raised the question of whether or not there would be room enough at the end of the world to stretch out one's hand.

Leucippos, for his part, imagined space as having a porous structure, the "pores" conveniently serving also as his "atoms," which were thus automatically guaranteed their indivisibility since they occupied no space. Aristotle, on the other hand, took the surprisingly modern view that where there are no bodies, neither space nor time can have any existence. And his favorite pupil, Theophrastos, went a step farther in concluding that "space is no entity in itself but only an ordering relation that holds between bodies and determines their relative positions." He may quite naturally have come to this soul-satisfying conclusion after considering the baffling problem of determining distances between moving ships at sea — or between the "wandering stars."

If space, however, proved inordinately difficult to measure, as Kepler was to demonstrate so exhaustively, time has turned out to be even more difficult. And not the least of its snags has been the near impossibility of visualizing it objectively. The Puri Indians, for example, knew only one word for "time" and modified it in tense by pointing backward for the past, upward for the present, forward for the future. Their "time" sense was also almost literally circumscribed by space in the fact that they looked upon the distant skyline of mountains that surrounded them as the outer rim of time, as much as it obviously was a boundary of space, and anything beyond this horizon seems to have been presumed drifting intangibly in a kind of forever.

More advanced peoples, meanwhile, ran into the more sophisticated difficulty of trying to devise a calendar that gave the year a fixed number of days, all neatly apportioned into regular months or seasons by the gyrations of a methodical but not otherwise any-too-reasonable moon. The fact that the earth actually takes 365 days, 5 hours, 48 minutes and about 46 seconds to go around the sun once and that there happen to be 29.5305879 days from moon to moon and 12.3682668 moons from year to year was too implausible to imagine, if not too ungodly to accept. Yet most peoples eventually had to admit the futility of trying to match the incommensurable sun and moon. The Egyptians, in fact, set up what was possibly the earliest, and certainly one of the most successful, calendars largely by ignoring the moon. Their year was timed instead by the sun, the stars and the flooding of the Nile. It had twelve months of thirty days which combined into three equal seasons: Flood Time, Seed Time and Harvest Time, each four months long. At the tail-end of this 360-day stretch they added a "holiday week" of five days to complete the calculated year.

The early Romans, on the other hand, struggled along with a carefree calendar of only ten lunar months, March through December, not bothering about the sixty-some days left over. "What earthly good are winter months anyhow?" But when the seasons kept getting twisted around, they grudgingly added January and February, plus, every other year, a thirteenth month, Mercedonius (literally "extra pay" for the legions), with whatever number of days the astronomers reckoned would best mollify the zodiac. The succeeding Julian calendar, decreed in 46 B.C. by Julius Caesar and modeled on the Egyptian with the refinement of leap year, was a vast improvement, even though its annual span averaged 11¼ minutes too long. This "negligible" amount, however, in the course of sixteen centuries, was enough to require a new adjustment of a week and a half when Pope Gregory XIII inaugurated our present Gregorian calendar (with its year still 26 seconds too long) in 1582—an adjustment which touched off unprecedented "calendar riots" among the tradition-bound people of England, Germany and Poland, some of whom risked death in frantic attempts to save "eleven God-given days" from being "forever lost."

Years, months and days have not been the only perverse units of time, however, for, curiously enough, the Roman hour (which survived until the mid-nineteenth century when Ponchielli, perhaps significantly, composed his "Dance of the Hours") was of flexible duration because its definition as a twelfth part of the sunrise-to-sunset day required it to stretch to about 75 modern minutes in June and to shrink to a mere 45 in December, varying also, of course, in different latitudes even on the same day.

Redefining the hour as a twenty-fourth part of the complete average day+night 360° of earthly rotation was the obvious solution and a big step toward reasonableness. But it left the hour still not exact enough for modern scientific purposes. Even though the earth's turning can now be clocked against remote stars to an error margin of less than one part in a quadrillion, a coarser uncertainty remains because terrestrial motion is itself by no means absolutely regular. In other words, the earth perceptibly speeds up and slows down like a live thing, probably mainly because of nutation (see page 424), the complex gravitational effects of sun and moon which are always influencing her equatorial bulge to sway a little farther northward or southward as they shift their declinations. And there must be a plethora of other slight tidal, magnetic, thermal, geodynamic and unknown influences. At any rate, it has been observed that the earth slowed down enough to lose about 27/100 of a second between the years 1680 and 1800.

During the nineteenth century, however, she picked up nearly $^{31}/_{100}$ of a second, then mysteriously lost a tiny bit between 1900 and 1920, since when she has been gaining again. There are shorter variations in her pace from year to year, of course, and also the inevitable longer-range running down of her spinning energy (see page 49), due mostly to over-all tidal friction, meteorites and erosion — all of them together slowing her by a whopping ten seconds each million years!

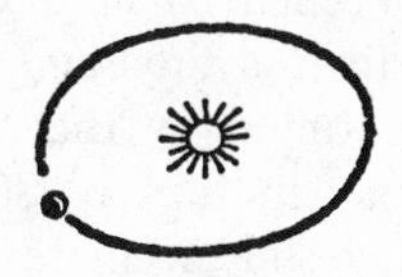

And if these are not enough fine points to consider, don't forget that the earth's nearly elliptical orbit alternately fattens and slims its form to its own very slow rhythm, the orbit now growing more circular as we approach the millennium of A.D. 26,000, when it is due to describe its most nearly perfect circle before slumping back toward the old biased symmetry. Every factor relating to the earth's energy or mass, indeed, has its effect, be it ever so subtle, on her rates of rotation and revolution — every little earthquake, storm or other shift of material alters her balance just so much — the melting of Antarctic ice, the Mississippi River carrying silt toward the equator, a new building rising up to augment the friction of the winds . . .

In view of the ponderable total of all these eccentricities of earthly motion, a man-made watch can be considered in essence an instrument for accurately interpolating between astronomical observations. If well made, it may run continuously for a quarter of a century ticking five times a second, 432,000 times a day, without replacement of a single one of its 180-odd parts, the smallest of which looks like a speck of dust and is actually a polished screw so tiny that a million of them will barely fill a demitasse. Its accuracy under normal conditions is often rated to within one part in 50,000, and I know of a navigation watch tested for 405 days in World War II, including travel in ships, planes, trains and submarine, that proved accurate to one part in 400,000, which means its errors averaged scarcely ⅕ of a second per day.

Mechanical watches and clocks are only one kind of timepiece, however. They were preceded by hour glasses and water clocks accurate to about one part in a thousand. Modern electric clocks, on the other hand, are timed by the alternating current of a power station, which may in turn be stabilized by the natural vibration rate of quartz crystals accurate to one part in a billion.

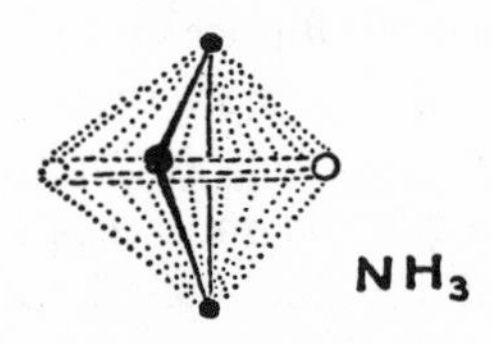

The first atomic clock, built in 1948, reached farther into the microcosm by using the ammonia molecule (NH_3) as a timer, taking advantage of the fact that ammonia's single nitrogen atom naturally flips back and forth through its triangle of three hydrogens at a steady frequency of 23,870 megacycles a second. And newer atomic clocks use the lonely electrons in single hydrogen atoms, which vibrate between their two possible positions (page 261) with such imperturbability that, when amplified by the new laser resonance principle and applied directly to the regulation of a clock, they can stabilize its movement to one part in a quadrillion or an error of less than a second each thirty million years!

Such constancy, of course, is not only much greater than that of the turning earth. It is even greater than that of the "fixed"

stars of the constellation of Horologium (the clock), which, like other stars, are admittedly moving around on unknown courses among themselves. So atomic clocks are a lot more regular than they need to be just to interpolate our celestial time, their more important uses being measurement of vibrations and rotations in basic atomic research and checking such questions as whether the microcosm runs on the same time as the macrocosm.

In practice, the standard to which ordinary clocks are set in the western world is now called UT2, a new "uniform time" determined by the United States Naval Observatory, which reconciles it with the earth's uncertainty by "occasional step adjustments . . . on Wednesdays at 1900 UT (7 p.m.) . . . when necessary, of precisely plus or minus 20 millionths of a second." This is comparable to Bach's introducing his well-tempered or compromise musical scale in place of the ideal, but impractical, diatonic scale that preceded it (see page 396).

But where does it leave our standard second, which formerly was defined as $1/86,400$ of a mean solar day? There are means and means — but where is the mean of a well-tempered day? To settle this problem, a new standard second was recently set up by international authority and defined as $1/31,556,925.974$ of the tropical (equinox to equinox) year of 1900. Time reckoned in such 1900-style seconds is called "ephemeris time." Thus, you see, we have already got ourselves a few different and conflicting time rates, officially recognized — and all right down there on one planet.

Some physicists are currently in favor of defining a second solely in multiples of atomic frequencies, these being the most constant natural time intervals we know of — and never mind if the earth spends different numbers of atomic seconds going through different astronomical centuries or its vernal equinox starts sneaking toward April or February. The physicists indeed are already using atomic wave lengths as their criterion of length — since an international Conference on Weights and Measures, meeting in Paris in October 1960, officially fixed the new standard meter at 1,650,763.73 times the wave length of an orange spectral line of krypton 86. So an atomic second, if also adopted as standard, would bring both space and time appropriately into the same basic definition.

But there would be serious complications. Atomic vibrations seem to have no vernal equinox, no reference point to set a clock by. If our hydrogen clocks should somehow stop, where would we set them after they got started again?

Also how could we check the constancy of light? That is no minor consideration in science, and to define both space and time in atomic vibrations is in effect like defining 186,282 miles as the distance light travels in one second, and one second as the time it takes light to go 186,282 miles. Under such a ruling, no matter how fast or slow light ever got to be, the distance it would go in a second would have to be 186,282 miles — no more, no less — because of the definition. We actually have evidence that a new ray of light starts off a little slower than full speed, warming up to its constant pace of 186,282 m.p.s. (c) only after its initial photons have in some mysterious fashion prepared or smoothed the way for later ones. But measuring such velocity differences would obviously be futile if our distance or time units should stretch or shrink in the same ratio as the speed of light.

Yet, believe it or not, just such a ridiculous stretching and shrinking of units — not merely by definition but in objective reality — actually turned up near the end of the nineteenth century and so profoundly baffled physicists that it took a drastic revolution, in the form of the theory of relativity, to clear the mental air.

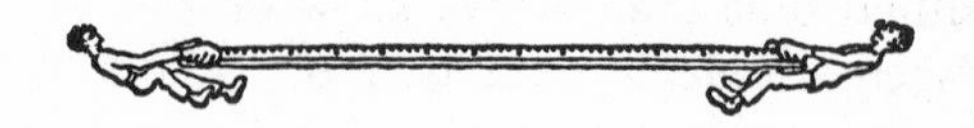

This was a scientific and philosophical cataclysm whose pressures had been secretly building for hundreds of years. It really started way back when physicists first began to wonder what is waving when a light wave "waves" through empty space. Sound, of course, had been pretty easy to visualize and diagnose as a vibration of molecules, and sound was found to travel faster in liquids than in gases, faster still in solids, for evidently the solider or more resilient the medium, the stronger its vibrations.

But light proved just the opposite. Its speed was found to be only three fourths as great in clear liquids (see page 422) and only two thirds as great in transparent solids as in air or a vacuum, ¾ and ⅔ being, you'll remember, the familiar harmonic ratios of the musical intervals of the fourth and fifth. And beyond these mysterious (perhaps coincidental) shades of Pythagoras lay the greater enigma of light traveling faster and faster through less and less, and fastest of all through an apparent nothing. If water waves wave water and sound waves wave air, what indeed could light waves wave as they speed from star to star?

Unable to believe that light radiates in absolutely nothing, scientists therefore postulated the seeming nothingness as actually a subtle somethingness. They called it *ether*, after the old Greek word for a supposedly similar ancient concept. Although intangible, invisible, presumably weightless and perhaps forever undetectable, the ether was assumed to fill the universe and to be present in all substances in greater or less degree. Nobody knew what it was, but the idea of it seemed so logical, even inevitable, that it quickly gained general acceptance as a basic material of creation. A preacher even "proved" its existence on the ground that it must be the reason for atoms gathering to make matter, since "where the carcass is, there will the eagles be gathered together." Maxwell's ensuing discovery that light is just one form of electromagnetic radiation, all of whose fields of force logically require a material structure, seemed to establish the ether more firmly than ever.

From here it was reasoned that, since the ether permeates all space and probably all things in space, at the same time binding all parts of the universe together, it must itself be integral and stationary — a sort of absolute frame of reference by which the motions of planets and stars might be judged. Did not the aberration of light, explained by Bradley in 1727 (see page 424), suggest just such a rigidity of space? Yet obviously, the astronomers could not expect to see the stars swimming through ether like goldfish through a bowl of stationary water, nor would a planet's speed be logically measurable by any sort of wind instrument of the airspeed-indicator type that shows motion in relation to a gas or matter. For an ether wind (not being made of liquid or gas or any tangible substance) must needs be an intangible wind and therefore it would have to be detected — if, indeed, detection were possible — by some much more subtle or abstract means.

This was a clear challenge to the ingenuity of nineteenth-century physicists, who very much wanted to know which bodies in the sky were really moving and, if so, how fast were they going and in what direction — and could there ever be, in a real sense, anything anywhere standing still?

One of the first attempts to measure the ether wind was made by astronomers who compared the refraction of light from a star when the earth was moving toward it with light from the same star six months later when the earth was moving away. In the first case, reasoned the astronomers, the eye is moving at about 18½ m.p.s. toward the light as the light at 186,282 m.p.s. approached its focal point in the telescope, while in the second case the eye is moving away at the same speed, and the 37 m.p.s. difference between +18½ m.p.s. and −18½ m.p.s. should make a noticeable difference in the nearness of the eye to the focal point. But no such shift in focal relations was ever detected.

The most plausible explanation offered for this failure at the time seemed to be Fresnel's old 1818 concept of the "ether drag," a theory that since ether is thicker in material bodies than in vacuous space, all objects tend to drag it along just as a ship drags water, the resulting ether eddies canceling out much of the measurable flow of the ether wind close to the surface of the moving object. To test this idea, a new experiment was clearly needed to measure light's velocity in some fairly dense material that was itself moving in relation to the earth and presumably dragging its own share of ether along with it. Such an experiment was performed by Fizeau in 1859 when he clocked light in both directions (upstream and downstream) through swiftly moving water, demonstrating that a difference in the motion of the water definitely does make a difference in the earthspeed of the light, a result regarded as supporting the idea that ether is draggable.

But ether was by no means yet completely established, for what evidence was there that the motion of water could not influence light even without ether's help? Or, if ether clung to the earth, how could aberration of light be explained? Clearly a much more sensitive experiment would be needed to prove the ether beyond any reasonable doubt, perhaps one in which the ether alone would be measured with respect to light. And for this purpose the ailing Maxwell, in the last year of his life, originated and proposed the most crucial experiment of the century — in some ways the most important test in all the history of science — an extremely delicate optical analysis of light's velocity in different directions in relation to the earth's motion through the postulated ether to see if some sort of "ether wind" could be detected.

Although Maxwell died too soon to carry out the experiment, it was actually performed two years afterward in 1881 by the American physicist Albert Abraham Michelson and again, with better equipment and the help of Edward Williams Morley, in 1887. On each occasion the velocity of light was timed both parallel and perpendicular to the earth's orbital motion by an ingenious arrangement of mirrors and light beams (see illustration), any difference in the light speeds to be revealed by the very sensitive interference pattern made by the two separated light beams upon rejoining each other. On the principle that a swift river must retard a swimmer's combined upstream and downstream speeds more than his cross-current speed, a large difference in the light speeds should show that the earth was moving rapidly through the ether, a small difference that it was moving only slowly, the apparatus being delicate enough to record discrepancies at least as small as one mile a second. But on each occasion, to the consternation of the world of science, the result came out almost completely negative. There was no evidence at all of any ether flowing past the earth in any direction.

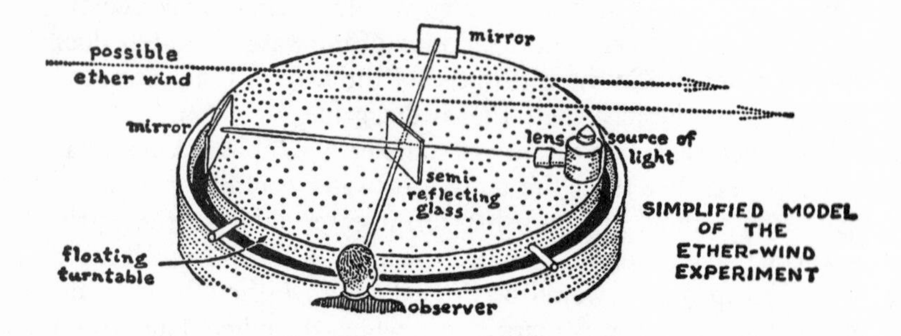

Michelson quite naturally interpreted this to mean that the local ether must be sticking fast to the earth, rotating and traveling along with it. If true, it would have been another vindication of Fresnel's "ether drag" hypothesis; such a rationalization, however, would still be up against the impasse of trying to reconcile itself to the long-established phenomenon of aberration of light, since any ether clinging to the earth would logically have to carry light with it (assuming light lives in ether), and there appeared no reason for such earthbound light to drift backward toward the earth's wake as aberrant light is always found to do.

Could the negative result, then, have meant that light in this case traveled not in ether at all but only in the stationary air, which hardly feels the earth's motion? Science emphatically said "no" — for although air has been found to have some effect on light transmission, light obviously does not require air as a medium for travel, nor is its velocity appreciably affected by it.

Yet something was certainly preventing the earth's motion through space from influencing the speed at which light strikes the earth. It seemed incredible that a velocity of 18½ miles a second in relation to the sun could not stir up the faintest whiff of ethereal breeze or other expectable effects of motion. It was like an airplane maintaining constant speed and direction over the ground regardless of all wind changes. There was little doubt about the fact, for many more "ether-wind" tests were made in succeeding years with new refinements in sensitivity — all to negative results.

In the first few years after the Michelson-Morley discovery, quite understandably, the whole thing seemed to most scientists to make no sense at all. It was only later that little by little some of the freer minds among them began to wonder whether the question answered by the experiment could really be the same question they had assumed had been asked of it. Come to think of it, now, perhaps they had been assuming too much. Should they have assumed that light is a wave motion in ether? Might they not better have been humble enough to ask the more elementary question: does light really move in waves, or in particles? Or in some other way?

Logically, if light is made of waves in ether, its speed of travel should be constant *relative to the ether*, while if it is made of particles shot from its source, its speed of travel should be constant *relative to the source.* Asked to choose between such alternatives, the Michelson-Morley experiment would definitely have had to pick the latter.

Yet there was something wrong with that answer also, for if light consists of particles, the particles shot out by two bodies moving at very different speeds would likewise naturally move at different speeds just as a bullet fired forward out of a jet plane goes at a different speed than a similar bullet fired from the ground. But astronomers well know that the light coming from the twin stars of a binary pair arrive on Earth at exactly the same speed regardless of which star is approaching and which receding or how great their disparity of motion. And the same with light from differently moving atoms in all stars. So starlight evidently disagrees with the particle concept of light.

The first explanation for the riddle that could withstand thoughtful criticism was pretty fantastic and with an Irish tang to it.

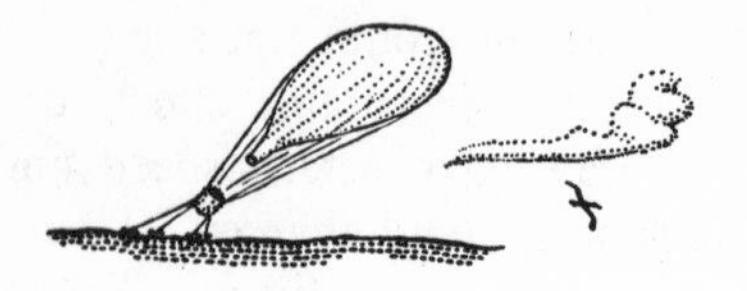

It was advanced rather quietly in the early 1890's by a professor named George Francis FitzGerald of Dublin. FitzGerald got to thinking about the flexibility of matter -- of how a rubber ball is forced into a different shape upon hitting a wall or how a tethered balloon is distorted by the wind — and he saw no reason why the pressure of an ether wind might not also in some way distort matter, contracting it just a little in the direction of its motion through the ether. Anyhow, such a contraction in a single dimension could be made to explain the outcome of the Michelson-Morley experiment quite nicely and precisely: the arm of the apparatus moving against the ether (parallel to the direction of the ether wind) would be shortened (by ether pressure) just enough to compensate for the slowing down of light (by the same ether wind.) In other words, the shortening of the light's path along the arm would naturally and automatically be proportional to the slowing of the light during its round trip, since both effects stem from the same source (the ether), thus making all light slowed-by-1-percent travel all distances shortened-by-1-percent in the same old time and therefore nominally at the same old speed. Of course, the arm placed crosswise (perpendicular) to the ether wind would not be shortened as long as it held that position, since FitzGerald's contraction applies only to the direction of ether motion. Objects change shape, therefore, only

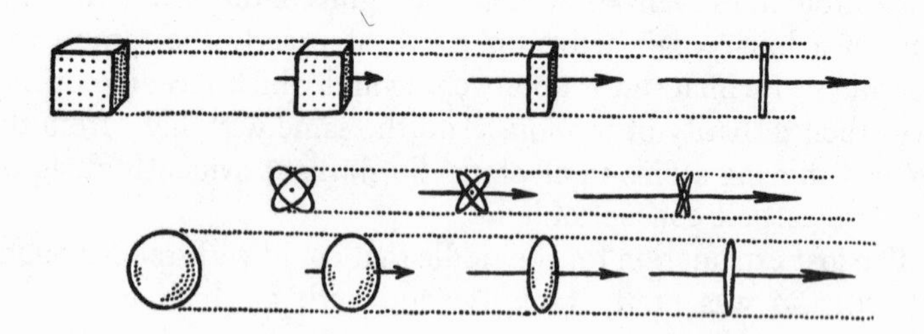

as they change ether orientation, cubes conforming to a rhomboidal foreshortening upwind-downwind, spheres becoming doorknob-shaped as their speed rises, all things always (no matter how they rotate) keeping their ether-shrunk dimension of motion parallel to the stream.

What could science make of this idea? Was it to be taken seriously? Or was it something dreamed up out of Alice in Wonderland, the recent best-selling fairy tale? Or just a bit of long-haired blarney?

Whatever it was, it certainly was ingeniously provided with a built-in booby-trap for the critics, because those who thought to disprove the reality of the contraction by measuring it soon realized that whatever footrules, eyes, lenses or other measuring devices they used would, according to the theory, contract along with everything else in the same dimension, thereby ensuring that each length of, say, 56.1 shrunken yards measured by a proportionately shrunken yardstick or retina would still number 56.1, revealing nothing as to whether there had, or had not, been any contraction.

More remarkable yet, FitzGerald's brash proposition was almost immediately up-graded toward respectability by the publication in 1893 of a similar but more complete theory by physicist Hendrik Antoon Lorentz in Amsterdam. Developed independently at about the same time as FitzGerald's, Lorentz's erudite hypothesis not only worked out the same one-dimensional contraction of moving matter in full mathematical detail (ratio of transformation $1 : \sqrt{1 - v^2/c^2}$, etc.) but explained exactly why it should happen in terms of the electric and magnetic fields inside molecules and atoms, even including distortions (from "spheres" to "ellipsoids") in the shapes of electrons, those still very mystical particles that had just been discovered by J. J. Thomson (see page 251) as kinds of negative beans in the positive soup of the atom. Electrical forces, pointed out Lorentz, are propagated like light, and "a moving electric charge should have a weakened field in the direction of its motion." For this reason, matter, which is held in shape by electrical forces between atoms, would naturally shrink in proportion to its velocity.

It was a theory, you may have noticed, based on old-fashioned absolute motion framed on a fixed empty space. Absolute motion was Newtonian motion, the common kind taken for granted in those days — albeit a motion that could, quite possibly, never be detected. Altogether the whole contraction idea was very peculiar, one cannot deny — yet temperature and pressure had long been known to contract matter, so why should not motion do it too? As no one arose to oppose the concept, it gradually came to be considered entirely plausible, even eventually to win a surprisingly general acceptance.

Only one man can I think of who definitely would not allow himself to swallow any such a contrived explanation — and that was a shy young patent clerk in Bern, Switzerland, who had been working up a few notions of his own on this and other aspects of science. But perhaps you may think such details of history not very important. So why bring this up? Who was this patent clerk anyhow?

His name — as the world was soon to find out — was Albert Einstein. He had been born to an ordinary Swabian family in Ulm, Bavaria, in 1879. Appropriately, perhaps mystically, like Newton's birth that came in the year of Galileo's death, Einstein's came precisely in the departure year of his great predecessor, Maxwell.

The sensitive little boy developed slowly, taking so long to learn to talk that his parents worried that he might be feebleminded. At a Catholic elementary school in Munich he seldom could think of the "right" answer in class and was unceremoniously crossed off as an "odd duck" by most of the other boys, not to mention being sidelined as a "sissy" during sports. He hated playing soldier and cried when taken to see a military parade. He made few friends, impressed no teachers and seemed to care for nothing much but fooling around on the family piano, dreamily making up little chants or sentimental hymns of an afternoon which he would hum to himself next day on the way to school.

But he eventually grew to enjoy reading and one day, playing at navigation with a magnetic compass his father had given him, he discovered that (as he later put it) "something deeply hidden had to be behind things." His father, an unsuccessful manufacturer and salesman of the strange new electrical goods just coming onto the market, sometimes let him play with sample batteries and switches or hinted upon the secrets of mysterious-looking dynamos and transformers. His uncle, an engineer, also told him one day about "a funny science called algebra in which, if you go out hunting and hear some animal in the bushes but do not know what kind he is, you just call him X until he is discovered." Albert was intrigued. And, having no reputation to uphold, he did not mind asking all sorts of impulsive questions, even if they made him look foolish, for by nature and instinct he seems to have well understood the old Chinese saying: "He who asks a question is a fool for the moment. He who does not ask a question is a fool forever."

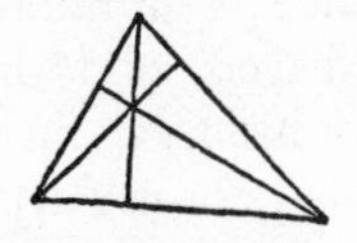

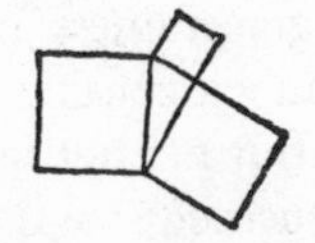

At twelve he got hold of a copy of Euclid's *Geometry* and puzzled over such mystic abstractions as the inevitable conjunction of the three altitudes of a triangle at a single point. It is remembered also that he was moody for days while pondering the Pythagorean theorem, of which, thirty years later, he was to recall, "It made me realize that man, by thought alone, can attain . . . order and purity." At thirteen he tackled Kant's *Critique of Pure Reason* and wondered what Kant meant about "space and time" being "forms of sense perception but not objective things." At about the same time, he discovered the rapture of playing Mozart sonatas on his violin.

But Albert was never happy in his gymnasium (high school) where he later remembered the teachers as being "like army lieutenants," always drilling with rigid formulas instead of whetting the students' appetites. At the age of sixty-seven in his autobiography he was to write that at this period he had indulged in "a positively fanatic orgy of free thinking coupled with the impression that youth is intentionally being deceived by the state through lies — a crushing impression. Suspicion against every kind of authority grew out of this experience — a skeptical attitude . . . which has never [entirely] left me." When his reaction to public education eventually caused one of the teachers to remark, "Your presence in the class destroys the respect of the students," he left the school and set off light-heartedly for Italy to join his bankrupt father, then trying to set up a shop in Milan.

He was enchanted with the carefree life of the Italians, the songs and jokes and gracefulness of villagers, and shortly renounced both his German citizenship and his legal membership in the Jewish religion. But neither in Milan nor in Pavia did his father's electrical shop succeed, so the young man increasingly felt the urgency of finding a profession.

Continuing to pursue the fundamentals of whatever and wherever his curiosity led him to, however, he did not bone up hard enough for his entrance exams to the highly rated Swiss Federal Polytechnic College in Zurich and flunked nearly every subject but mathematics. Although he managed to make the grade the following year, he remained rather indifferent to approved channels of thought and smiled indulgently over such unquestioned doctrine as that "in the beginning God created Newton's laws of motion." He seldom attended lectures but browsed freely in the library and daydreamed in the laboratory. Even after graduation as a physics major, he produced such a poor impression in the academic world that he was dismissed from three teaching jobs in succession and soon reduced to a hand-to-mouth existence. Nevertheless, he often found himself arguing far into the night about abstruse problems in logic with a quiet Serbian girl named Mileva Marec he had met in college and who, though studying to be a mathematician, was adventurous enough to accept his bashful and far-from-promising proposal of marriage.

The following year, with a baby son to support, young Albert tried harder for jobs in other fields — applying anywhere that some technical training might be of use — and counted himself lucky to land a berth processing patent applications in the Swiss capital. He had long been fascinated with new-fangled gadgets anyway, and here he was at the age of twenty-three, actually getting paid for examining and classifying them. Besides, his leisurely schedule allowed him a good many hours a day for browsing in the esoteric realms of space, time and energy — and almost every evening he could be seen wheeling a baby carriage along the street, halting now and then under a lamppost to jot down a row of mathematical symbols.

What was he thinking about in those days? It is hard to say, for Einstein himself not only could not remember afterwards but often admitted in later years that he found himself a poor source of information on the genesis of his own ideas. Even in his last important interview, only two weeks before he died at seventy-six, he insisted that "the worst person" to document any sequence of discovery is "the discoverer" himself. But we know that in Bern he must have meditated deeply upon the mysterious elusiveness of the "ether wind" and presumably explored ways its absence might be explained. Although he was later to become a close personal friend of Hendrik Lorentz, who "meant more than all the others I have met on my life's journey," at this stage Einstein had neither met Lorentz nor read any Lorentzian work more recent than his paper of 1895.

Instead, Einstein evidently was engrossed in his independent research, which was as much philosophical as scientific, wracking his brains to grasp what nature could possibly be trying to say through making the speed of light the same for all viewers regardless of motion. At one point, he told a colleague in the patent office that he did not think he could keep up such a hopeless search much longer. Yet somehow his innate faith in "the rationality of the universe" kept him going, and he never quite quit. Years later, reminiscing about his youthful quest, he spoke emotionally of "this huge world that exists independently of us human beings . . . that stands before us like a great, eternal riddle" — the "contemplation" of which perpetually "beckoned like a liberation."

The main difference between Einstein's thinking and the thinking of Lorentz and other scientists of the day seems to have been that Einstein instinctively hungered harder for the fundamentals of things, for the large view, the broad perspective, the simpler hypothesis. So deeply did he yearn for basic truth, in fact, that he resisted the corrosion of prejudice more successfully than anyone else and overcame the strong human temptation to take things for granted, to assume something true because everybody "knows" it to be true.

When all available evidence and logical reasoning forced him to conclude that "motion is never observable as 'motion with respect to space'" despite Newton's common-sense laws of motion based on long-accepted absolute space, he firmly decided that, having "no basis for the introduction of the concept of absolute motion," one no longer has the right to believe in "absolute motion."

It was an epic resolution of independent mind and spirit, an act of rare integrity born out of the inscrutable detachment of genius that Einstein himself would one day describe as a "solitude . . . painful in youth, but delicious in the years of maturity."

But it was only a beginning — for to renounce absolute motion was one thing. To replace it with some truer breed of motion was another. And how to make a new motion stick in the face of such tests as Newton's famous "absolute-space" experiment: his hanging of a bucket of water upon a twisted rope so that, as it spun around and around, the water's rotating concave surface (pressed outward by centrifugal force) showed that "absolute space, in its own nature, without relation to anything external, remains always similar and immovable"?

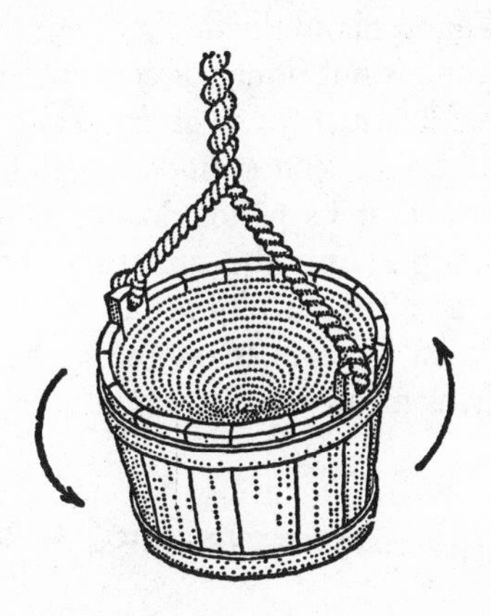

Almost from the first, young Einstein, meditating over his baby carriage, was instinctively drawn to the idea of relativity — the concept that the only possible motion is relative motion. But it seemed elusively nebulous in the beginning. "Relative motion" —

relative to what? Would two or twelve or seventy thousand relative motions be relative only to each other without any fixed standard to judge them by — without anything anywhere really at rest in a frameless world? How disconcerting! And who would ever accept such a theory?

Einstein had not yet worked up much appetite for history but he must have been aware from reading Kant and other great thinkers that relativity was not a new concept. In point of fact, relativity had been a principle in mechanics for a long time. And was it not Heracleitos of Ephesos who opined that "the way up and the way down is one and the same"? And Empedocles who, pointing to the setting sun, remarked, "The sun does not really set. It is the earth that brings night by coming before the light"?

Even if such observations seem to some readers today as little more than witticisms, in the fifth century B.C. they were great leaps forward in human imagination. And some of the later Greeks, notably Zeno, maintained that, contrary to the fluid feeling it gives us, time is not liquid or continuous but, like a piece of wood, has a durable structure or grain. Zeno would have been heartened if he could have seen a modern watch, especially under a microscope to show that its hands do not move steadily but in tiny unavoidable jumps. And the quantum theory could hardly have failed to be congenial to him. In China also, at about Zeno's period, a Taoist logician is quoted as having said, "There are times when a flying arrow is neither in motion nor at rest."

Are such quantum thoughts anything less than probes toward relativity? By the end of the Middle Ages relativity ideas could be expressed more convincingly by Nicholas of Cusa, greatest philosopher and scientist of the early fifteenth century, who not only ventured to suggest that "all religions are essentially one" but introduced the "cosmological principle" of spherical symmetry in space with the argument that wherever in the heavens anyone may be placed, it naturally appears to him that he is in the center of the universe. Galileo brought the idea closer to home by pointing out that "uniform motion in a straight line has no discoverable effects," while Newton ruled that "the motions of bodies included in a given space [such as a ship's cabin] are the same among themselves, whether that space is at rest or moves uniformly . . . in a straight line."

Other thinkers echoed the relativity concept, from Leibniz's concession that "there is no real motion" through Huygens, Berkeley and Kant to Maxwell's balanced summary that "all our knowledge, both of time and space, is essentially relative." Such consistent conclusions could hardly have arisen from disconnected, subjective fancies but obviously reflected, instead, objective observations of nature over a long period of history. Certainly Einstein himself understood this by the time he modestly declared in 1921 that his theory of relativity "is not speculative in origin" but "owes its invention entirely to the desire to make physical theory fit observed fact as well as possible" — that "we have here no revolutionary act but the natural continuation of a line that can be traced through centuries."

Accordingly, it was almost inescapable that Einstein would have to begin where Maxwell left off. And, as Maxwell's greatest work was in the field of electromagnetism, quite understandably the very earliest example of relativity published by Einstein was not mechanical but electromagnetic. It appeared on the first page of his famous paper of 1905, *On the Electrodynamics of Moving Bodies*, and described "the reciprocal electrodynamic action of a magnet and a conductor."

Einstein had noticed that an electrically charged body (say, a piece of amber freshly rubbed with wool) was generally considered in a static condition as long as it remained "stationary." That is, when not moving in relation to the earth or some other sizable frame of reference, it merely retained a charge of "static electric-

ity" (surplus electrons) without any current flowing. But if it were noticeably moving, the included motion of its electrons could be considered a flow of electricity and would, by virtue of its motion, create a magnetic field. Thus, as Einstein observed, there was a presumed "sharp distinction" between the conditions of rest and motion. Yet he realized — in what is said to have been his first clear conception of relativity — that if a man walks past a "stationary" piece of charged amber, the amber is given a relative motion in respect to the man and therefore acquires an electric current with a surrounding magnetic field *from the man's point of view.* In other words, electricity and magnetism are essentially relative along with their fields, and a magnet may be a magnet for one observer but not for another (see page 353), or a conductor may carry almost any strength of current, or none, depending solely on how you look at it.

Of course, this does not mean that such electrodynamic phenomena are imaginary or merely mental effects, for the man walking past a piece of amber may not sense its excess of electrons or be aware that it is amber or even that it exists. Its existence is really beyond and in spite of him. Which is to say that electrodynamics is, in a sense, an objective concept, yet it is also essentially relative and has a subjective aspect in that it changes according to the motion of whatever frame of reference it is observed from or compared with — in the same way that heat is a kind of statistical relative motion of molecules (see page 266) or the motion of a rocket in space is measurable relative to the moon, earth, sun or whatever you choose.

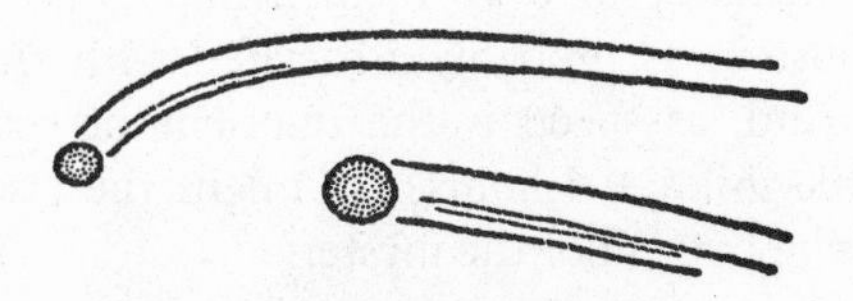

And this electromagnetic relativity was also separately suggested by a kind of ether-wind experiment in its own realm conducted after the Michelson-Morley experiment and known as the Trouton-Noble experiment. The apparatus here consisted of a strongly charged condenser delicately suspended so it could turn freely and easily with any magnetic field on the presumption that it might align itself perpendicularly to the earth's course through the ether. Yet it showed no such tendency in the slightest degree, a negative result that only added to the death knell of the already discredited *absolute* motion.

So Einstein was confirmed in his conclusion "that the phenomena of electrodynamics, as well as of mechanics, possess no properties corresponding to the idea of absolute rest." The only kind of rest or motion anywhere in the world, then, simply had to be *relative* rest or *relative* motion — rest and motion being, in this new sense, exactly the same thing from different viewpoints — and this as true in the microcosm as in the macrocosm, a universal law reaching from inside the atom to beyond the Milky Way: Einstein's great discovery that he introduced quite characteristically as "this conjecture, the purport of which will hereafter be called the Principle of Relativity."

It would have been a simple enough idea, too, once you got used to it, except that it did not seem to be consistent with the stubborn factual evidence of the constancy of light. For if all motion is relative and anything can be considered at rest from some frame of reference, why is light's speed through space fixed at 186,-

282 miles a second in relation to every body? Why cannot light be at rest relative to something? Or why cannot its speed be reduced at least from 186,282 miles a second to 186,272 miles a second as measured from a rocket moving along "with" it at 10 miles a second?

The queer constancy of *c* at 186,282 m.p.s., no matter what, appeared to Einstein as probably associated with the very meaning of speed. And, as speed is the quotient of space and time (25 m.p.h. = 100 miles ÷ 4 hours), so might the essence of space and time be at the bottom of the mystery.

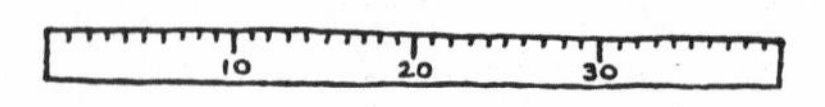

But what is the essence of space? Of time? How does one go about finding out such things in a relative world where nature has the viewpoint of no one in particular? Young Albert wasn't sure he knew, but he was curious and eager and he decided he would have a good go at it anyhow.

First of all, he determined to reject as much prejudice as possible from his thinking about space and time, including all ordinary common-sense and obvious mathematical assumptions like Euclid's famous axiom that "a straight line is the shortest distance between two points" or that a man walking at the rate of 2 miles an hour downstream upon a barge drifting 2 miles an hour down a river must be moving at 4 miles an hour in relation to the earth.

"Who can judge what is really a straight line?" he asked himself. "And who can say just when or where 2 and 2 will make 4?"

Almost any dimension is mysterious when you get right down to it, he reflected — even plain, old-fashioned miles on a map. What does distance really mean? How long, for example, is the west coast of Europe from Gibraltar to the North Cape? A good first guess might be four thousand miles — and that would be quite correct as measured on a small map by leaving out the Baltic Sea and making judicious short cuts over most fiords and river mouths. But five or six thousand would do just as well, merely by using a more detailed map and zigzagging in and out of more bays and inlets. Eight or ten thousand miles could easily be covered by choosing local maps of the coasts, twenty thousand by including sand bars and some of the larger rocks, a hundred thousand by tossing in individual stones with or without barnacles according to need, even millions or billions of miles by descending to grains of sand, and then on down to molecules or atoms. Thus, one must admit, distance is quite a flexible dimension and a material coastline can be of almost any length one chooses — without deviating an iota from reason or reality.

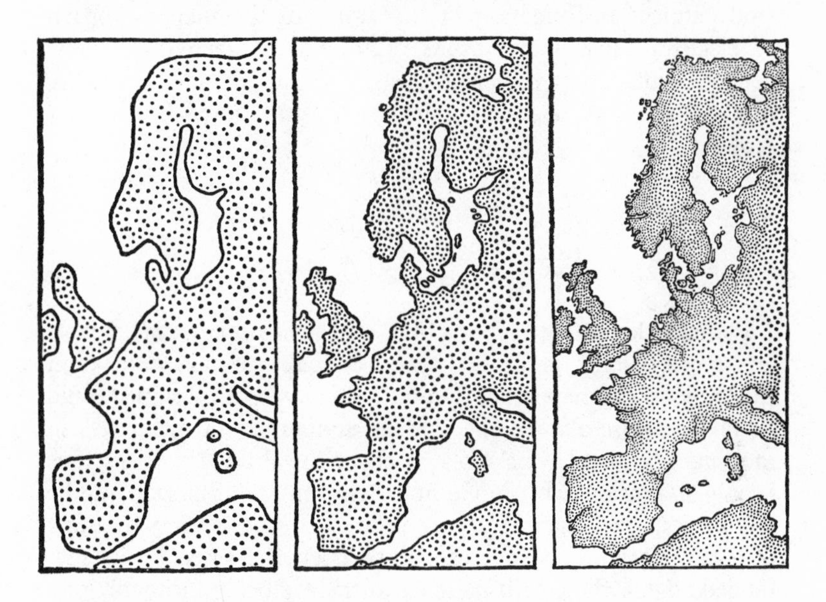

Time, likewise, seems to be a flexible dimension and closely associated with space. For, as clocks in Berlin tell time that is an hour "later" than in London, so do every fifteen degrees of earthly longitude equal an hour of earthly time. And just as minutes and seconds of longitude (which are angles) grow shorter (in miles) toward the pole, why may not minutes and seconds of time shrink proportionately from some correspondingly changing point of view?

Groping for a brainhold in the midst of such hard-to-visualize abstractions, Einstein found that the language of mathematics was much better than German, English or French for expressing and evaluating the concepts. And in working out his own solution to the Michelson-Morley discovery of the nonabsoluteness of motion and rest, what should he come up with but the identical, the inevitable, formula of Lorentz: that the dimension of motion of any moving object must be "shortened in the ratio $1 : \sqrt{1 - v^2/c^2}$"! It was the only answer that would agree with the experimental results. Yet Einstein could not be content in attributing such a contraction to anything so limited as electrodynamic transformation in the atom, to the mere distortion of electrons' shapes and orbits. Rather must there be a more general explanation somewhere deep in the nature of relative motion — of shifting many-angled space and time. But where? How?

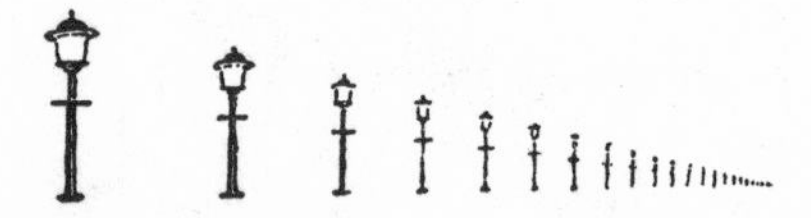

Perhaps the young father looked at the lampposts and people and houses along the street beyond his baby carriage and began to wonder if their apparent diminution in size with distance could be in any sense akin to the evident diminution of all matter in motion. He may have asked himself, if the shrinkage of distance is only an illusion, why should not the shrinkage of motion also be an illusion? Is one more real than the other? The dwarfing of a lamppost by distance can be precisely measured by its subtended angles, the area it presents to the eye being invariably proportional to the inverse square of the distance. If such an optical transformation of a lamppost, drastically obvious from the first glance, is illusory — how much more illusory must be the negligible shrinkage of an object in motion, such a diminution being normally too slight for any known technique of measurement, so slight in fact that the shrinkage of the whole earth's diameter at its solar velocity of 18½ miles a second totals scarcely 2½ inches!

It is quite possible that Einstein was fully conscious of the affront to common sense in such a comparison of the abstruse theoretical transformation due to motion and the patently illusory shrinkage of things receding into the distance, the latter a phenomenon every toddling child soon discovers does not *really* happen. But Einstein's nature just could not take things for granted, even though he saw how much easier it is to believe an error one has heard a hundred times than a truth one has never heard before. So he stubbornly determined not to let himself be taken in by any denials of people in the distance that their bodies had grown smaller than nearer people, not even if they pulled out tape measures to "prove" their bigness — for he just knew that the tape measures had shrunk along with the people. In a few cases he had even seen people go so far away that they became almost infinitely tiny — so tiny that they literally vanished from sight. What size *really* were these people? And how could they possibly protest any judgment made about them?

In a sense, Einstein might be said to have retained something of the direct simplicity of a child in his mental processes — for, like the little boy on his first airplane ride who asked, "Mama, when do we begin to get smaller?" he could accept the diminution of distance as having its own kind of legitimate reality. He could re-examine this so-called illusion and test it in the light of what no one considered an illusion and find the two essentially of a kind.

Einstein later personally inquired into the question of where or when a growing child acquires his concepts of space and time, and at least one psychologist, a Swiss named Jean Piaget, responded by demonstrating that a baby's first idea of space is "centered on his own body and on the location of successful actions." At this stage the baby has developed no clear sense of time or sequence, nor does any object have much continuity to him. When a six-month-old reached for Piaget's watch, for instance, Piaget flung a handkerchief over it, whereupon the infant withdrew his hand as though the watch had been taken away. Then a year-old baby, who had learned to look under the handkerchief for the watch, easily found it but, after seeing the watch subsequently placed in a box across the room, looked for it again under the handkerchief and was astonished that it was not there.

"The baby believes in an object as long as he can localize it," explained Piaget, "and ceases to believe in it when he can no longer do so." This attitude is amazingly similar to that of the modern sophisticated physicist who will accept a meson or a neutrino as real only when it is proved to occupy a known localized position.

Correspondingly, a baby's first conception of velocity appears to be based not on any comprehension of distance or time nor on the relation between them, but rather on the more primitive awareness of order or priority. To his newly budded brain, the object that arrives first is the faster, regardless of where or when it started. Nor does continuous travel necessarily carry the child mentally ahead on the map, as was dramatically expressed by the tired four-year-old who entered his third identical superhighway restaurant in one day, protesting, "Gee, Pop, we've been going all morning and all afternoon — and here we are again."

Even an old woman can retain enough naïveté, under some conditions, to fail to comprehend the simplest modern locomotion. This happened quite strikingly when the mother of Tenzing, the Sherpa who surmounted Mount Everest, was given her first train ride, from Jaynagar to Darjeeling, at the age of eighty-four. A few minutes after the train started, she suddenly asked her son in alarm, "Tenzing, where is that tree I saw in front of the window by the station?" After Tenzing laughingly explained to her what a train is, she sighed with relief, saying, "Never in all my life have I seen a whole house moving like this from place to place."

More common, however, is our normal confusion as to which thing is *really* moving when two things are seen changing their relative positions. "Is that other train starting to move now?" "Or is it us?" Looking down from a bridge at one's reflection in

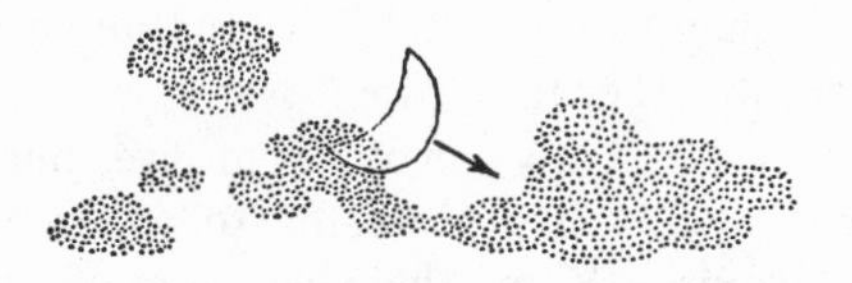

a gentle brook, it is easy to get the impression that the water is standing still while you yourself are drifting through it. And the same holds for the moon sailing between thin clouds or a high tower toppling against the mist. The optical rule is that when a smaller object moves in relation to larger or surrounding things, no matter which is actually moving or which standing still relative to the observer, the smaller or enclosed object is what normally *appears* to be moving while its enveloping "frame" seems absolutely stationary.

Although this fact was established in a long series of visual tests by one J. F. Brown at the University of Berlin only in 1927, about a quarter of a century after the dawn of relativity theory, it sheds significant light on man's traditional notion of fixed space, accounting substantially also for the almost universal acceptance of Newton's postulate of the absolute — an assumption that Newton evidently made for practical reasons without much expectation that it could ever rigorously be proved true, but an assumption which nevertheless was in exact accord with a very general human illusion.

The persistence of such an illusion of the absolute through the era of railroad building last century obviously explains the apprehension many people expressed then that, even if passengers inside the new "steam cars" could be shielded from wind pressure, the mere fact that they would be moving at 30, or perhaps even 40 or 50, miles an hour might prove too much of a strain for the human constitution to withstand. Motion in relation to the earth's surface, in other words, was unthinkingly assumed to be absolute motion without regard for any other motions the earth itself or its part of the universe might be involved in. Where, indeed, were any astronomers or philosophers with perspective enough to point out that two blinks of a human eye, blinked one second apart, may be separated by something else than a second or by something other than time? For it was not yet realized that space too is inevitably involved in eye blinks. In fact, space is involved in all physical events to whatever degree any viewpoint may require. From the viewpoint of our moon, for instance, the earth's populated surface is spinning at about a quarter of a mile a second, so that two blinks of an average human eye would occur not just a second apart but literally a quarter of a mile apart. From the sun's view, the much higher orbital velocity of the revolving earth would stretch the length between blinks to about 18½ miles. While the views of remote stars, necessarily involving the sun's motion in relation to their own, could separate the blinks by hundreds of miles or, if far beyond our galaxy, by many thousands. Indeed, one might reasonably speculate that, from the extreme viewpoint of the opposite side of the universe (if the universe has sides), any second of time separating two earthly eye blinks could be literally transformed into 186,282 miles.

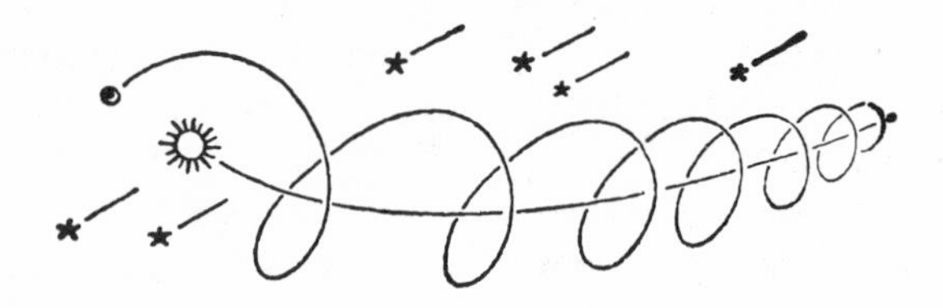

You may call this fantastic. It *is* fantastic from a commonsense earthly view, of course. But Einstein, pushing his hungry baby through the streets of Bern, could not afford to limit himself to common sense or even an earthly view. When a man explores such abstractions as space, time and the speed of light, it would hardly seem appropriate for him to confine his thoughts to any single star or planet. Besides, the very word *fantastic* in such a context could well be expected to unfurl a new dimension of its own meaning.

So Einstein let his thoughts soar freely into both space and time, disciplined only by the most universal and abstract logic. Ele-

phants were no longer "large" to him. Ants no longer "small." For that would have been looking at them through provincial human eyes, as if a mere man had a right to set an absolute standard of size. Instead, Einstein looked out of the sun and saw that elephants could be tinier than germs. And he looked out of the atom and saw that an ant could spread across the sky like a Milky Way! He taught himself that human words like *long, near, bright, hot, soft, loud, strong, soon* are all relative and therefore meaningless unless you assume some standard viewpoint, which need not be a human viewpoint. The same lamp that is *bright* by night is *dim* by day. The same full moon that now rises in the east before my right shoulder at 6 o'clock in the evening is setting simultaneously in the west behind your left shoulder at 6 in the morning on the other side of the earth. Neither space nor time, therefore, has any fixed absolute meaning, and Newton's "absolute, true and mathematical time" is just as outdated as his "absolute space."

Of course, you can point out that the different times of morning and evening on opposite sides of the earth are really different aspects of one time system and easily transposable from time zone to time zone, from longitude to longitude, angle to angle. And you can calculate that the exploding star we see now on Earth is not really still exploding *now* from its own viewpoint, since it has taken the light of the explosion more than a century to travel from the star to us. Yet the transposition of times and spaces is not nearly so simple as it first appears — for there is always the question of the relative motions of the two bodies. Where was the star, when a particular photon departed from it, in relation to the earth at the moment the same photon arrived? Such an interval, you must realize, is not simple. It is not just a distance. Nor is it just a period of time. Rather it is a compound containing factors of both space and time. It is what normally separates two *events* as distinct from two *places* or two *instants*. On the universal scale, neither places nor instants turn out to have meaning in themselves but only when combined into events. Thus there is really no such time as *now* except in some place. And there is really no such place as *here* except at some time.

Nebraska as a place, for example, means something to people now because the *now* makes it more than a place. It makes it an event. But if you throw away all ties with time, Nebraska evaporates into nothing. Where were the rocks and soil of Nebraska a couple of billion years ago? They certainly did not compose any river and hill skeleton we could recognize as Nebraska, and most of their elements must have been spread about the earth in liquid or gaseous form as parts of the primordial seas or steaming swamps or lava flows of that hardly imaginable period. Many of Nebraska's present atoms indeed had not then even arrived from the sun or from the space clouds that feed it. So Nebraska, in the cosmic view, has to be a time as well as a place if it is to mean anything. Strictly speaking, it is an *event* in earthly history.

That is part of why increasing remoteness from the *here* and the *now* add unsuspected distortions that dawned on Einstein only in irregular spasms, that led him to realize that neither space nor time alone is as universal, fundamental or constant as it seems — that, in fact, as the world must increasingly realize, space and time are only aspects of things. They are abstract points of view — dimensions of events — and, although related in subtle ways that are almost always taken too much for granted, each establishment of space or time is literally a separate function — an aspect that cannot be assumed to be part of the same system as any other space or time establishment.

Einstein indeed seems to have been the first human being ever to suspect that simultaneity between two events is a provincial illusion. Pointing to the significance of simultaneity in his 1905 paper, he wrote, "We have to take into account that all our judgments in which time plays a part are always judgments of *simultaneous events*. If, for instance, I say, 'That train arrives here at 7 o'clock,' I mean something like this: 'The pointing of the small hand of my watch to 7 and the arrival of the train are simultaneous events.' "

He admitted that observing such simultaneity was a reasonably accurate way for a person holding a watch to tell the time of an event happening next to the watch, but insisted that, on principle, the method could not be relied on for timing events far away from the watch, especially by someone moving in relation to the other things involved. The question, of course, has little to do with visibility or lack of visibility of the watch, which can be of any size from a vibrating atom to a whirling galaxy, but rather hinges on the relative nature of simultaneity itself—simultaneity being no more absolute than motion or fixity in space since, as Einstein noted, two events that are exactly simultaneous in one frame of reference may literally be far from simultaneous as envisaged from a differently moving standpoint.

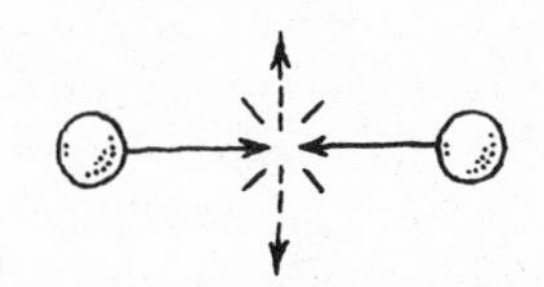

COLLISION OF TWO OBJECTS

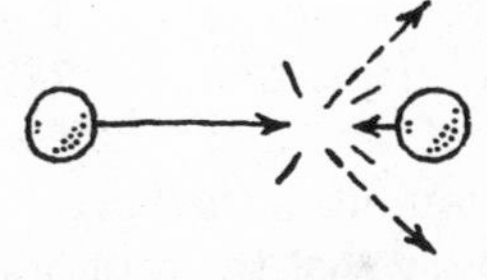

THE SAME COLLISION SEEN BY A MOVING OBSERVER

COLLISION RELATIVITY

THE SAME COLLISION SEEN BY A FASTER MOVING OBSERVER

I do not think there is any better illustration of this elusive truth than the story of the astronomically long freight train, traveling at celestial speed, which was suddenly struck by lightning at both ends. Although the damage proved slight, somehow a troublesome argument arose as to whether the two hits had been made "simultaneously" or not. First a reputable scientist, riding at the exact middle of the train, swore he had recorded the light of both bolts of lightning (with aid of mirrors) exactly simultaneously. Then an equally reliable observer, who had been standing on the ground at the exact mid-point between the strikes, proved he had recorded the flash from the rear of the train a sizable fraction of a second earlier than the forward flash reached him.

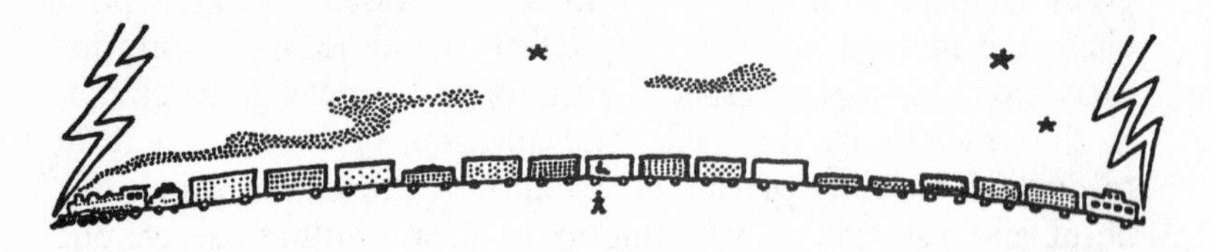

Most earthbound persons, if asked to settle such an issue, probably would incline to the view of the observer standing on earth who insisted the two lightning strikes could not possibly be simultaneous, their "simultaneity" as seen by the train witness being only apparent, not real, since the train had carried him closer to the forward flash by the time he received their "simultaneous" light from points that were (by then) no longer equidistant from him. However, as those who can accept the broader-minded relativity of motion will realize, it may be just as true to say that the train was standing still and the ground speeding by it as vice versa — so that the ground observer could have been "carried" beyond the equidistant point as easily as could the train observer. From Einstein's celestial view, in fact, the lightning flashes (along with their mid-point) are free to be considered an integral part of the train, or a part of the solid earth, or of any other frame of reference. Every observer may choose his own viewpoint. Was the observer on the train really moving toward the *place* the lightning had struck at the front of the train? Can you decide whether, after the lightning strike ended, its *place* of striking (the source point of its light) moved along with the train or stayed with the earth or, more reasonably, disintegrated into both these subjective components? Where, after all, can any *place* be if it is not where someone thinks it to be? Any viewpoint is true and right. No view is wrong.

Hard as it is to "get" such relative simultaneity through one's head, one somehow just has to take off from the accustomed runways of one's mind and fly one's imagination into the problem if one wants to understand relativity. For simultaneity is one of the first keys to the space-time relationship and a prime clue to the hidden subjective nature of these elusive abstractions that compose it. Just stop and figure it out now: if the answer to the question of whether lightning strikes the train twice at the same time or twice at two different times truly depends on whether the lightning is regarded as part of the earth world or as part of the train world, then simultaneity must be relative. And if simultaneity — a coincidence in time, involving space — is relative, obviously time and space are relative too. If this is true, it indicates that time and space have a deceptively subjective character — that somehow they flow and wimple and warp individually, fast or slow, this way or that, according to whether they are associated with a train or the earth — or something else. They are sensations of dimensions, indelibly but subtly related to your frame of motion or nonmotion, consider it which you wish. They are aspects of things, like angles of view or lampposts in Bern that you can make grow or shrink to any desired size merely by walking toward or away from them. Come to think of it, they are remarkably like Lorentz's collapsable atoms that are squeezed in line of motion by the "ether wind."

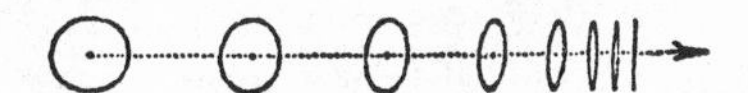

This very thought — or something closely comparable — when it flashed upon young Einstein in Bern surely must have given him a moment of profound exaltation, for he saw then that it could explain the Michelson-Morley mystery of "where is the ether wind"? — sounding the final knell of absolute space and time. This it may have been that he referred to half a century later when he explained to an eager young student in Princeton, New Jersey, "The mind can proceed only so far upon what it knows and can prove. There comes a point where the mind takes a leap — call it intuition or what you will — and comes out upon a higher plane of knowledge."

At any rate, Einstein suddenly realized that the formula of Lorentzian transformation, in the form $c : \sqrt{c^2 - v^2}$, was none other than the beautiful Pythagorean theorem on which he mused as a child, only with c here signifying the hypotenuse as a path of light, and he knew it applied not just to the shape of matter or to the orbits of electrons but to the very nature of the space and time that enclose them. What did this mean? It meant that motion must have a shrinking effect not only upon moving objects but upon all the space involved in their motion — and upon all the time involved also. In other words, it was a very fundamental effect.

But how could time shrink? Could it literally contract in speed of flow proportional to its relative motion — motion relative to any observer — just as space evidently could contract in line of motion? This was an eerie, unsettling idea, of course. But it added up mathematically. Its logic clicked. And it explained the case of the train struck by lightning as well as it cleared up the ether doldrum. It meant not only that a yardstick moving lengthwise must be shorter than a stationary yardstick but that a moving clock must run slow as compared with a stationary clock. If anybody should ask, "Stationary in relation to what?" the answer would be, "In relation to ANY thing or ANY observer — it makes no difference what." In a way, then, it was an illusory effect, Einstein realized, a kind of new perspective on time-space, yet as real as most fundamental things in this world, including matter itself. Certainly it was just as real as the lampposts he could see shrinking into the distance before his very eyes.

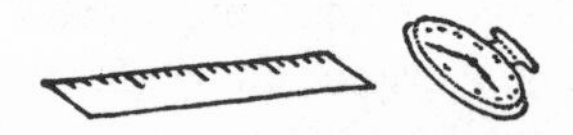

Let us try to *get* this relativity idea now — as straight and simple as it is possible to get it. It is really not so frightful once you catch on. It doesn't demand any mathematics, even though mathematics helps to prove it. It is elementary. If anything, *too* elementary. All you need is enough imagination to shake free from the earth and from your old notions about what is standing still and what is moving, from your comfortable assumptions that clock motion or the passing days are in any way inevitable.

The main thing to *get* is that somehow motion transforms things: it foreshortens their lengths and slows down their tempos. The foreshortening was dramatically revealed by the Michelson-Morley experiment last century, while the slowing down was directly proved in 1936 by H. E. Ives of the Bell Telephone Laboratories, who found that the very steady vibration frequencies of hydrogen atoms are invariably retarded by their speed in exact accordance with Einstein's equations. But this still adds up to be an illusory and relative effect, not something absolute that can be pinned down once and for all, no matter what. I mean that there is a kind of uncertainty principle about motion itself, and that no one can really decide what is moving and what (if anything) is not — so that there is deep doubt also about what is transformed and what (if anything) is not. In other words, while one yardstick is foreshortened by its motion relative to another, the other must also be foreshortened by its motion relative to the first, making them each symmetrically shorter than the other from the other's viewpoint. Which is like two men walking away from each other until each appears smaller to the other than to himself.

Something similar occurs in the time transformation: each of two relatively moving clocks running slower to the other than to itself. Which again is like two men walking so far apart that the legs of each appear slower in the other man's eyes than in his own.

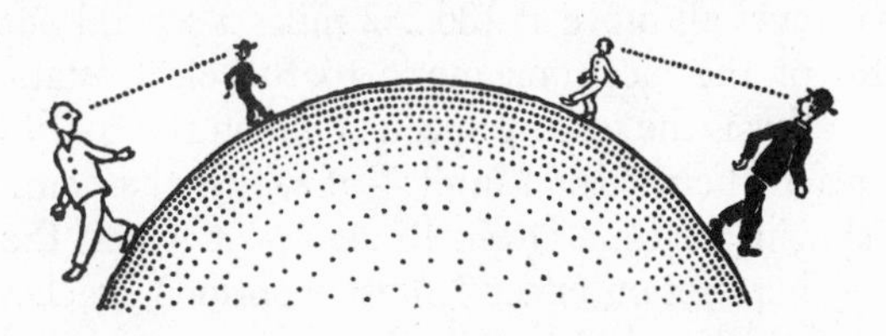

Thus the whole Lorentz transformation of motion can be exemplified by two rather arrogant men striding past each other, each of whom disdains the other as "smaller and slower" than *he* is. And this curious, symmetric relativity quite simply and logically solves the riddle of the missing ether wind. It solves it not only more fundamentally but more satisfactorily and much more elegantly than either FitzGerald or Lorentz could solve it. According to some of the more sensitive commentators, even the mathematics of it is beautiful, relatively speaking, and quite convincing — if you can be convinced by mathematics — in the way of showing you how the shortening and slowing effects of the earth's orbital motion, in relation to her other motion, exactly equate with Michelson's two perpendicular beams of light.

Even the queer constancy of the speed of light is clarified by the fluidity of relative time and space, for the constant *c*, of course, is not an *absolute* but a *relative* constant, being defined in miles and seconds that themselves must vary with relative motion. If a light flash, for instance, is sent out from a place where two men are passing each other, every point on its spherical wave will have advanced 186,282 miles from *each* man after a second by *his* watch even though the men are no longer together — this "miracle" of equidistance being possible because of the individual transformations of the miles and seconds of each man, which are functions of his relative motion. It is not, therefore, the fact that it is *light* that gives light its miles-per-second constancy to all observers. It is only the fact that its velocity is the *highest relative velocity of the transmission of influence in the material universe.* For if sound (or smell or touch) could travel 186,282 miles a second, its effects would presumably also reach each recipient at that same constant speed regardless of which way, or how fast, he was going. Indeed, not alone light, but electricity, heat, radio, x-ray and gamma waves all move at 186,282 miles a second because this is "the ratio of the electromagnetic to the electrostatic unit of electricity" — this is the relative speed at which the transformations of motion reach their natural limit — at which the contraction of space and slowing of time attain to zero — at which the increase of mass (another motion effect I must mention) reaches infinity. When space has thus (relatively) shrunk to nothing, when time has ceased to flow, and when mass has exploded to infinity, how can speed be further increased? Obviously, if you admit those limitations, there can be nowhere left to go nor time to go there in — quite aside from the nebulous question of whether any infinite force could ever have the motivation or capacity to accelerate any infinite mass. Indeed, without the very coordinates of speed (space and time), what in the world could any additional speed be made of?

And as for the increase of mass with motion, like the shrinkage of space and slowing of time, it has been amply proved by experiment, even before Einstein explained it — proved in the astonishing massiveness of the fastest (about .8 *c*) electrons emitted by newly discovered radium and uranium at the turn of the century. This mass increase means, in essence, that energy from the motion has been added, since mass and energy have turned out to be as much two aspects of one quantity as parallax and distance are two ways of describing the range of a star. Of such did Einstein abstract his $E = mc^2$.

2 + 2 = X

It is all pretty logical, you see — even if highly unearthly or, shall we say, a trifle uncanny. And it is decidedly explicable in the HOW sense, though perhaps you'd better not ask WHY. The WHYs of the most fundamental questions, as I've said before, are not satisfactorily answerable in this phase of life. Things work certain ways, according to the laws of nature as we find nature — just because things are the way they are or were the way they were. . . . What more may one say? Is there a "now" that is not "here"? Or were the Puri metaphysicians right in assuming that things farther away in space must also be farther away in time?

Concepts that are way down deep under our thoughts must be accepted almost blindly, it seems, like an axiom, on intuition or faith: a straight line that is the shortest distance between two points, the combining of 2 and 2 to make 4, or the objective reality of matter — things you feel in your bones, including even your skull bone, and which (if bones may be trusted) are true. Axioms, of course, cannot be proved. If they could, they would be theorems, not axioms. But that should not bother anyone, since axioms are deeper than theorems or thought, being made essentially of heart, not mind — of basic feeling, the very bedrock of reason.

This at least is a way of saying that relativity is not logically so hard to understand, but that you must build its logic upon the right axioms if you want it to add up to something credible. You must somehow attain a certain minimum breadth of perspective as a foundation, which you cannot possibly do on reason alone. Rather, you must let yourself be swept off your mental feet, so to speak, in a kind of faith. For what is the use of learning the logical steps of relativity if you are not persuaded that relativity exists? In a word, the least of your problem is to *understand* relativity. The most is to *believe* it!

It may help here to reflect on the evolution of your own personal beliefs about the nature of the world. You can hardly be expected to recall much of how your intuition grew and changed as you discovered your environment — but surely your intuition *did* change. As a baby looking out from your crib, your life was almost timeless. You could not notice motion on the clock. Mother was just Mother. You had never seen her younger nor had you any reason to imagine her older. Neither had space acquired much meaning until you realized one day that things getting bigger were things coming nearer. You did not yet distinguish right from left except that the wall on one side always had a window and the wall on the other a door. If you had had more capacity for generalizing, you might have assumed that right walls had windows and left walls doors.

Later, when you grew big enough to crawl around and notice new angles, you suddenly got a fresh intuition: the walls with windows were sometimes on one side of you, sometimes on the other. Like magic, right walls turned into left walls and you could make them do it. Little by little, this became easy. Then, sometime long later, you learned that a more reliable way to explain a right wall is to call it an east wall. The idea of compass directions in fact endowed you with an exciting new intuition that reached out all the way to the horizons.

At this phase, more than likely, your world was still flat. East was toward that hill over there and west was down that road, both directions continuing straight on as far as you could see. Up was up and down was down — unchangingly, inexorably, absolutely.

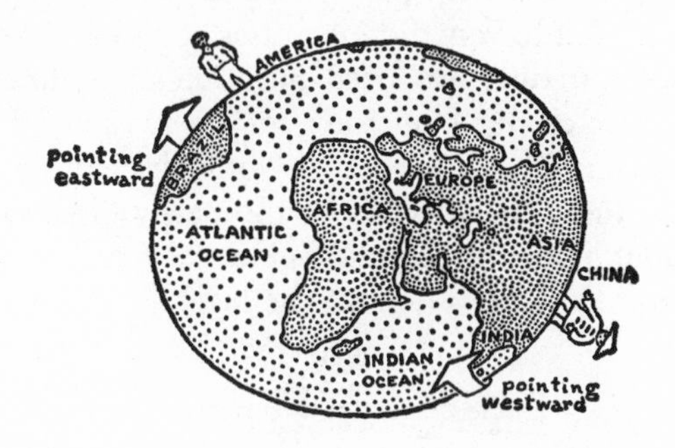

So it was a day of some shock and surprise when you finally caught on to the earth's being really a ball — which meant that "up" to you is the same way as "down" to a Chinese, and that "east" in Brazil points straight "west" in India. In fact, if you had lived in New York and had been given a thirtieth-century subterranean telescope that would let you look through the earth at Moscow, you might not have believed your eyes when you saw the Kremlin standing "horizontally" out of the "wall" of Europe as you gazed "upward" at it from "below."

The modern round-world concept in intuition, we must realize, was not only very little understood on Earth before the days of Magellan, but has taken centuries getting established since then, being still by no means accepted without question among all peoples. You see, it takes a lot of conscious effort to acquire such an intuition, particularly as the older, simpler intuition of a flat earth has not been finally disproved as wrong. When you come right down to it, the earth *is* flat — at least for a little way around you on the average. That's just as obvious as absolute space and time — which are also true intuitions in your immediate vicinity. Intuition thus turns out to grow in concentric spheres —

successive approximations of reality — consecutive awakenings to wider awareness . . . from Ptolemy to Copernicus to Kepler to Newton to Einstein. . . .

The earlier of such intuition stages, however, have seemed to be within the reach of average minds — not too steep for creeping credulity. I mean stages no further off than the round earth or the encircling moon and the planets. But when we get to Einstein's new leap upward into very distant worlds at speeds that drastically distort space and time, we need intuition of a different order. We need to stretch our homespun comprehension far enough to take in something wholly beyond any ordinary earthly experience. We need to study the workings of the newly discovered laws of nature without pride or prejudice until they make some logical sense — even if the foundations of the logic seem to be floating in emptiness.

I do not know whether, in attempting this here, we will err more in confusing the greenhorns of relativity or in boring the veterans, but the veterans at least can afford to skip over the next few paragraphs, leaving the greenhorns free to stumble onward in their full unmonitored innocence.

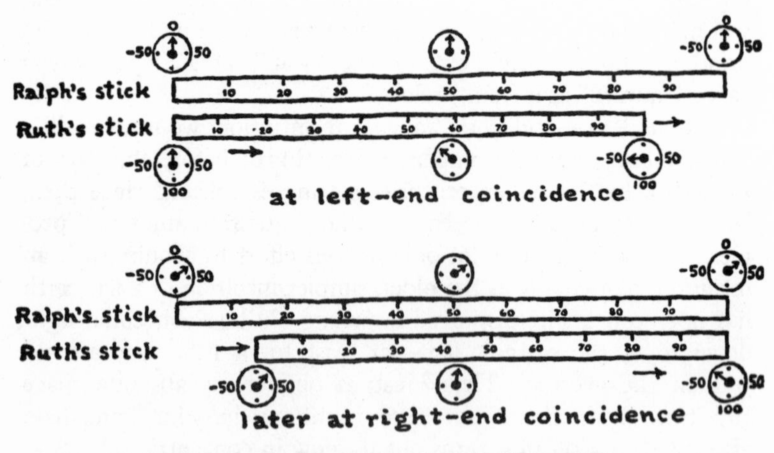

FROM RALPH'S VIEWPOINT

The Lorentz transformation, to take a clean-cut example, may be demonstrated by two meter sticks equipped with special clocks and moving lengthwise in relation to each other at half the speed of light ($c/2$), which calls for a relativity foreshortening of 14 percent. Let us say the upper stick belongs to Ralph and is at rest relative to him, while the lower stick (moving to the right) belongs to Ruth and remains at rest relative to her.

The first picture shows the two sticks from Ralph's point of view, first as their left ends coincide at the zero instant ($t = 0$) and next as the right end of Ruth's apparently shortened stick *later* reaches the right end of Ralph's — this occurring, curiously enough, at what turns out to be $t = -26$ by her nearest clock but $t = +28$ by his. You will notice that all Ralph's clocks agree with one another here, since they all represent his single, instantaneous view, but that Ruth's clocks differ (in his viewpoint) all down the reduced length of her moving stick. To him, this picture represents the essential facts of simultaneity.

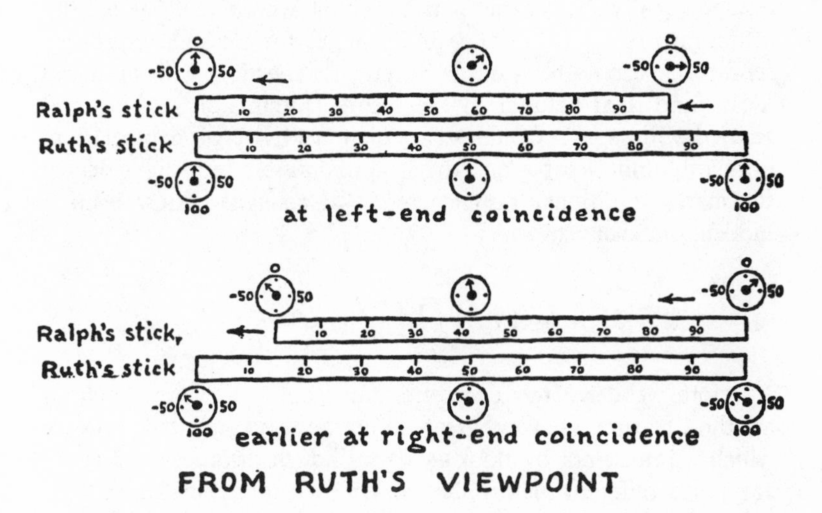

FROM RUTH'S VIEWPOINT

Ruth's viewpoint, shown in the second picture, is distinctly different. Although the left ends of both sticks still coincide at his $t = 0$ (which is her $t' = 0$), she regards his moving stick as the shorter and all his clocks as unsynchronized while hers are in perfect accord. Trusting her own view of time, she believes the right ends of the two sticks passed each other *before* the left ends met, the time of right-end coincidence being $t' = +26$ on his nearest clock and $t' = -28$ on all of hers. Thus, although both viewpoints agree that the other one reads clocks almost correctly, they almost never can admit that the other's clock is right.

I must point out that the two series of clocks could as easily be set to agree at the right ends as at the left ends, since the relationship is symmetrical. But that would not alter the more significant and fundamental fact that, when two bodies move past each other, each one sees part way into the other's past at the other's forward end or correspondingly into the other's future at the other's rear end, or both! Naturally, this has nothing to do with common sense and would have been regarded as completely daffy before Einsteinian relativity, yet it is now supported by a wide range of experiment and experience. And it even embellishes our previous analogy of Lorentzian transformation as resembling "two arrogant men," each of whom looks upon the other, in passing, as "smaller and slower," for the men must also tend to see each other's heads in the past (possibly with outdated ideas) and their behinds in the future (presumably sagging and gnarled with age). Could some such relativity of depreciation, ordinarily unnoticeable on Earth, be responsible in a slight degree for man's traditional disparagement of foreigners? Or even a hidden source of war?

Before we delve too deep into philosophy, however, there is another feature of very great importance in relativity theory which we must not overlook — that although distances and times vary with different viewpoints, they also combine naturally into the quantity called interval, which is the same from every view. Interval, you will recall, refers not just to space and not just to time but to the ubiquitous combination called space-time that actually separates events all the way from collisions of electrons to exploding stars. It is hardly possible to explain how a length of space and a length of time can each appear differently to differently moving observers, while their combined space-time interval remains the same for all, except by pointing out a peculiar mathematical fact: that the interval squared always equals the difference between its space squared and its time squared, a difference that is constant and unaffected by shifting observers' viewpoints or the relative proportions of space and time involved.

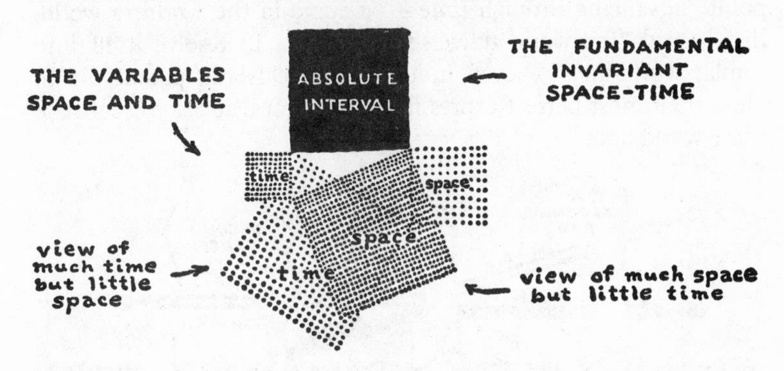

Which amounts to saying that the separation between two exploding stars is an absolute space-time quantity, regardless of whether you view it (from some relatively "slow"-moving planet) as composed more of space than time or (from a "fast"-moving rocket) as composed more of time than space (see page 585).

This Pythagoreanlike discovery appears to have stemmed from the exuberant mind of Einstein's old mathematics teacher, Hermann Minkowski, who made a stirring address in Cologne on September 21, 1908, promoting relativity into a system of "world geometry." In his third sentence, Minkowski boldly proclaimed that "henceforth space by itself, and time by itself, are doomed to fade away into mere shadows, and only a kind of union of the two will preserve an independent reality." It was his way of saying that the *interval* is the sole objective physical relation between events, the mathematician's fundamental invariant, the prime ingredient of world texture and probably one of the few absolutes left in our fathomless new ocean of relativity.

Incompatibly moving observers could thus look forward to ultimate agreement upon any interval as a whole, however they must differ about contingent space and time. From this rock, Minkowski made so bold as to associate a point in three-dimensional space with a corresponding point in one-dimensional time in order to establish a four-dimensional *world-point* as the theoretical cornerstone of the interval. That is how he created what he called "an image, . . . the everlasting career of the substantial point" advancing through time — "a curve in the world: a world-line." And "the world universe" was "seen to resolve itself into similar world-lines" — even inducing physical laws potentially to "find their most perfect expression as reciprocal relations, between these world-lines."

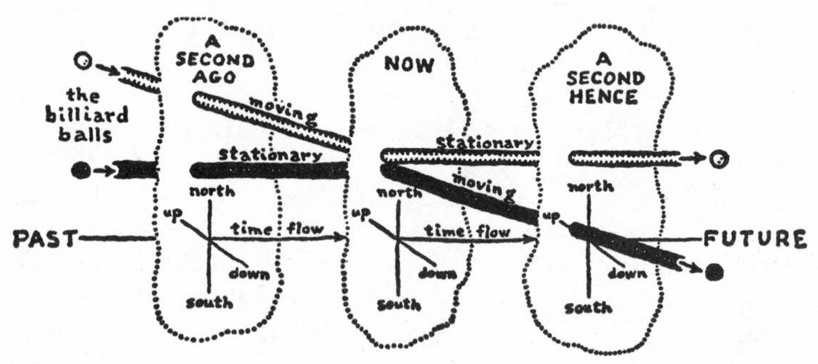

WORLD LINES OF TWO COLLIDING BILLIARD BALLS

Was this four-dimensional world of Minkowski in any sense *real* — something one could get one's teeth into (real teeth, I mean) — or was it a kind of mathematical trick to comfort the theorists? Einstein himself had his doubts at first and is said to have remarked that "since the mathematicians have invaded the theory of relativity I don't understand it myself any more." Yet after due reflection, he realized that the traditional three-dimensional space and separate one-dimensional time made such a complicated frame for relativity that he could not visualize how to generalize it further, as he wanted to. So he enthusiastically accepted Minkowski's contribution as just what relativity needed to synthesize its inspirations into a simpler whole. And, along with space and time, he let electricity and magnetism have their correspondingly compound dimensions also, which fused these hidden, subjective forces into a new objective, four-dimensional electromagnetic "tensor" permanently embedded in four-dimensional space-time.

And of all the aspects of relativity that made themselves known before World War II, the fourth dimension, as a fresh conception of time, was what most profoundly got under the world's skin. For the fourth dimension not only seemed to ask strange questions but more than hinted at still stranger answers. Plotting out time alongside space as a basically similar dimension, for instance, inevitably led to its division into "past" and "future" and raised the question of the relation between them. This relation seemed to be a causal connection involving all sorts of "forces" that insured a continuous flow from earlier conditions to later ones. Yet a "force," as Bertrand Russell pointed out, is something like a sunrise — just a convenient idea for explaining something. In larger perspective, a "force" no more "forces" anything to happen than the sun literally "rises." To say that one thing causes another almost surely means only that the one thing naturally precedes the other. Spring weather produces eggs in birds' nests. How it does it is not necessarily a causal flow in one continuous direction, because, for all we know, causation works both ways. Your receiving a letter logically "compels" its writer to have written it to you — just as surely as his having written it to you "compels" you to receive it. Perhaps somewhat more so, for I can imagine a letter written without being received more easily than I can imagine a letter received without having been written.

A physicist might try to pin down causation further by experimenting with a causal "force" like gravity, which "causes" the earth to revolve around the sun according to Newton's law of the inverse square of the distance. I have in mind a thought experiment in which the distance acted upon by gravity can be shortened all the way down to zero. But there, at the actual place of contact between mass-point and mass-point, the "force" (gravity) must finally explode, according to the law, for the inverse square of zero is infinity. Does this mean, then, that causation itself must disintegrate in an "infinitistic catastrophe" upon the theoretical meeting of ultimate matter with ultimate matter?

Argue as they would, and testing every available theory, the scientists do not seem to have established much more of an absolute difference between past and future than between right and left or back and forth. Thus they have practically confirmed the reasonableness of the fourth or time dimension — the fourth way in which length can be measured — the path of change — the direction in which Minkowski's *world-points* move along *world-lines* — the natural time track that seems to be just as much relative as any spatial dimension. I mean by this that earlier and later do not necessarily correspond to past and future, for an event may as easily be later as earlier in your past, or as easily sooner as later in my future, and the vastly different tempos of existence in the universe (almost inversely proportional to size) enable worlds as diverse as atoms and grains of sand and stars and galaxies to coexist harmoniously. Also time is something like the sideways continuation of a great wave that breaks and plunges upon a shore, ceasing to exist here, then ceasing there, yet progressively rolling shoreward farther and farther along the beach through its ceaseless unfoldment in the lateral dimension. The remote past in its way appears just as dark as the remote future — possibly darker, as it seems to be going away — yet it is not certain that the two directions are perpetually opposite or, if they are, that they may not change places. In the present century, in a sense, earth time is extending in more than one direction, for the scope and memory of history increase backward as well as forward — particularly where archeologists dig up the past (say, in the upper Nile valley) in desperate competition with politicians just as eagerly shaping the future.

Another kind of time relativity is suggested by the viewpoint of water molecules in a river. To them, one might say, the frothy tumble of a mountain brook represents the past and the broad, salty ocean the future, while time progresses downstream. Yet to a spawning salmon swimming upstream out of his ocean past toward a future brook, the river's time is diametrically reversed. There is a symmetry to time, you see, which — as with motion in space — offers full choice of viewpoints.

Still another example may show how completely mirrorlike time-space can be. Many years ago a Turkish artillery officer, condemned to die, was given the hard choice of jumping off a high tower or being shot out of a huge cannon. Either method would kill him instantly: one by the shock of the abrupt change from high speed to zero speed, the other by the change from zero speed to high speed, these being completely reciprocal. It is said that the officer was actually killed — believe it or not — by *both* methods — which, in the understanding of relativity, are the same — for he managed to jump from the tower in such a way as to land inside the muzzle of the cannon, which killed him just as if he had been shot upward by it in reversed time!

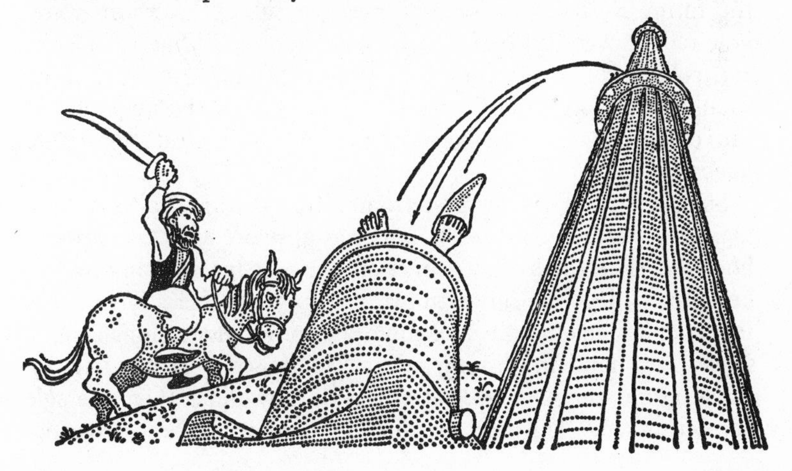

Philosophers have long debated such relativity of time, often confessing with Saint Augustine, "What is time then? If nobody asks me, I know. But if I were desirous to explain it to one that should ask me, plainly I know not."

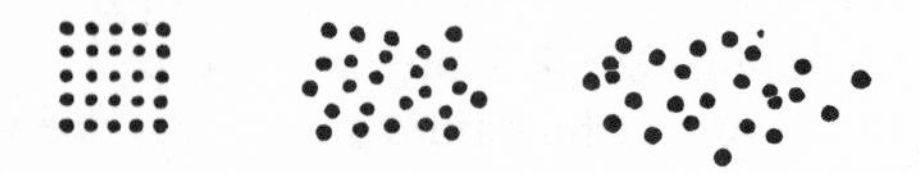

The most nearly reliable distinction between past and future on Earth has had to await the theories of modern science — in particular, the second law of thermodynamics concerning the natural drift of matter toward relative disorder or entropy (see pages 214, 267). That discovery made it possible at last to define a kind of temporal arrow or general direction to universal time flow so that, if something such as a time-lapse movie film, say, of some strange planet's geological history should ever miraculously appear, scientists could tell its beginning from its end. It would explain why astronomers could not seriously think any creatures who might be living on a withershins moon (such as Saturn's Phoebe) would uneat their food or unbreathe their atmosphere just because their world was revolving oppositely from most of the solar system. It also might suggest why we cannot always calculate the past we have seen, although we can often calculate the future we have not seen. I mean, a tub of lukewarm water *now* tells us very little about the combination of buckets of hot and cold water it was composed from *a minute ago.* Yet if we should mix those same buckets of known temperature *now,* we can easily calculate the temperature they will create *a minute hence.*

Such a way of telling past from future — which takes a thermometer instead of a clock — is not as absolute as it may appear, however. The catch is: it works only in our macrocosmic world, because it depends on entropy. Entropy, in fact, is just the tendency of large numbers of molecules or atoms to shuffle themselves into random disarray. It is a statistical effect, and no single atom can have an entropy state any more than a single bee can swarm. This may explain why modern physicists consider the atom timeless or at least ambiguous in its past-future-past relations. For nothing has been found in the laws of physics to prevent time from running "backward" as well as "forward" in an atom. Nor probably even in a large molecule or a crystal lattice. Macrocosmic vibrations back and forth from state to state, like those of a plucked harp string, of course cannot realistically be called timeless, for we know they do not quite reverse or duplicate themselves. They could be likened to successive equinoxes on Earth which only approximately repeat former equinoctial states — states that are distinguishable enough from one another to be called *events*. Yet there is no evidence that atomic states really are events, any more than that anything else which exactly repeats itself again and again can avoid eventually being considered uneventful.

Thus there may be said to be a lower limit of time, perhaps at the bottom of the macrocosm, just as there must be a lower limit of ordinary heat somewhere below the molecular level of size and motion. And this idea seems to have been corroborated by mathematical physicists such as Eugene P. Wigner, who in 1932 showed how "the reciprocity theorem" can be applied to molecular and atomic collisions — how, for instance, the reaction of two hydrogen molecules ($2H_2$) with an oxygen molecule (O_2), creating two water molecules ($2H_2O$), liberates energy in the form of the two water molecules flying violently apart, while two

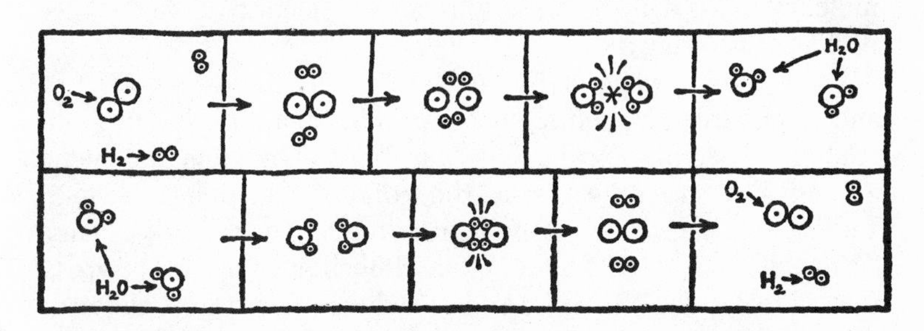

water molecules colliding with that same amount of energy can decompose the water right back into hydrogen and oxygen in exact reciprocity. Nor does such time alternation stop at the atom, for, as you may remember (see page 326), Dirac and Carl D. Anderson discovered a similar reciprocity among elementary particles about the same year. And the phenomenon has since been successfully attributed to nuclear collisions.

Trying to prove time reversal in our macroworld, however, is something else again, and has usually turned out rather like that report in Coeur d'Alene, Idaho, of a mysterious car that for several hours was seen "tearing around the town in reverse." The case was only cleared up when the teen-age girl at the wheel explained to a cop, "My folks let me have the car, and I ran up a little too much mileage. I was just unwinding some of it."

Although we have no indication that any celestial orbit has ever been unwound by a reversal of direction, it appears impossible to prove that our world can *never* return to one of its earlier states, particularly if you consider the universe as enclosed or finite. Some philosophers have even pointed out that a returned state of the world, if complete, would logically include all records, instruments and memories also "returned" to the same state, which would make the return undetectable—so that if such "time loops" are for some unknown reason occurring right now millennially or daily, or on any other frequency, we would have no way of knowing it.

Furthermore, what is the difference between going forward and backward? Of course, one can tell what is "forward" in relation to familiar local events but, once away from the home ground and current contexts, the criterion grows less obvious. Indeed, whichever way the earth turns we naturally name "forward" and so define it. The Phoebeans (if any) would undoubtedly feel their world revolves "forward"—different though their word for it might be. And considering time itself as nothing but the sequential dimension of events, then time also can be made to flow in any "direction" just by events—and its "direction" is, of course, illusory under the relativity principle, being conceivable as "forward" or "backward" only in the sense of "in tune with" or "opposite" what we are used to.

What actually could constitute a backward flow of time through our consciousness anyway? It is worth stopping to think about this. If we began to ungrow and life's processes were reversed, mouths evacuated and rectums fed, and friction and gravity took on the opposite of their present tendencies, would that amount to a reversal of time? We might reasonably assume so, especially if every aspect of life were reversed consistently. But, if anything at all were not reversed, we would likely think there had only been a strange shift in certain laws of nature. If the stars alone continued their motions and rhythms of brightness, for instance, we might conclude that the earth's nature had shifted rather than time itself.

Yet, in the light of the success of the relativity theory and the demotion of NOW, we should realize that earth time is not rigidly geared to star times — that there is a measurable leeway in our celestial relations, making the earth chronologically individual and giving a curious fluid texture to the whole temporal dimension. This is an aspect of, and closely related to, the condition in space known as curvature (which we will soon come to) and which, in combination with distorted time, makes up the characteristic wavy "grain" of space-time.

The theorists have found that the geometry of space-time thus tends toward hyperbolic forms, its basic unit being roughly (and appropriately) representable as an hourglass of which top and bottom spread out indefinitely into the past and future while the narrow middle is the local present, the here-now, laterally encircled by a spatial wedge of elsewhen-elsewhere. On two-dimensional paper it can look like the diagram on the next page and may be worth your serious contemplation.

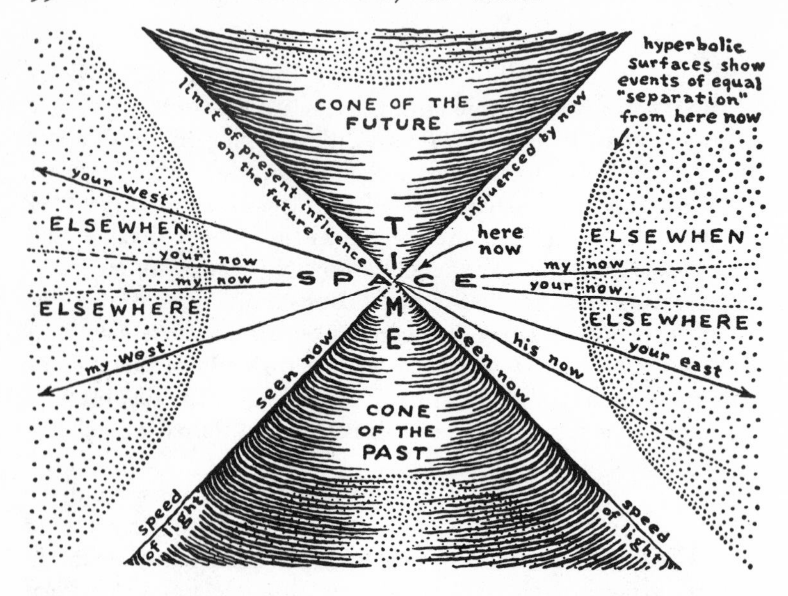

If you are surprised to see that the *past* meets the *future* only *here*, not *elsewhere*, just remember that Einstein has quite thoroughly exploded the myth that simultaneity prevails throughout the universe. And this means that the only definite location of *now* is *here*. In fact, every man's *now* is *here* (the "*here*" meaning "where he is"). While beyond *here, now* becomes more and more a matter of viewpoint, of relative motion, as in the case of the train struck by lightning. In terms of simple geometry, *now* can be represented as a line — not necessarily a straight line — of continuous simultaneity. Part of everybody's *now* line touches *here*, where everybody's past and future sometimes meet, but the rest of the *now* lines' lengths stretch out at all possible angles, according to their individual relative velocities, forming the circular wedge of other space that is not *here* and not (for everyone) *now*, nor definitely *past* nor *future*.

The meaning of this weird collar of temporal ambiguity might be clearer to you if you sent a flash of light or radio waves to Mars and it were reflected back to you, arriving, say, half an hour later. For anything that happened to you during that half-hour (from a universal range of viewpoints) could be neither definitely *before* nor definitely *after* nor definitely *simultaneous with* the flash's arrival at Mars. This is not mere theory, either. In the near future there are bound to be lonely ladies on Earth praying for loved ones "out there" on expeditions to Mars and some of them will yearn for the consolation that "he is thinking of me now," presumably at a prearranged moment. Yet "he" will have to think of "her" continuously for a half-hour to be (universally) sure their thoughts mesh as planned. And if he were on Neptune, it would take him a tedious eight hours. If on a nonsolar planet, at least several years. And all because, as Eddington put it, "the *nowness* of an event is like a shadow cast by it into space — and the longer the event the farther will the umbra of the shadow extend."

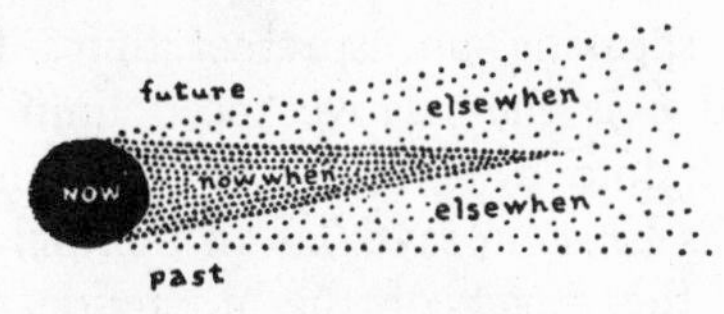

That would seem to limit practical two-way communication, such as a conversation (and possibly any sort of mutual telepathy), to a single planetary neighborhood — for even radio conversations between Earth and moon are annoyingly sluggish with a 2½-second pause before the snappiest possible response to anything said. The basic limitation here, of course, is the speed of light, which is also the maximum rate that any kind of influence can be propagated relative to any material thing in the universe. It is a limitation that not only seems to prevent any sort of a signal or causation from going anywhere except into the future, but also defines the meaning and surface shape of that same future through the presences and absences of those very same causing influences — if you see what I mean. Another way of saying it is that one event is only definitely *before* another if it can influence that other in some way. Thus, to get into somebody's future, you must be moving slower than light in relation to him — which explains why we, who never pass each other as fast as light, are forever getting into each other's futures. It also suggests why light itself is timeless and therefore aloof from the effects of motion (compounded of space and time) as evidenced by its baffling constancy of speed, no matter whether you add or subtract your own speed to (or from) light's by going with or against it. Or you might figure out light's timelessness by the Lorentzian formula of transformation, which says clocks must slow down to zero and stop altogether in relation to anything moving by them at the speed of light.

Is then light an aspect of time itself, since a light beam has no motion through time? No, one cannot quite say that light as light is, properly speaking, an aspect of time. But its speed is. I mean its speed *c* is the relative speed limit in the material universe and therefore a natural seam of time, "playing the part physically," as Einstein says, "of an infinitely great velocity." Indeed, *c* holds the secret of the mysterious composition of velocities whose sum can never exceed 186,282 miles a second no matter how many of them there are or how they collate to the velocity of light. So *c*'s constancy truly must be of the essence of time.

Doubtless that is why science holds no hope of our ever being able to overtake a flash of light in this world. However fast we go, the light must still go away from us at 186,282 m.p.s. according to the strictest measurements — even though, paradoxically, the measurements are deceptive. I mean, if a scientific observer could be sent off in a rocket at, say, 161,282 miles per second (earthspeed) in pursuit of a flash of light, the light would seem to witnesses on Earth to be going only 25,000 m.p.s. faster than he was. But he would inevitably measure the difference as 186,282 m.p.s. — because the relative foreshortening of his measuring scale would make his miles only as long as half-miles on Earth, while the retarding of his clocks would turn his seconds into double-seconds of earth time. These two distortions together would stretch his estimate of the light's speed to 100,000 miles a second (really half-miles per double-second). Then by adding on the discrepancy between his *now* line and the earth's *now* line, his different clocks would be warped out of synchronization with each other (from an earthly view) just enough to bring his reading of the light's departure to 186,282 miles a second. Easy, isn't it? And I trust it's twice as clear to you as it is to me!

Anyhow, while you are still quietly digesting this rather sticky conserve of mulled abstractions, may I gently suggest still another curious, and perhaps more helpful, thing about time — that, if any object moves at 186,282 miles a second, the reason it cannot progress through time is that time is essentially a relation between things and themselves — as distinct from space, which is a relation between things and other things — and 186,282 miles a second of relative speed has the very extraordinary effect of stretching a body's *world-line* out until it is so unbelievably tenuous that the body can no longer have any relation to itself. World-lines, remember, may have *any* kind of shape, depending on how they are regarded and on the motion of the frame you look at them from — so they may seem to be going straight like the light of a star or spiraling like the earth's track around the moving sun or, perhaps, vibrating radially in and out like a throbbing atom or rocking, spinning, tumbling, yawing or even possessing some sort of multidimensional twist not yet imagined.

If this thought does not convey much of a picture of time, it may be only because time is far too abstract for easy visualization, its direction of motion being elusively "perpendicular" to all three dimensions of space at once. Furthermore, time also seems to have its own multiple dimensions or frames of reference, which react upon each other. I mean, a reel of movie film can serve nicely as a model for flowing time. But what is this "time" flowing through? Some framing medium is obviously needed. So the *motion* of the reel constitutes a framing model that reveals a second dimension of time. And the *clock* that times the motion of the reel, a third dimension; the *rotation* of the earth that regulates the clock, a fourth. And so on.

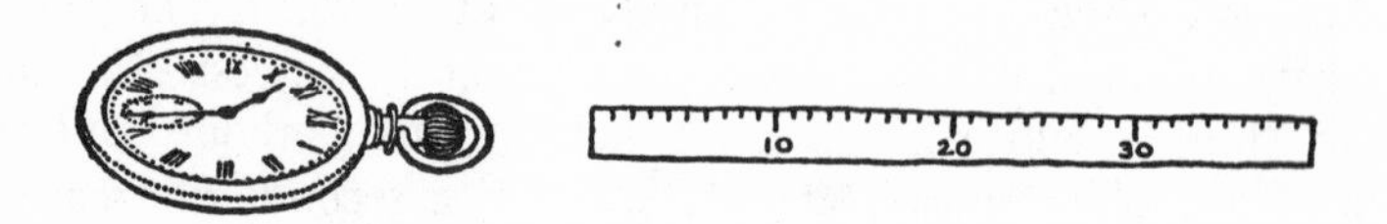

Then there is the curious linkage between time and space which is so familiar we are rarely conscious of it. We use the word "times," for instance, to mean "multiplied by" because time multiplies a three-dimensional amount, such as space, into a four-dimensional product, such as space-time. That is how a hen's daily production of one egg, repeated seven times — or *times seven* — becomes her total product for a week — a nice example of what Aristotle must have meant when he said, "Time is the number of motion."

In a hundred ways thus daily, without thinking of it, we acknowledge our continuum by combining space and time. We write down the *temporal* order of the spoken word in the *spatial* order of visible letters. We inspect the surface of a target to see where bullets passed from their known past into their calculable future. We dive under the sea — holding our breath to a depth as readily measurable in fathoms of space as in seconds of time. And the light-year is a unit, not of time but of distance.

So is it any wonder that mere places evaporate while signals pass between them? That the whole abstraction of *place*, which we once learned to trust, turns out to be nothing but a viewpoint that is different from every side? That, as Einstein declared, "it is neither the point in space, nor the instant in time, at which something happens that has physical reality, but only the event itself"?

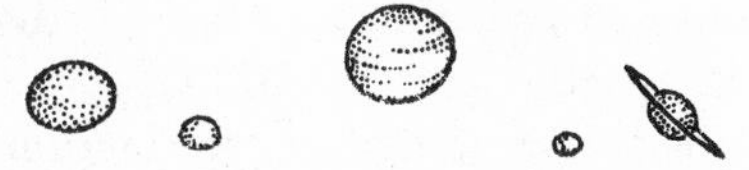

I think it was about here in his thinking that Einstein must have realized that Newton's great law of gravitation, for all its beauty and simplicity, needed to be reconsidered —

> That very law which moulds a tear
> And bids it trickle from its source,
> That law that makes the earth a sphere
> And guides the planets in their course.

It was a staggering task, but logically unavoidable. For Newton had neglected putting *time* into his inverse-square equation and, if this world really was the space-time continuum that relativity had already shown it to be, any major force such as gravity acting through *space* simply had to act also through *time*.

This is not to say that Newton was wrong in taking *time* for granted. He was a hundred percent right for the seventeenth century, and much more than ninety-nine percent right even for our own twentieth century. He was as right as anyone could be who assumes that the distance from Earth to moon is a definite distance at any particular instant. And his rightness had been well tested in thousands of working mechanical applications, in practical engineering formulas, in special eighteenth-century expeditions to Lapland and to the equator to measure the gravitational flattening of the earth toward the poles — even in seemingly miraculous astronomical predictions — during three centuries. Yet Newton's caliber of rightness could no longer be quite enough. For a new principle had evolved. Relativity had just demonstrated that space as such cannot be absolutely definite in amount. Not only does it shrink for relatively moving observers but the time at one end of it can no more be the time at the other end of it than the *now* that is *here* can be the *now* that is *there*.

So Einstein set about broadening his relativity theory in the direction of gravitation. He did not see any reason why it could not encompass gravitation if it could be generalized enough. Up to now, he had been dealing only with rather special cases, such as the relative motions of two trains or two measuring sticks, moving at steady, uniform speeds past each other — special cases which had neatly explained Lorentzian transformation and the ether doldrum. He could logically call his conclusions from these cases the Theory of Special Relativity.

But from here on he must consider other motions, more usual and general motions, specifically unsteady and changing ones — in short, *accelerations*. And he would call his findings about their relationships the Theory of General Relativity.

You may wonder at first what all this has to do with gravity. But its appropriateness will become clear if you can visualize Newton's famous apple as it fell to the ground. For apples fall with natural acceleration, and Newton certainly recognized such as the acceleration of gravity, as had Galileo before him.

However, Newton quite reasonably concluded that the apple, in falling, was being *pulled* downward toward the earth. That is, he felt himself standing stationary upon solid ground while the apple approached him. And it evidently never occurred to him to look at things from the apple's point of view.

But Einstein, being a relativist and keenly aware of the importance of nonhuman and exotic viewpoints, wanted to ask the apple what *it* had to say about Newton, the earth and its own motion. And he may have posed his unheard-of question in somewhat this vein: "Tell me, apple, how did Newton look to *you* when you fell toward him?"

"Fell toward him?" retorted the astonished apple, by this time feeling not only bruised but sour. "Fell toward him — nothing! I'll swear it was Newton and that damned earth who fell toward *me!*"

"Well, well," Einstein must have thought, and probably with a grateful twinkle, "I wouldn't be surprised if you've got something there." For why shouldn't the earth fall just like everything else?

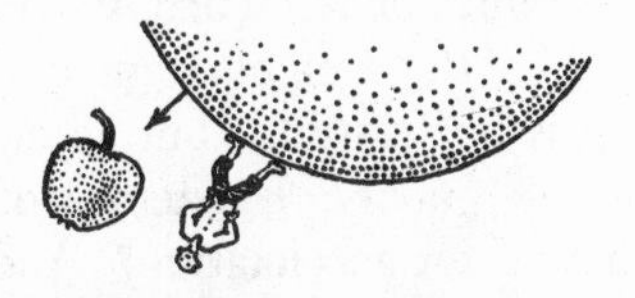

Why could it not be as true to say that the earth falls *up* to meet things in the sky as that things in the sky fall *down* to earth? Is not that simple relativity? Or am I too full of apple juice?

It *is* relativity — yes, he checked himself. But perhaps it is not so very simple. For although the apple is regardable as accelerating downward, can one . . . could one possibly justify a claim that the earth is accelerating upward just as much?

On the face of it, upward terrestrial acceleration would seem quite impossible for more than a very small portion of the earth, since upward, globally speaking, is not *one* direction but *every* direction. In fact, if the earth's surface is accelerating upward all around the earth, the earth as a whole must be exploding. Which it obviously is not. Yet, come to think of it, maybe the earth *is* exploding in a sense. What about its atoms and molecules, gyrating like mad in all directions? Is not that a kind of explosion? Certainly the hammering of molecules against Newton's feet were a more cogent cause for his upward acceleration, even according to his own laws of motion, than anything he could claim as sufficient cause for the apple's acceleration downward.

But perhaps standing on the earth's surface is too special a situation in the universe to be a fair example of universal gravitation — so, as Eddington once suggested, let us consider Newton as standing at the center of the earth, where there should be neither any noticeable gravity nor preponderant molecular pressure pushing him toward the apple that he might see (by subterranean telescope) "falling" toward him from high above. Here the gravitational circumstances of the converging man and apple are equivalent. Neither is being forced by anything visible or tangible toward the other, and each feels *he* is standing still while being fallen upon.

It is a case of two frames of reference, moving relatively — symmetrically — exactly like the freight train and the earth except that here the motion is not constant but accelerating. Could this be the true essence of gravity, Einstein wondered? Is gravity really just another name for acceleration? And if so, why should it attach itself so steadfastly to massive bodies like the earth?

Instinctively, he felt the answer must somehow be bound up with the still cryptic, scarcely emergent nature of space-time, as had been the answer to the riddle of the ether wind. But where to approach it for a real understanding? How to aim one's mental scalpel into the gravitational marrow?

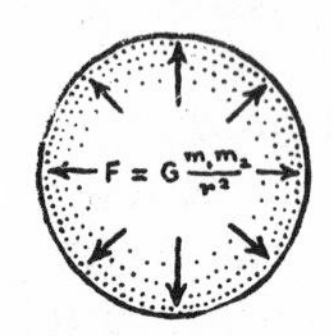

As it turned out, more than ten years were to pass while Einstein struggled with this profound problem — "years," in his own words, "of anxious searching in the dark, with their intense longing, their alternations of confidence and exhaustion" — while around him the most powerful nations of Europe steadily and stealthily aligned themselves into bitter rivalries that eventually ignited the conflagration of World War I.

Sometimes during his unprecedented trial, as the war clouds glowered and later the newspapers were filled with reports of horror and mass destruction, Einstein clutched at dim and fleeting geometric notions of space-time, trying to visualize dimensions and again dimensions. Sometimes he pondered the thoughts of Newton and reviewed the rival gravitational theories of the distorted medieval days — recalling Descartes' celestial vortex of swirling atoms (see page 313) that had been rejected because it contended that things fell only when their surfaces were pushed from above, an argument that did not account for gravity's demonstrable proportionality to the *masses* instead of to the *surfaces* of gravitating objects. The whole vortex idea indeed was about as fantastic as Lorentzian transformation. Yet

Lorentzian transformation had worked out successfully after all, even though for a reason its originator did not foresee. So assuming Lorentzian transformation really was true, Einstein wondered, could it somehow be possible that Descartes' vortex idea was also basically right — although perhaps the hypothetical vortex itself might have to be replaced by some still unsuspected ethereal pressure inherent in the geometry of emptiness?

About at this stage, Einstein happened to look one day at a Mercator map of the earth. You will remember Mercator's projection as the one that spreads all lines of longitude out straight and parallel so that the farther a country is from the equator the more absurdly swollen it must appear in size. And it occurred to Einstein that if anyone who thinks the earth is flat should use such a map he would logically also believe Greenland to be as big as all Europe, because a flat earth is accurately representable on a flat map. But then, if Europeans traveling across Greenland should invariably report that distances there seem to be much shorter than in Europe — that is, "appear" much shorter than they "really" are — the flat-earth believer might well conclude that

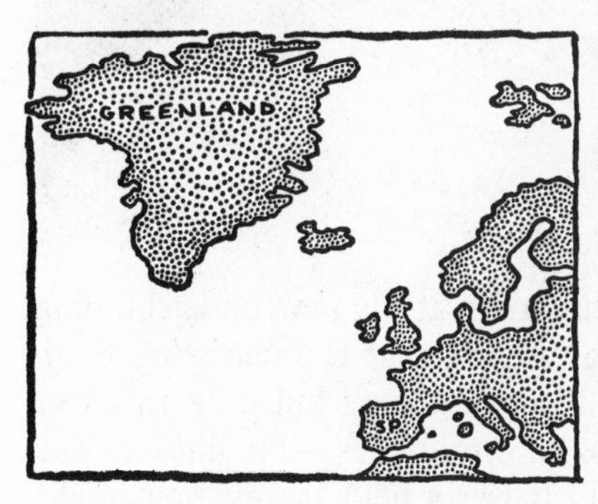

Greenland affects people's minds, somehow hypnotizing them so that they do not realize the "true" passage of time, or perhaps actually slowing down their clocks and foreshortening their courses of travel as in the Lorentzian transformation.

Could this not be another analogue of relativity? If so, it would work equally well if Greenlanders drew their own Mercator maps on a different base line, projecting a reciprocal Spain bigger than Greenland — and giving this Spain the mysterious power to transform the space and time of traveling Greenlanders.

It would work in a slightly different way also if two ships steamed side by side across the equator, heading exactly north. If their captains believed the earth were flat, for instance, they would logically assume the two ships could continue parallel to each other along parallel lines of longitude as far as the sea extended. So when, after a few days, they noticed that the ships were gradually coming nearer together and, no matter how perfectly they steered north, that the convergence was steadily accelerating, one of the captains might postulate a mysterious force of attraction to explain it, a force proportional to the inverse square of their distance apart — in short, a law of gravitation hardly distinguishable from that defined by Newton!

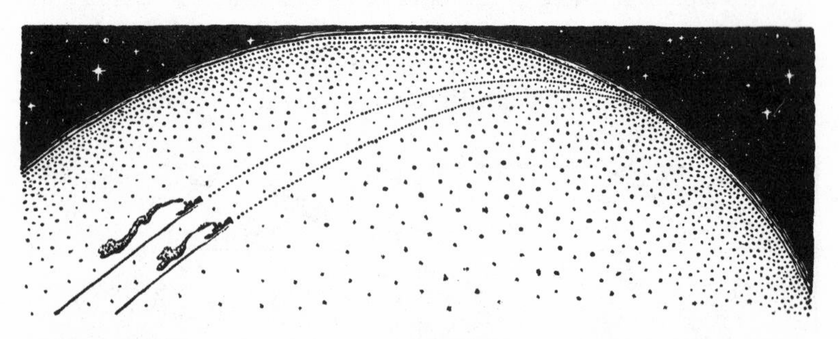

In similar manner did Einstein arrive at his new understanding of gravity, at his revolutionary awareness that the persisting tendency of things to fall is not really a force at all but only an effect of geometry — a nonparallelness of world-lines — an illusion due to the "curvature" of the world. It was a field theory, you might say, for it avoided the "action at a distance" implicit in Newton's old law by attributing falling to the natural response of objects to their immediate or field surroundings. That is, free matter in space literally takes the easiest and most leisurely path, following whatever sort of curve of space-time it finds itself in — a track that has come to be known as a *geodesic*. This, in a sense, is a reversion from Newton back toward Descartes and his vortex pressures. But, much more, it is the discovery of a deeper meaning in the ancient Greek observation that "God always geometrizes."

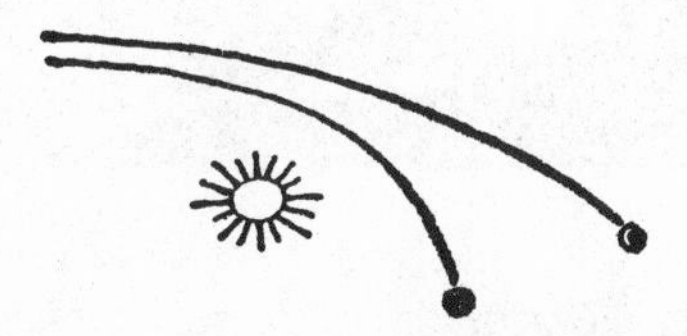

If this still seems vague and mysterious, at least it can be tested by fairly simple experiments. For it is a remarkable fact that, although no one is known to have suggested it before Einstein, artificial gravity can be created in a small enclosure — say, an airplane or an elevator or this space ship — that is so exactly like natural gravity that no one can tell them apart. This is the basis of the G unit of acceleration (equal to gravity's strength at the earth's surface) that the Air Force now manipulates at will to create gravitational "forces" dozens of times stronger than normal or even to "banish" gravity altogether in a zero-G trajectory for the better part of a minute. This is the reason an airliner in a cloud may bank in a turn so tight its wings point vertically up and down without the passengers becoming aware of it. Few pilots and fewer spacemen can be found to hold brief any longer for the outmoded notion that there is any fundamental difference between gravitation and acceleration. And it is naturally the earthbound folk who cling hardest to the old materialistic concepts of gravity, for they know gravity as something that can be felt, something that never relaxes its relentless tugging-tugging-tugging toward the earth. It is hard for them to remember that the sensation of weight in truth is not gravity itself but only the resistance of the earth to their natural response to it. If they could only get away from the earth somehow and be free of their constant bombardment by the tireless molecules of pavements and floors and chairs and beds, they would be elated at the ethereal liberation of floating at large undisturbed — of *answering* gravity instead of *fighting* it — of conforming gracefully to nature's orbit, drifting in peace upon a world-line, accepting the lovable abstraction of the geodesic.

Suppose, for example, that you were born and brought up (or down) in a falling elevator or, more realistically, in my orbiting space station. In such a vehicle, an apple does not fall and, if you had never heard of the phenomenon of falling, it would take a great effort of imagination to visualize and understand the strange world of a planet's surface where a wholesale molecular bombardment is always pushing "upward" so hard that loose things like apples seem to want to go "downward." It would sound absurd to be told that in fact the only reason your apple inside the vehicle does not "fall" is that the whole vehicle including yourself is already "falling." Yet from outside we know that this is true — at least relatively speaking. We can see at last that the state of falling is the normal state of bodies in the universe — of all free, whole bodies at any rate — and that it is only the denser subregions (like a planetary surface) containing tiny captive crumbs (like you) that must undergo the frustration of not being allowed their full response in unhindered falling.

To make the point clearer with an extreme case, consider your space vehicle as orbiting close to the very heavy dwarf star, Sirius B, which, as we shall presently see (page 585), has turned out to be extremely important in relativity theory. Since the density of this body is sixty thousand times greater than that of water, its gravitational field must be of such unbelievable ferocity that, if you could be left standing upright on its solid surface, you would instantaneously collapse in a splash upon the ground. Yet in falling freely through the same gravitational field in your vehicle, even though your acceleration "downward" exceeded what it would be if you were fired from a cannon, whipping you around Sirius B like a feather in a tornado, you would feel no force and would continue to float in weightless relaxation as if you were a trillion miles away in utter emptiness.

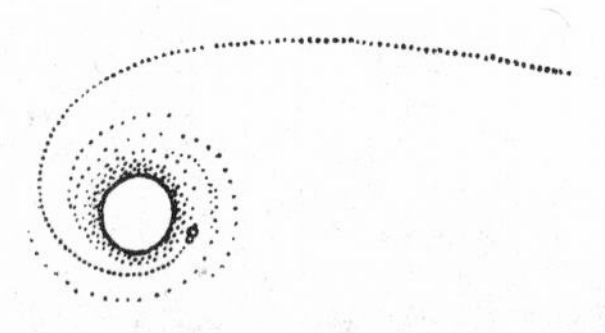

Thus is manifest the outer mystery of gravitation, the paradox of acceleration — two equivalent aspects of a single phenomenon as defined in Einstein's now well-known Principle of Equivalence, a principle in which the "force" (of gravity) that flattens the standing body on Sirius B is revealed to be identical with the "force" (of acceleration) that whips the vehicle around it — both being illusions of geometric space.

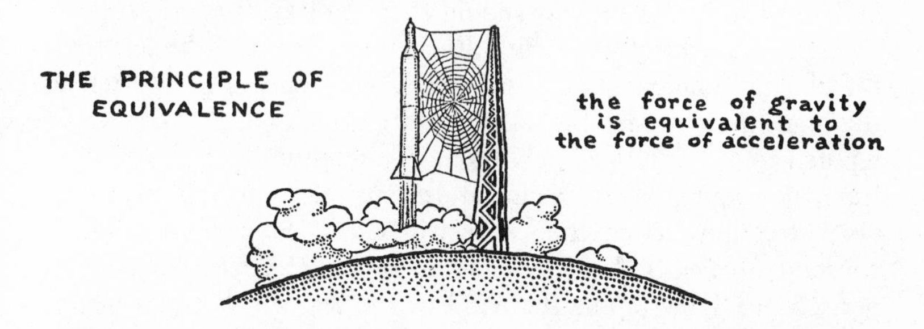

Einstein eventually became quite aware of the shock his free-ranging concepts would inflict upon unprepared classical minds, and so he began — at least on one occasion — by begging his eminent predecessor, "Forgive me, Newton," when he introduced his startling new ideas on gravitation, from there on gently but confidently working into a discussion of the abstruse "geometric property" that makes things "gravitate." By the time the derivative called "acceleration" turned up as another face of the same property, however, some of his listeners had been left behind while others, more discerning, were relatively gaining — among them Sir Arthur Eddington, Sir James Jeans and, significantly, Bertrand

Russell, who went so far as to denounce acceleration as "a mere mathematical fiction, a number, not a physical fact."

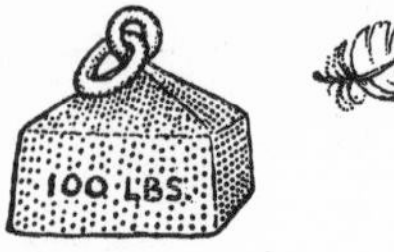

But there remained still another face of the new property to be dealt with: inertia — which, you remember, was defined by Newton as the tendency of a resting body to remain at rest or of a moving body to continue moving uniformly in a straight line at constant speed. Inertia had long been regarded as a rather passive, feminine force with a quiet predilection for complementing the more active, masculine force of gravity, as in the case of the earth's motion around the sun where centrifugal inertia nicely balances solar gravitation. Yet how distinct in fact are these apparently individual forces? And why should they so often merge into one?

After only a little pondering, Einstein got a strong hunch that the two are fundamentally inseparable. He seems to have reached this conclusion under the inspirational aegis of Galileo, who is reputed to have dropped pebbles and cobblestones simultaneously from the leaning tower of Pisa to demonstrate that these weights, no matter how much they vary, will always fall with the same constant acceleration — later determined at 32.2 feet per second per second. How unlike the wind or the sea, which treats heavy stones so differently from light ones, was this curious impartiality of gravity! Its force must act in strict proportion to the criterion of individual mass. But, come to think of it, that is exactly how inertia also acts — a fifty-pound stone requiring ten times as much force to accelerate it or slow it down as a five-pound stone.

It was too much for Einstein to believe that two really independent forces could both obey such an odd rule by pure coincidence, so he began to look for gravity-like influences that might be hidden within inertia — until suddenly he recalled the speculation of Ernst Mach of Austria, that the stars may be mainly responsible for inertia — that the water spreading to the outside of Newton's

famous spinning bucket (see page 521) was probably not so much clinging to "absolute space" as gravitating toward "the masses of the universe," where "all bodies, each in its share, are of importance for the law of inertia."

From here on, Einstein set about the task of incorporating inertia as the *universal* aspect of gravity, along with the reciprocal goal of treating gravity as the *local* aspect of inertia, both being relative geometric properties of the integral space-time continuum — so that the earth no longer need strain centrifugally or schizophrenically against the tension of solar gravitation but instead could float in complete relaxation along her natural track, her private geodesic line through space-time.

Step by step, you see, geometry was growing, maturing, its previously unimaginable abstractions enveloping in turn space, time, motion, acceleration, gravity, and now inertia. Practically all the mechanical sciences had been influenced if not consumed by geometric angles, coordinates and curves. Even electromagnetism, light and the atomic world were apprehensive as to their own approaching days of reckoning. For, amazingly, the metric of space — essence of geometry — seemed able to explain anything.

Eddington, with characteristic lucidity, compared medieval humanity to a race of flat-fish living in an ocean of two dimensions, where fishes in general swam in straight lines except for one area where they all seemed bewitched and could not help going in curves. Some thought

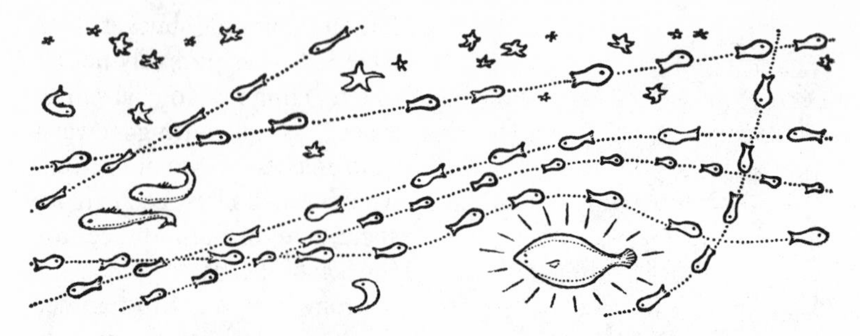

there might be an invisible whirlpool there until a bright fish named Isaac pointed out the "real" cause of the trouble: a very large fish — a sunfish — lying asleep in the middle of the region, who attracted all the other fish in proportion to their sizes. This adequately accounted for most of the peculiar curves, so nobody bothered about the lesser attractions of a small moon-fish circling near by or the great numbers of fixed starfish twinkling in the background. And the only discontentment left was the carping of a few carp who did not see how the sunfish could exert such great influence from such a distance, though they presumed his influence must spread forth somehow through the water.

Then a very unusual fish called Albert got interested in the problem and began analyzing it — studying not only the "forces" involved but, more particularly, the *courses* being swum. And suddenly he arrived at a surprising explanation: there must be a mound where the sunfish lay! Flat-fish could not sense it directly, nor understand it, because they were two-dimensional, but any fish who swam over the slopes of the mound — no matter how straight he aimed — got turned somewhat "inward" toward the mound, just as a piece of cloth over your right shoulder will tend to hang more or less leftward.

Thus there was no longer any need to assume "influence at a distance" in accounting for curvy swimming, since the curves were now built right into the two-dimensional world — even though they were almost impossible for a flat-fish to deduce — thereby making it essentially a three-dimensional world despite the fact that the fish themselves were denied access to the third dimension.

This analogy gives a pretty fair idea, I think, of how a hidden curvature in the world can impart the illusion of an attracting force such as gravity. And of how hard it is to comprehend the curvature while we are enclosed *in*side of the same dimensions it is *out*side of. Our real world, moreover, has mounds not only of space but is teeming with mountain ranges of space-time, not to mention canyons, funnels, gyrating hyperbolas and all manner of indescribable, multidimensional and illusory shapes which are aspects of its grain or structure.

On the subject of shape, I might mention that mere *shape* is always relative and therefore largely illusory — bitter though this may be for the young or beautiful to swallow. It works out that *order* is the intrinsic relation in world geometry, and order is a function of the absolute *interval* that binds the order. But space and time which compose the interval are free to shift and distort, relatively, along with any shapes formed by their dimensions. Thus a square turns into a diamond if you look at it obliquely, and the form of a woman is essentially given her by her observers — which, obviously, I mean not solely in a psychological way.

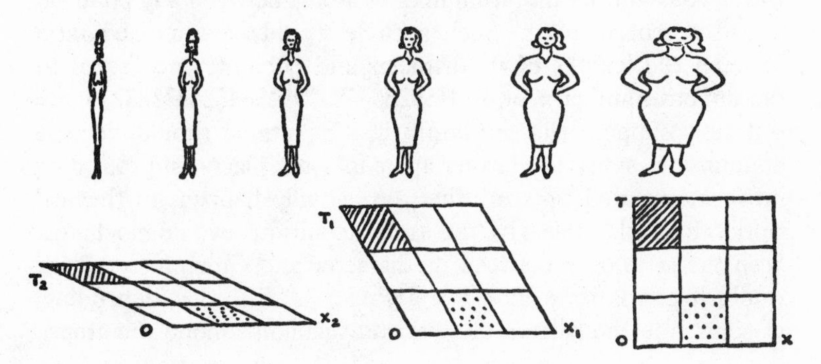

This is why space-time curvature, while very important relatively, must retain the illusory quality that makes it so hard to grasp in the real world. You can visualize a gravitational field around a star or planet, for instance, as a kind of invisible dimple in the skin of space-time, a pucker in the cosmic cloth of at least four dimensions — but that will be an ambiguous picture at best, for it must vary with every frame of observation. Nor is it in any way simple. Indeed even if you knew the G values at hundreds of points around a space-time pucker you could scarcely comprehend its geometry since, in Eddington's words, "what determines the existence of the pucker is not the values of the G's . . . It is the way these values link on to those at other points — the gradient of the G's and, more particularly, the gradient of the gradient."

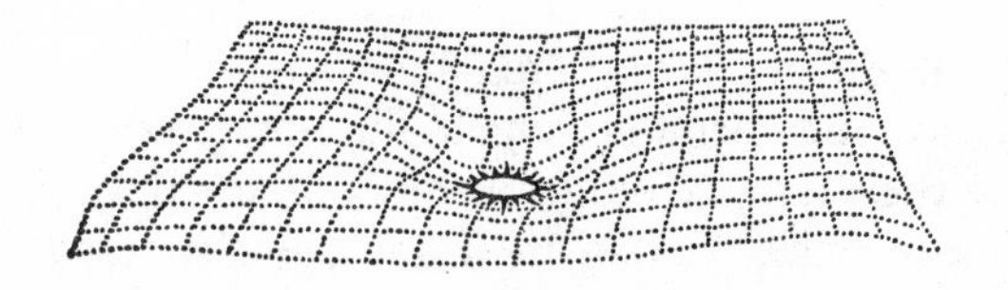

A better way to picture curvature in the cosmos might be to think of space as uniformly studded with floating atoms so that any spatial distance is proportional to the number of atoms it encompasses. A circle in such space could be defined as a curve with a constant minimum number of atoms between any point on it and a central point. Such a circle would have a fixed ratio between the lengths of its diameter and circumference equal to the famous number π or 3.14159265358979323846264338327 9 . . . But now suppose the uniformly floating atoms should become nonuniform, spreading farther apart in some places and crowding into concentrated clots in others, as actually happens in the real world. Then a circle (by the same definition) would no longer keep the ratio of π between its diameter and circumference, nor would it continue to look like a circle, its distortion depending, of course, on the pattern of clots and vacuums among the atoms.

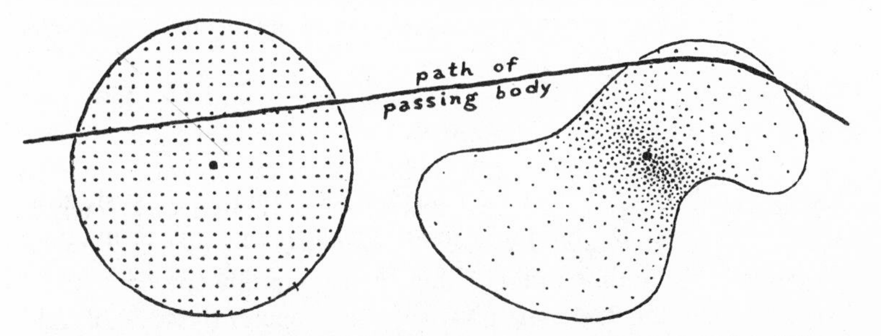

CIRCLES IN UNIFORM AND NON UNIFORM SPACE

Considering this an analogy of actual space, the atom clots representing the regions of stars, now suppose some object should move through the space, penetrating the edge of one of the clots. It would obey the natural law of least resistance, automatically following the path that passes fewest atoms. Although when the space originally was studded *uniformly* with atoms this path was a straight line, now in *nonuniform* space the object naturally must curve around the dense clots, staying far enough away to avoid atom crowds yet not so far that its lengthened route will add unnecessary hermit atoms, either. In this way, the object actually describes a track concave toward stars in proportion to their masses, which makes it appear to be attracted by them — like the flat-fish crossing the sunfish's mound.

But the attraction is not what it seems, for the object is really just drifting along a geodesic, its private line of least resistance, which also (under the influence of Lorentzian transformation) is the track where its time flows fastest, or (which is the same thing) the track that takes the longest time. At least that is what actually happens, according to all the evidence now available. And it explains why gravitation commonly amounts to a repulsion as well as an attraction, sending comets away from the sun about as often as toward it or, teamed with inertia upon a rapidly rotating asteroid, making stones "fall" upward into the sky. It may even suggest how gravitation can be only the effect of looking at things from an accelerated point of view (as from an upward-pushing earth) — or only the effect of the nonparallelness of matter (as in the case of Newton's apple, which fell when its geodesic and the earth's were no longer parallel).

In any case, we cannot go on assuming, as did Newton, that material bodies naturally *want* to go straight. For stars distort space as birds distort air. And straightness is an ideal of almost no practical meaning in the real world, even though *any* track is straight *relative* to *some* frame of reference. By which I mean (as shown in the illustrations) that the supposedly curved orbit of the earth around the sun, drawn on an apparently straight grid of Newtonian rectangular space, can just as well be considered

EARTH'S ORBIT REPRESENTED AS A CURVED LINE ON THE STRAIGHT GRID OF NEWTONIAN SPACE

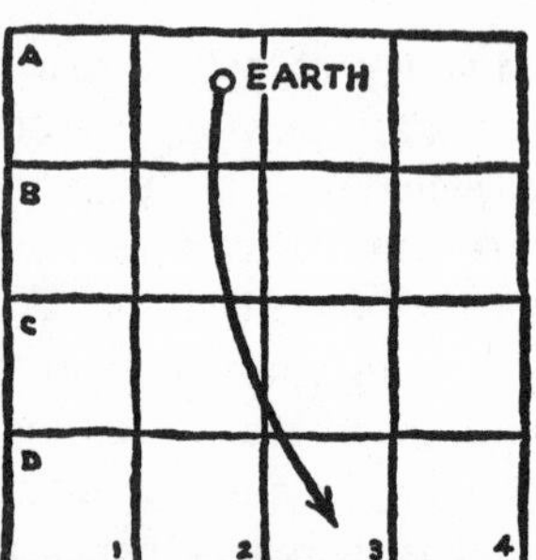

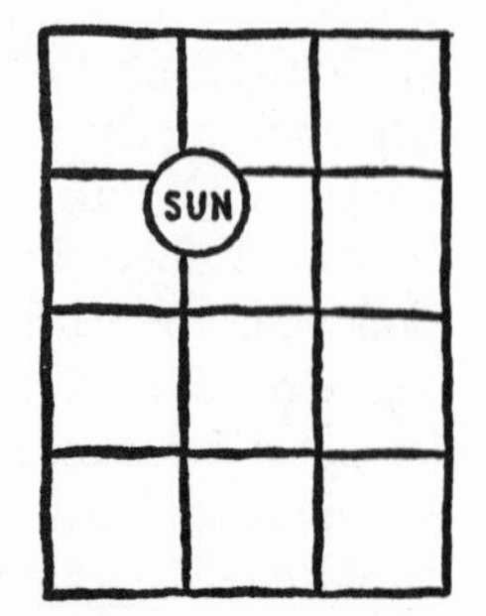

a straight orbit or geodesic, drawn on a curved grid of Einsteinian distorted space. And the latter, newer interpretation is actually a little better suited for interpreting *all* the known facts, as will (I hope) soon be apparent.

An almost obvious test of curvature is the measuring of angles to see if geometric figures come out according to the ideals of

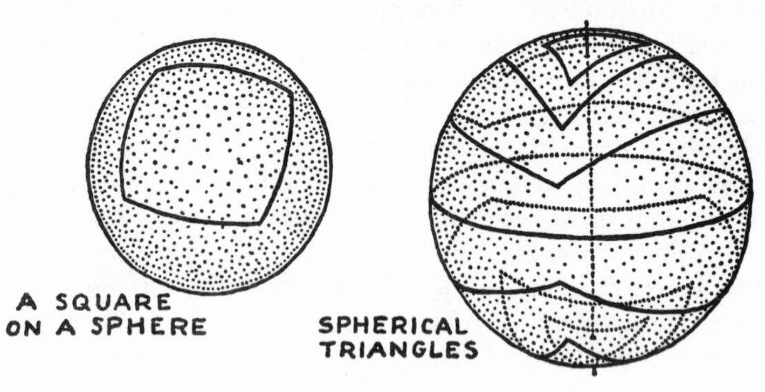

Euclid, whom you presumably studied in school. For it is evident that a "square" with sides one foot long, drawn on the surface of a sphere, must contain more than one square foot of area — and that a "triangle" drawn upon earth, with great circles for sides, will have angles that total more than 180° or two right angles, particularly if such a "square" or "triangle" is big in relation to its sphere. The terrestrial "triangle" formed by the Greenwich

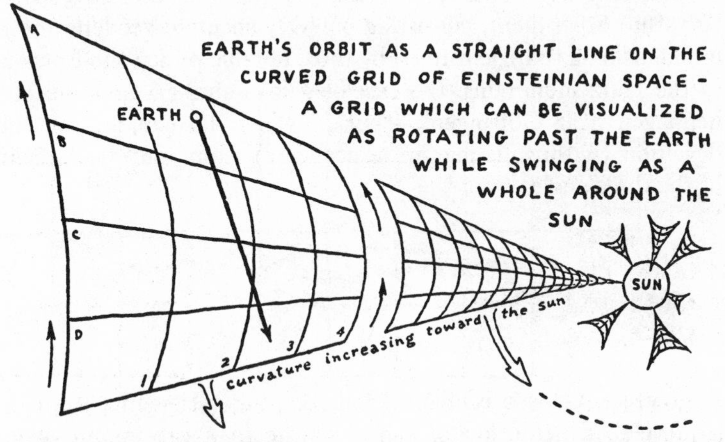

meridian, the 90° west meridian and the equator, for example, has three right angles. Still bigger "triangles" have still bigger angles — all the way up to a total of six right angles when the "triangle" has stretched out into the form of a continuous great circle. And if you want to push the "triangle" beyond this extreme, in effect turning it inside out, its three angles can approach an ultimate total of 900° or ten right angles!

But this, you may protest, is merely going from plane into solid or three-dimensional geometry, and does not basically conflict with Euclid. Nevertheless what is true of a triangle on the blackboard is not necessarily true of a triangle too big for the blackboard. For the blackboard represents a straight, flat, two-dimensional ideal that does not exist in nature — a nature where added dimensions turn up apparently willy-nilly and may not be gotten rid of permanently just by sweeping them under the mental rug.

The truth in this seems to have been first realized by Gauss, the prodigious nineteenth-century German mathematician who was seriously questioning the foundations of Euclidean geometry at the age of twelve and a few years later created the "metric theory of surfaces" that described *positive* (spherelike) and *negative* (saddlelike) curves in terms of the angles between their p and q coordinates resembling latitude and longitude lines. Such a study almost inevitably led him to perform the first space-curvature experiment, consisting of "very accurate geodetic measurements on a triangle formed by three notable peaks: the Brocken in the Harz mountains, the Inselberg in Thuringia and the Hohenhagen near Göttingen" — but in which he could detect no deviation (within his margin of accuracy) from 180° in the sum of the three angles.

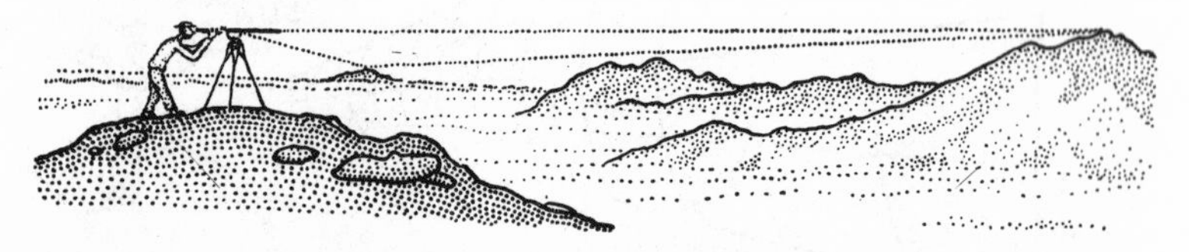

However, Gauss's brilliant pupil, Bernhard Riemann, went on to develop a definition of the curvature in a continuum of an indefinite number of dimensions — a feat which half a century later, as if by accident (in the oft-noted way of true history), served Einstein in the development and perfection of his theory of general relativity. For naturally Einstein, having discovered a new *explanation* for gravitation, needed a new *law* of gravitation at least as precise as Newton's inverse-square law which it was to improve on. And this law had to say something about curvature.

The simplest form of spatial curvature, however, was so far beyond normal human experience that it was not really visualizable even to Riemann. The nearest he could do, in fact, was probably to imagine a series of concentric spheres of greater and greater radii — of which, although geometricians used to think their surfaces had to be proportional to the squares of their radii, the surfaces of the outer ones began to be a little less than this criterion called for, then lesser and still lesser (relative to the squares of their radii) until a maximum possible spherical area was attained, beyond which the areas became smaller again until finally they disappeared into a point, the antipode of space — a definite limit that could no more be called a boundary than could the antipodal point opposite you on the surface of the earth.

If that model tried his imagination too hard, Riemann might have chosen some other one such as Jeans's analogy of spindle-shaped bodies, many of whose pointed ends meet at A (as in the illustration), occupying all space surrounding A, but whose other ends (like opposite magnetic poles) can somehow simultaneously stretch and distort themselves into converging, surrounding and meeting at another single point B — thereby demonstrating the spatial curvature between A and B.

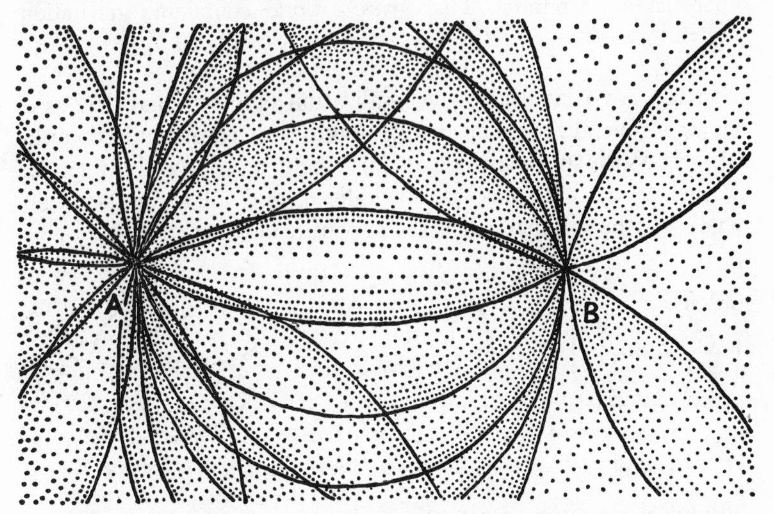

And the curvature of space-time was still more unimaginable than that of space alone, being a four-dimensional manifold embedded in what Eddington describes as "as many dimensions as it can find new ways to twist about in." In reality, as Eddington goes on to say,

a four-dimensional manifold is amazingly ingenious in discovering new kinds of contortion, and its invention is not exhausted until it has been provided with six extra dimensions [three electric and three magnetic ones] making ten dimensions in all. Moreover, twenty distinct measures are required at each point to specify the particular sort and amount of twistiness there. These measures are called coefficients of curvature. Ten of the coefficients stand out more prominently than the other ten.

Einstein's law of gravitation asserts that the ten principal coefficients of curvature are zero in empty space.

If there were no curvature — that is, if *all* the coefficients were zero — there would be no gravitation. Bodies would move uniformly in straight lines. If curvature were unrestricted — that is, if *all* the coefficients had unpredictable values — gravitation would operate arbitrarily and without law. Bodies would move just anyhow. Einstein takes a condition midway between: ten of the coefficients are zero and the other ten are arbitrary. That gives a world containing gravitation limited by a law.

Such a law is as much a generalization of Newtonian law as a hen is a generalization of an egg — and with comparable complications. With subtle and beautiful logic it reaches and influences all matter, all radiation, all world-lines, and all action — action which may be summarized as the curvature of the world.

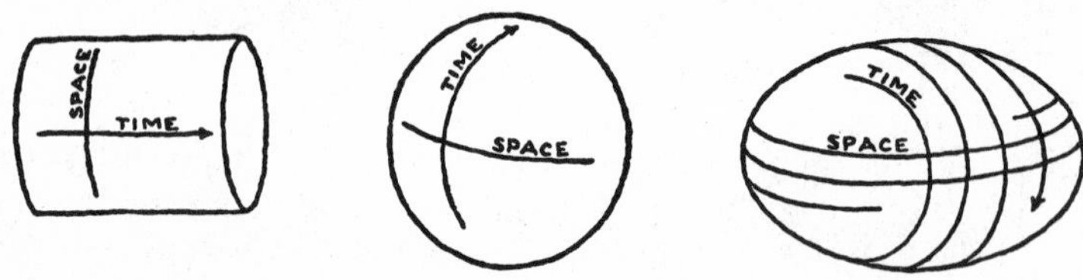

HOW DO SPACE AND TIME COORDINATE ?

Present indications as to the galaxies in our universe (whose numbers seem to be increasing faster with remoteness than do the cubes of their distances away) suggest that cosmic curvature tends to the negative — something like a four-or-more-dimensional saddle, the surface of a doughnut hole or, if you prefer, a negative sphere. Such curvature may, however, exist only in the observable regions of the universe, for there is nothing known to prevent a reversion to positive globular curvature in regions beyond the horizon of present knowability, possibly where matter changes to anti-matter or where some other little-understood seam of symmetry takes dominance over polarity or handedness.

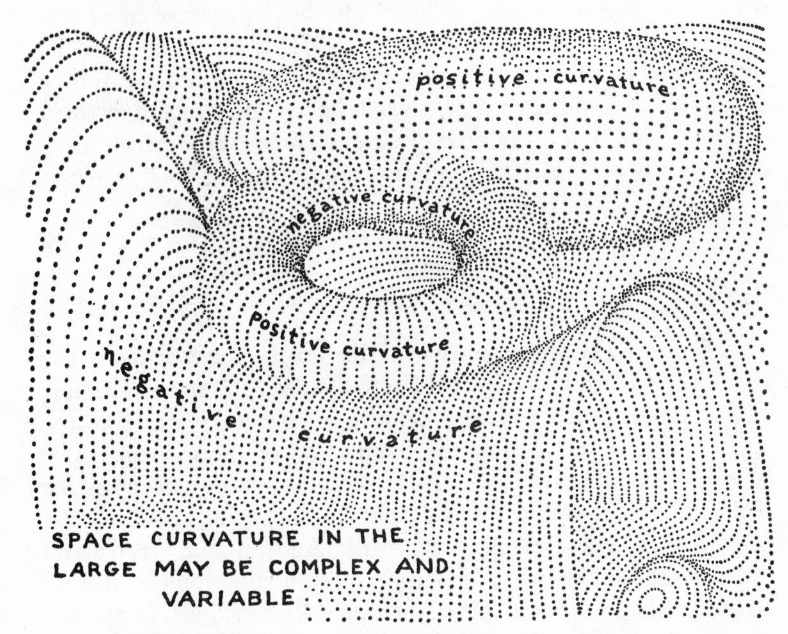

As for the effect of the curvature of space on light and vision, we literally *see* along the gravitating geodesics of light and along the actual orbits of photons without consciously realizing it — in fact, we see around all light's curves and twists. Even though the curvature of light is much too slight to be noticeable to our senses, it has been proven that light actually drops, like other objects on earth, with a constant acceleration that carries it 16 feet in the first second, 48 feet in the next, and so on. And it is only the fact that light also moves forward 186,282 miles during each second that keeps us from noticing the drop.

The direct and very precise measurement of "falling" light on Earth, however, has turned out to be too delicate and ambiguous to be convincing as a proof of general relativity, so Einstein suggested instead "weighing light" on the sun, where gravity is about 27 times stronger than on earth. This could be done by very accurate calibration of star positions just outside the rim of the sun, as seen during a total solar eclipse, then comparing them with the same stars' normal positions to see how much each star beam is bent by the gravitational curvature of solar space.

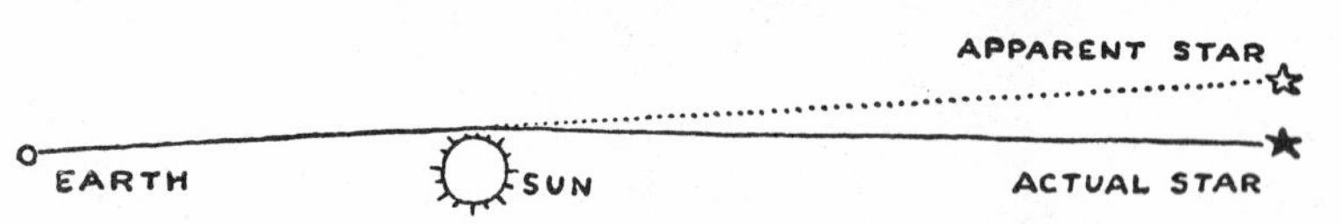

It happens that May 29, when the sun is lined up with the Hyades star cluster, is what Eddington called "the most favorable day of the year for weighing light" and that 1919 was the first year of relative peace after the publication of the General Relativity Theory in 1917—so it seemed something of a divine happenstance if not a miracle that a total eclipse of the sun "chanced" to occur on May 29, 1919. And on that day several astronomers, including Eddington, substantially confirmed Einstein's prediction that a star beam would be bent by 1.74 seconds of arc in passing the edge of the sun. This amounted to practical proof that the equation for gravity should not be simply Newton's $F = Gm_1m_2/r^2$ but, more specifically, Einstein's $F = Gm_1m_2/r^{2.00000016}$—which meant that the abstraction of invisible curvature was real.

At this point it would be only reasonable to begin to wonder, since light is bent by space which in turn is bent by gravitational masses, just how massive a star would have to be to bend its own light completely backward against itself. Answering the question, Eddington replied that a star the size of giant Betelgeuse (just about big enough to contain the orbit of Mars), if it were as dense as the sun throughout, would strain space around it so severely that its "light would be unable to escape and any rays shot out would fall back of their own weight." In fact, Eddington ventured, "the curvature would be so great that space would close up round the star, leaving us outside — that is to say *nowhere*."

Which, if true, would suggest that our existence *somewhere* — yours and mine — is just possibly one of the reasons why Betelgeuse is so vacuous. And quasars so enigmatic . . . ? (page 213)

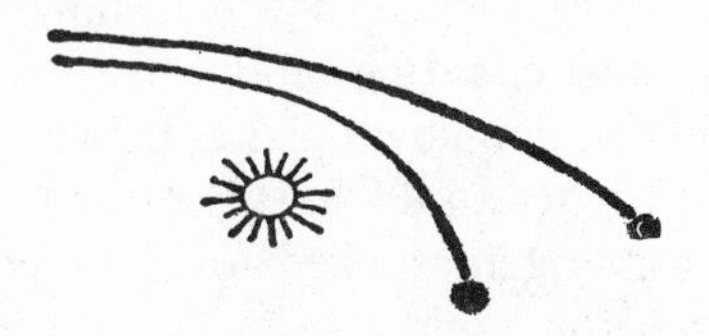

A second difference between Newtonian and Einsteinian gravitation is that Newton's laws are based on a planet's orbit's being a simple stationary ellipse, such as Kepler discovered was true in the case of Mars. But under Einstein's theory a planet's orbit works out instead to be a rotating ellipse that does not quite repeat — that precesses like a wobbling top. Part of the reason

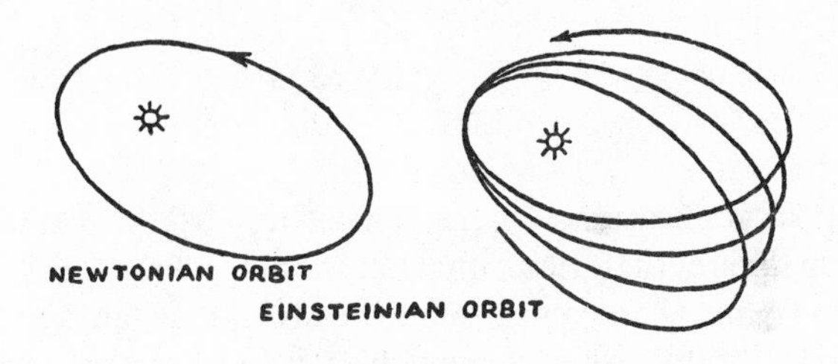

for this, as you may have guessed, is that every time the planet nears the sun it must accelerate a little and therefore, by Lorentzian transformation, its mass increases and its track foreshortens, distorting the surrounding space just enough to carry the orbit a tiny way beyond where Newton would have placed it. The difference is so slight that, in the case of the earth's nearly circular orbit, it precesses only 3.8 seconds of arc per century and a whole precessional revolution would take 34 million years. For this reason, planetary precession would have been a difficult phenomenon to prove were it not for the existence of a particular and convenient planet of very elongated orbit and possessed of unusual swiftness: the planet Mercury. Mercury, indeed, had been known for a long time to be precessing about 43 seconds of arc per century more than its perturbations warranted, and Leverrier, famed for his accurate prediction of Neptune (page 99), showed in 1845 that this amount of orbital shift could be produced if Mercury were being further perturbed by an unknown planet between it and the sun — a planet he felt so sure would be discovered that he named it Vulcan. And it was eagerly looked for and speculated about by astronomers all over the world.

Yet Vulcan never showed up. And the mystery remained unsolved until the general theory of relativity was applied mathematically to test the precession of Mercury and the result turned out — to nearly everyone's relief — to be 43 seconds of arc a century — a proof that has also been found to have its counterpart in the precessing "rosette" orbit of the electron.

The third verification of general relativity has to do with Einstein's prediction that time must flow more slowly not only in a moving frame of reference but logically also in an accelerating frame, which would be equivalent to a field of gravitation. Thus the larger the gravitational mass, the more sluggish its clocks must be, though not necessarily by enough to notice in ordinary life. In fact, Einstein calculated that "a second of time on the sun should correspond to 1.000002 earth seconds."

How could such a slight discrepancy be detected? The answer found was that the discrepancy need not remain so slight, for another star could be used where the time retardation would be much greater than on the sun. The handiest such star turned out to be the well-known dense dwarf Sirius B, with a diameter only a thirtieth of the sun's but a density 25,000 times greater.

Nor would it be necessary to send a clock there, since atoms vibrate reliably enough to regulate the best of clocks, and the spectral colors from the star must clearly reveal the frequencies of its atoms, hence (by comparison), their rate differences from Earth. And so, as very precisely measured in 1925, time on Sirius B was slipping steadily behind earthly time by more than a second every five hours — in accord with general relativity theory.

On the basis of such exotic yet sound evidences of the reality of general relativity, physicists can now describe with considerable confidence the basic time disparities that will appear in a future high-velocity space-ship voyage to some such star as Vega, whether or not it is then still fixed (from an earthly view) in the constellation of the Harp. If the traveling space ship accelerates to a speed close to that of light and uses its motors to reverse its direction fairly quickly at Vega, it may return to Earth in sixty years as recorded by earthly clocks and calendars, yet with only sixty months (of earth time) elapsed upon the space ship itself. I mean of course that an actual disparity in time-flow will have taken place between the space ship and the earth — not just an apparent difference — and that the returning space travelers will find most of their friends and relatives dead and, like as not, some of their own grandchildren already older than themselves.

Whether disparity of this kind is literally possible in any degree has been debated a good deal in this century, with such a noted physicist as Herbert Dingle insisting that two synchronized clocks, no matter how differently or drastically they move through the universe after being separated, will remain symmetrical because "nothing is in question except their relative motion. Hence they cannot show different times on reunion." Yet the leading theorists in general relativity reply that there is an "absolute distinction" between the time of earthly clocks and the time of clocks on space ships that are accelerating through space-time. The distinction lies in the different amounts of acceleration. While the earth merely drifts along on its line of least resistance, on its geodesic of longest time, the space ship, as long as its motors are accelerating or decelerating it (two views of one thing), must inevitably experience a slower passage of time. This was basically established by the Sirius B experiment in 1925, again demonstrated by a tuning experiment between hot and cold crystals of the isotope iron 57 at Harvard in 1960, and is now being conclusively tested by actual clocks in space.

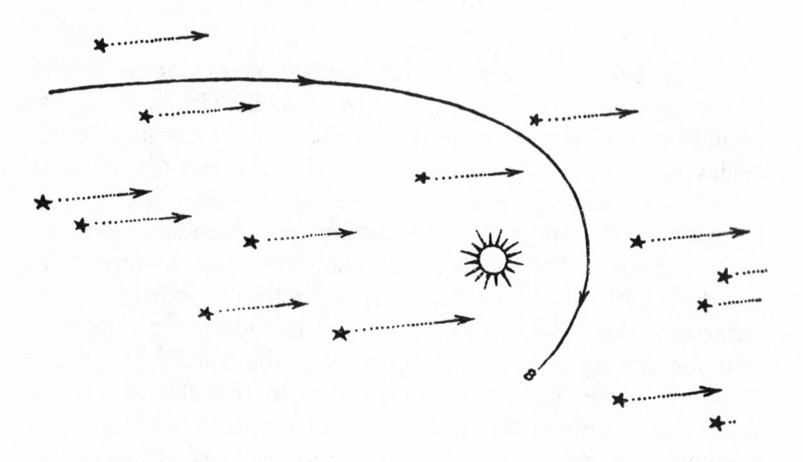

There is also the strange abstraction of gravitational potential, the potential energy created in raising any mass relative to a gravitational field through which it may fall. Every time you go upstairs, for instance, your gravitational potential increases and so does the potential energy of the earth (including you) by the same amount. But the strangest thing about gravitational potential is that it remains directly proportional to the relative slowing of time upon any accelerating object, so that when the space ship swings (under terrific acceleration forces) around Vega at a range giving it vast gravitational potential toward the earth, its clocks and heartbeats are going proportionately slowly as compared to those on Earth. The relative time field of the earth then, you might say, is flowing past the space ship at an alarming rate so that, by the time it returns home, it finds itself some two generations behind.

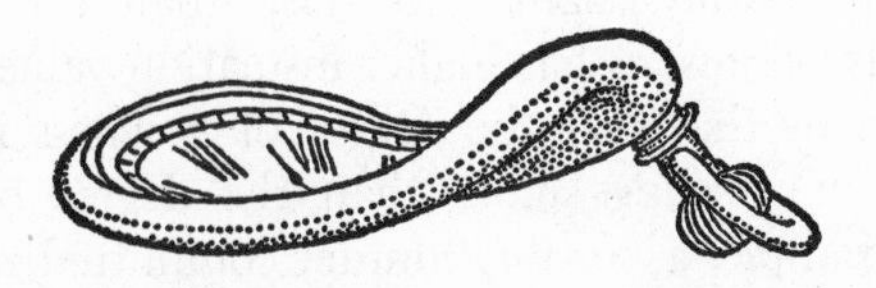

Such unearthly temporal distortions as these, of course, are way beyond our present intuition. Yet we can deduce that long space journeys to the stars are theoretically possible within reasonable portions of the travelers' expectable lifetimes if we can only achieve velocities close to that of light. If it were possible for a space ship to move exactly as fast as light, freezing the flow of time altogether and foreshortening space down to nothing (in effect making the speed infinite), obviously it would seem to the voyagers to take no time at all, except for the months of acceleration and deceleration at each end. Indeed, the approaches to such a speed might be considered one of the seams of the universe, where the physical and mental worlds meet. At least the speed of light as viewed subjectively, by a mind theoretically keeping up with the light, would equal the speed of thought.

We could go on to such questions of cosmic relativity as "how much of anything must be put into a given volume of space to make it absolutely full?" which is a variation of the basic question: does "to be" necessarily mean "to be somewhere or somewhen?" Or to what extent is existence possible outside space-time? Almost any answer to this question can lead to the curious concept of *space without location or distance* — a seeming, and perhaps semantic, absurdity — a space which, if it really existed, could mean that the distance between things might increase while the things themselves remained at rest relative to one another, as in the case of the fictional fugitive from justice who "puts space between" himself and his pursuers. This is not known to happen literally on Earth, of course, yet it does make some unearthly and uncommon sense in explaining the constancy of the speed of light regardless of the approach or recession of its source.

Oddly enough, the time aspect in such unfamiliar considerations is almost always harder to grasp than the space aspect, perhaps partly because it has higher insulation value in consciousness. I mean by that that the degree of *now*ness in time is apt to influence our feelings more than the degree of *here*ness in space. For example, a horrible murder committed right *here* fifty years ago seems more remote, therefore less shocking, than an equally horrible murder being committed right *now* fifty miles away or even fifty thousand miles away — the shock of *now*ness presumably resulting from our subconscious realization that it might be possible to fly fifty thousand miles (by rocket) to the scene of the murder, thereby seeing or interacting with it approximately *now*, while no known vehicle could possibly carry us back fifty years to become involved in a murder *here* half a century ago or even a month ago.

Another curious thing about time is its particular application to bodies or selves of common identity. For, as I have already suggested, time seems to be essentially the relationship between things and things that have common identity — between, say, a boy and a man grown from that boy — while space is the relationship between things and things without common identity — as between a man and another man. Thus the deepest time-space distinction seems to me to be one of identity and, as long as — and only as long as — identity survives, the time aspect may be distinguished from that of space. But just as identity may never have been born inside the time-free atom, so it may die almost anywhere else — because identity is relative, not absolute, and tends to fade into vagueness as perspective recedes, at some level of abstraction ceasing to be determinable at all — even as we saw (page 556) the collar of here-now fading into the aura of elsewhere-elsewhen beyond the hourglass of space-time.

Consider how, when a man is alive, there is an identity between his body (there then) and his body (here now), indicating he exists in time. But after his death, the identity decreases between his body (there then) and his disintegrating body (whenever wherever), indicating that his body at last is loosening its bond with time. His mind also may well escape time before his body — as do abstract (timeless) geometric figures and possibly occurrences inside the atom or outside the galaxy.

I sometimes wonder whether humanity has missed the real point in raising the issue of mortality and immortality — whether perhaps the seemingly limited time span of an earthly life is actually unlimited and eternal — in other words, whether mortality itself may be a finite illusion, being actually immortality and, even though constructed of just a few "years," that those few years are all the time there really is, so that, in fact, they can never cease. Indeed, if time is the relation between things and themselves, how can time end while things exist? Or how can time have ever begun, since either a beginning or an end would logically and almost inevitably frame time in more of itself?

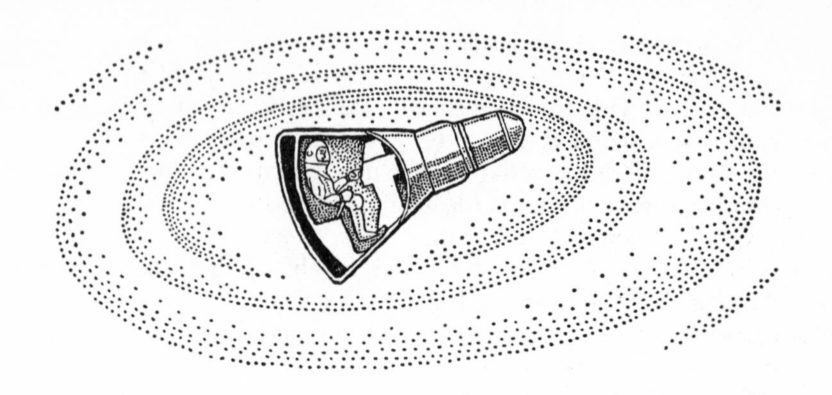

14. the sinews of reality

IT IS ABOUT TIME (Earth time), I think, for me to wind up this volume with some kind of a summary of conclusions as to the nature of the worlds, macro- and microcosmic, as they look from this detached but, I hope, not too dispassioned station. This implies that I should be able to arrive at a decision concerning at least some of the basic laws of nature. Yet I admit that such is, by the same nature, too tall an order for me — and perhaps for any mere man.

The traditional view is that the laws of nature remain constant and more or less hidden, awaiting discovery. But how do we know that the laws of nature are constant? Most things change. Why should laws be an exception? Or, for that matter, what assurance have we that there are any laws of nature at all? Could it possibly be that the men who think they are discovering them are really only thinking them up? This in actual fact was Poincaré's considered view.

It might be helpful here to look at some typical cases in natural law. Before Niccolò Tartaglia's book on ballistics (Venice, 1537), for instance, it had been generally believed that the trajectory of a cannonball consisted of three parts: the straight line of its firing, the curve of its adjusting to gravity and, finally, the vertical straight line of its fall enforced by gravity. While of gravity Leonardo said, "The desire of every heavy body is that its centre may be the centre of the earth."

But soon after Tartaglia and Leonardo, the various laws of nature implied in these beliefs were almost completely swept away before the strong wind of Newton's simpler inverse-square law of gravitation, of which Newton wrote, "The main business of natural philosophy is . . . to deduce causes from effects till we come to the very first cause, which certainly is not mechanical." There is little doubt that Newton had in mind God as his "first cause." And on at least one occasion, he expressed fear that the wheeling planets, moons and swooping comets would in time derange the solar system so badly that nothing short of divine intervention could bring it back into balance.

Then a few years later, Laplace demonstrated mathematically that the perturbations of the planetary bodies upon each other are "not cumulative but periodic" — so the system, after all, must really be capable of self-regulation.

The law that these successively unfolding revelations revealed thus turns out to be a moving, growing thing — a thing even Einstein could not expect to stabilize for very long. And the same may well go for all other laws of nature, it being found generally true that no matter how great the discoverer of anything is, or how small his successors, those successors will inevitably climb upon his high shoulders and from them see farther than he saw.

If this is, in a way, paradoxical, it is also characteristic of nature, which seems almost brimming with paradox — as if her main objective were to baffle our minds or perhaps, as I think more and more, to spark our spirits with wonder. Surely there is wonder in the paradox that, despite the newly proven quantum, nothing is quite sharp or exact in the true world. There is wonder in the facts that the speed of light cannot be determined (even in principle) closer than a few hundred feet per second — that, in any particular case, it has only a probable value; that if you put material into a box, no matter what the material is or what the box is made of, some of the material must escape the box — maybe only a tiny percentage of the elementary particles — but they leap right through it in a ratio of inexorable probability. And there is wonder that the atom is a perpetual motion machine — the very monad that young Democritos had in mind so long ago when he conjectured that "the love and hate of atoms is the cause of unrest in the world."

And, as if that were not enough, matter is both discrete entity and continuous field, at once particle and wave — a wave that is abstract, its illusory nature exemplified in the fact that part of every ocean wave, every sound wave, perhaps every light wave, is always moving backward. Is this also the stuff I am made of and you are made of — the material of life whose very existence is a challenge to the second law of thermodynamics? In speaking of life, of course, we are at the threshold of a whole new subject — but it also impinges on our present discussion and only slightly alters the mixture of our already-rich fuel for thought.

Where does a candle flame go, I might ask, when I blow it out? Is it dead? Is it a hole in nothing? Is it gone forever? Or does it still exist in some other form? The question makes about as much sense as "how fast is the earth moving through space?" It depends on one's viewpoint and various relative factors, including the concepts of identity and continuity. One can logically call a continuous flame "a single flame of one identity," for it is much like a single life. And it is in essence as much an object as the flaming, gaseous sun is an object. Yet, after it is blown out, it is easy enough to light it again if conditions stay about the same — and who can prove the new flame is not the same flame as the old? Thus, in essence and in relativity, death need not end life — not even materially.

This ties in with the basic abstract nature of the world — with opinions such as Jeans's that light waves as such do not really exist — that, when you get right down to it, they are actually only "waves of knowledge." But whatever they really are, the melodies in intelligence are legion when we listen intensively to the world — when we really harken to its inner and outer harmonies!

Since you have probably forgotten most of them, may I briefly review some of the principal kinds of music of the spheres touched upon in this book? First, the harmonic intervals of the 7 notes of the scale explored by Pythagoras, which have certain correspondences to the intervals of the 7 planets of antiquity, and perhaps to the 7 shells of the modern atom, not to mention the 7 octaves of the piano keyboard and the 7 octaves of the Periodic Table of elements — or, in lesser degrees, Bode's Law and Balmer's Ladder, the table of particles, the moon ratios of Jupiter (Io + 2 Ganymede = 3 Europa), the rings of Saturn and the nodes of the Trojan asteroids. Next there are the wave ratios that shape and govern sea waves (group vs. single waves), dunes, snowdrifts, tides, molecules, atoms and other forms of flowing energy — the harmonics and nodes of oscillation in pendulums, springs, and all manner of elastic, chemical, electromagnetic and living bodies. Then the radii of atomic shells, proportioned to the rule of single squares 1, 4, 9, 16, 25 . . . , and the numbers of electrons in atomic shells to the rule of doubled squares 2, 8, 18, 32, 50 . . . , the 32 symmetry groupings of crystals, the 230 distributions of identical objects in space, the various progressions and means, arithmetic, geometric, harmonic — and so on . . . and on . . .

The meaning of this seemingly endless music, to me, is that not only does the atom sing but it sings in tune with the stars and with its whole surround in space-time, resonating in significant harmonies with other atoms and with all sorts of energies known and unknown. In truth, reality seems to be a kind of blossoming limit imposed upon budding possibility by the harmonics of energy in motion — and that limit gives form and value to all things, gives concreteness to abstraction and actuality to potentiality.

This is shown in myriad ways almost everywhere I look — in every field of grass, of stars, of art, science and thought. From the geometric philosopher, R. Buckminster Fuller, for example, I learn that the molecular miracle of "synergy" expressible as $1 + 2 = 4$ is at the bottom of many of his greatest discoveries in construction engineering. As he himself explains it, "one equilateral triangle hinged onto 2 others can be folded into a 3-sided tent whose base is a 4th triangle. The inadvertent appearance of this fourth triangle is a demonstration of *synergy* or the behavior of a system as a whole unpredicted by its parts."

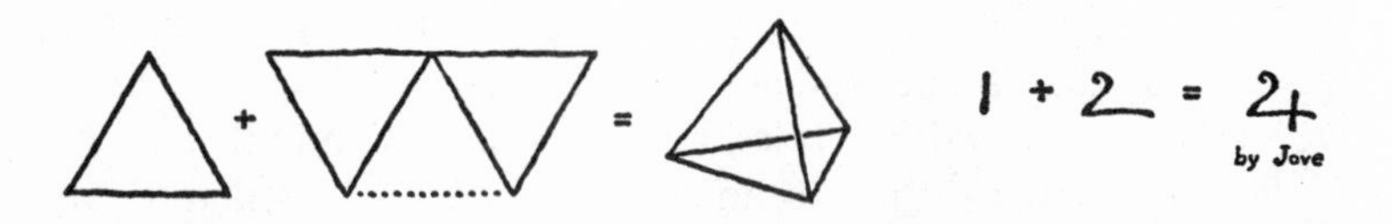

Continuing in similar vein, I consider a folding chair. Does the chair exist when it is folded up and put away in a closet? Probably most philosophers would admit that it still exists, even though in a different state — a state of diminished usefulness. If this is so, then, by the same principle, a dismantled watch is still a watch — at least it is a potential watch — and a sperm and an ovum that are capable of meeting and conceiving are, in essence, the creature they will conceivably be. In this sense, you exist before conception and, by the same reasoning, you still exist after decomposition. You ARE, therefore — regardless of time-space. You ARE, I might venture, perhaps partly in the way Jesus meant when he said, "Before Abraham was, I AM."

This is one of the points at which science and religion meet, I think. For if science and religion are opposed at all, they are probably opposed only as a thumb is opposed to fingers. And, as God arranged the hand, a thumb is just the opposition that enables fingers most effectively to grasp things. Furthermore, the *grasp* of things is a kind of created something beyond the members that do the grasping. Which in its turn may help explain why an electron alone is unobservable and therefore, from a practical, objective viewpoint, does not exist as an entity in itself. It is only as united with other particles that the electron, in fact, comes into being for us — that the whole is born of its parts. Thus the abstraction of *addition*, of electron + proton + neutron . . . is what matter IS. It is the abstraction that adds up to and thus creates our world.

In about every important sense, you see, the world is profoundly abstract. And not only the world but also the world's beings are made substantially of abstraction. A being is something like an idea or a song or an organized system. Where was the telephone in A.D. 1800? Where were you and I? None of us had yet been born. All the elements of our future compositions were in the world but not organized into integral systems. And just as the right combination of thoughts and actions produced the telephone and developed it as a system of communication, so did the right combination of motivations and germ cells produce you and me. Thus an organism of life is basically much the same as an organism of systematic ideas — an abstraction, a new combination, a larger reality . . .

This brings up the elusive question of what reality *really* is anyhow — of whether a cubic yard of material taken *out* of a highway is as dangerous or as real (to passing traffic) as a cubic yard heaped *upon* it. If a wonderful world that doesn't exist, for example — say, a mirror world or a thought world — has all the same laws of nature that our "real" world has, how should physicists go about testing it for actuality? Of course, the unreal stars in this unreal world would emit unreal light which could be seen by unreal eyes and recognized in unreal brains. Even the physicists themselves, if they were in such a world, might have to be unreal — but they could still be physicists as distinct from nonphysicists (from, say, sociologists or haberdashers). And there just might be a way for them to test world reality, in a thought experiment, by defining consciousness as reality and somehow determining whether the world in question included consciousness — which, if it was detectable, could (by definition) be real.

Obviously, reality does not need location — which explains how a lens with a focal length of ten feet can be packed in a box only two feet long. And reality does not need to have been experienced — which explains how the shadow of the moon upon Cornwall on August 11, 1999, is already in the world of inference. And reality does not even need to be "true" — which explains how James Watt could build a real and workable steam engine from calculations based on a wrong theory of heat.

Yet reality in some indefinable way does need some degree of existence. I wonder, for instance, about the difficulty of visualizing life on Earth very far beyond one's own birth or death. Could there be a deep reason why one cannot imagine nonexistence? Perhaps the real resolution of such a paradox could be found in the same cosmic positiveness that reveals coldness to be only a reduced heat and darkness but a lesser light. For why should not existence itself, like time and life and love, be relative and capable of infinite range? Who can decide where some degree of being really borders on the not-being? Or, by simple semantics, when the "not-be" will, if ever, begin to not be?

There may even be an uncertainty principle connected with being or reality, as Einstein suggested when he distinguished reality from certainty, saying, "As far as the propositions of mathematics refer to reality they are not certain, and in so far as they are certain they do not refer to reality." In which case, nature probably has a subtler symmetry than most human minds can yet understand — the human mind having hardly yet emerged from the jungle primeval.

As for my own jungle mind, I only hope my mirages turn out to be real mirages, not just picture mirages — and that my mental pictures become real pictures, not merely pictures of pictures. For, as Bertrand Russell once expressed it, how things seem to seem is not enough. We must somehow discover how things really seem.

This of course calls for a certain greatness in perspective. When you look at a small table globe representing the earth, for example, you must never forget that, on the same scale, the "sun" is as big as a barn and a mile away, the "nearest star" as far as the moon, and "other galaxies" way beyond comprehension — even on that small scale. Now if you look out upon the actual sky above your head and beneath your feet, as I do here, you discover it to contain not only much but ultimately everything, not excepting yourself nor all sorts of unbelievable things — not excepting milk from the Milky Way (the earth, incidentally, is part of the Milky Way) — and not excepting much more that is humanly inconceivable, in fact literally all there ever was and is — and perhaps all that might be — entire worlds and galaxies of worlds and clusters of galaxies of worlds and endless, showering ebullitions of all these and more, pouring and whirling ever outward into all space — containing among them unnumbered worlds more enlightened, more beautiful in spirit than our undistinguished speck — where ideas unfathomable and joys undreamable grow and flower and fulfill themselves in actual, breathing heavens beyond our reach yet, literally, right before our eyes!

Is there nothing then but illusory space-time between us and Kingdom Come? Naturally, I cannot reach out and touch it with my hand but I can imagine it someway with my mind and feel its potentiality in my heart. And I can see beauty and order there — and most especially the elements of music. I can hear, in a real sense, the music of the spheres.

Speaking of elements, by the way, one cannot help but be reminded of the restraint of nature in forming elements on such a planet as the earth. Suppose that, instead of a hundred chemical elements, there were a hundred million of them. In such a case, almost every pebble on every beach and every mote of dust would be of an unknown substance, and it would be next to impossible to develop any science of chemistry. But that is not the real way of nature. Nature has a beautiful simplicity of order. And the intuitions of Pythagoras and Leucippos and some of Plato's are proving substantially justified.

The world is made of abstraction with sinews of perspective — and music. Its waves gather knowledge and instruct the universe. Its melody is more than notes, its poetry more than words. Its stars are as much seeds of distance as earthly acorns are seeds of time — for if acorns unfold into oak trees after many years, so may stars "planted" in space be considered spores of mysterious potency — space seeds that we cannot fathom from here but which, if the unraveling of space allows, may soon draw near and blossom into our ken.

And as I float — in good company but alone — I must ask once again when is "now"? And where is "here"? As did Sappho while she sang:

> The moon is gone
> And the Pleiads set,
> Midnight is nigh:
>
> Time passes on
> And passes; yet
> Alone I lie.

For this black space has more dimensions than any man can know. It is the page on which I write, the harp on which I play. It is the high vantage out of which I seek new questions for old silences — new presences for old absences.

And I can still remember my youth on the rolling earth and see again the mountains that I saw. And ever must I remind me that when I see those mountains down there, those great and breath-taking mountains of the wrinkled crust, that there are even greater, even stronger, even more beautiful mountains somewhere far behind them — and atom worlds within their atom worlds — and oceans beneath earthly oceans — stars above the stars, constellations beyond the constellations, even "universes" outside "the" universe.

I am obviously still much too small and feeble and young to see all these things clearly. But somehow I know. I know that they are there.

index

This full index, made by the author, is especially suited for browsing. It lists hundreds of illustrations and abstract concepts and is liberally crissed and crossed with suggestive and interrelating references. Part One contains pages 1 through 226; Part Two contains pages 227 through 600.

A CATALOGUE OF SELECTED DOVER BOOKS IN ALL FIELDS OF INTEREST

A CATALOGUE OF SELECTED DOVER BOOKS IN ALL FIELDS OF INTEREST

CELESTIAL OBJECTS FOR COMMON TELESCOPES, T. W. Webb. The most used book in amateur astronomy: inestimable aid for locating and identifying nearly 4,000 celestial objects. Edited, updated by Margaret W. Mayall. 77 illustrations. Total of 645pp. 5⅜ x 8½.
20917-2, 20918-0 Pa., Two-vol. set $9.00

HISTORICAL STUDIES IN THE LANGUAGE OF CHEMISTRY, M. P. Crosland. The important part language has played in the development of chemistry from the symbolism of alchemy to the adoption of systematic nomenclature in 1892. ". . . wholeheartedly recommended,"—Science. 15 illustrations. 416pp. of text. 5⅝ x 8¼. 63702-6 Pa. $6.00

BURNHAM'S CELESTIAL HANDBOOK, Robert Burnham, Jr. Thorough, readable guide to the stars beyond our solar system. Exhaustive treatment, fully illustrated. Breakdown is alphabetical by constellation: Andromeda to Cetus in Vol. 1; Chamaeleon to Orion in Vol. 2; and Pavo to Vulpecula in Vol. 3. Hundreds of illustrations. Total of about 2000pp. 6⅛ x 9¼.
23567-X, 23568-8, 23673-0 Pa., Three-vol. set $26.85

THEORY OF WING SECTIONS: INCLUDING A SUMMARY OF AIRFOIL DATA, Ira H. Abbott and A. E. von Doenhoff. Concise compilation of subatomic aerodynamic characteristics of modern NASA wing sections, plus description of theory. 350pp. of tables. 693pp. 5⅜ x 8½.
60586-8 Pa. $7.00

DE RE METALLICA, Georgius Agricola. Translated by Herbert C. Hoover and Lou H. Hoover. The famous Hoover translation of greatest treatise on technological chemistry, engineering, geology, mining of early modern times (1556). All 289 original woodcuts. 638pp. 6¾ x 11.
60006-8 Clothbd. $17.95

THE ORIGIN OF CONTINENTS AND OCEANS, Alfred Wegener. One of the most influential, most controversial books in science, the classic statement for continental drift. Full 1966 translation of Wegener's final (1929) version. 64 illustrations. 246pp. 5⅜ x 8½. 61708-4 Pa. $4.50

THE PRINCIPLES OF PSYCHOLOGY, William James. Famous long course complete, unabridged. Stream of thought, time perception, memory, experimental methods; great work decades ahead of its time. Still valid, useful; read in many classes. 94 figures. Total of 1391pp. 5⅜ x 8½.
20381-6, 20382-4 Pa., Two-vol. set $13.00

THE SENSE OF BEAUTY, George Santayana. Masterfully written discussion of nature of beauty, materials of beauty, form, expression; art, literature, social sciences all involved. 168pp. 5⅜ x 8½. 20238-0 Pa. $2.50

ON THE IMPROVEMENT OF THE UNDERSTANDING, Benedict Spinoza. Also contains *Ethics, Correspondence,* all in excellent R. Elwes translation. Basic works on entry to philosophy, pantheism, exchange of ideas with great contemporaries. 402pp. 5⅜ x 8½. 20250-X Pa. $4.50

THE TRAGIC SENSE OF LIFE, Miguel de Unamuno. Acknowledged masterpiece of existential literature, one of most important books of 20th century. Introduction by Madariaga. 367pp. 5⅜ x 8½.
20257-7 Pa. $4.50

THE GUIDE FOR THE PERPLEXED, Moses Maimonides. Great classic of medieval Judaism attempts to reconcile revealed religion (Pentateuch, commentaries) with Aristotelian philosophy. Important historically, still relevant in problems. Unabridged Friedlander translation. Total of 473pp. 5⅜ x 8½. 20351-4 Pa. $6.00

THE I CHING (THE BOOK OF CHANGES), translated by James Legge. Complete translation of basic text plus appendices by Confucius, and Chinese commentary of most penetrating divination manual ever prepared. Indispensable to study of early Oriental civilizations, to modern inquiring reader. 448pp. 5⅜ x 8½. 21062-6 Pa. $4.00

THE EGYPTIAN BOOK OF THE DEAD, E. A. Wallis Budge. Complete reproduction of Ani's papyrus, finest ever found. Full hieroglyphic text, interlinear transliteration, word for word translation, smooth translation. Basic work, for Egyptology, for modern study of psychic matters. Total of 533pp. 6½ x 9¼. (Available in U.S. only) 21866-X Pa. $5.95

THE GODS OF THE EGYPTIANS, E. A. Wallis Budge. Never excelled for richness, fullness: all gods, goddesses, demons, mythical figures of Ancient Egypt; their legends, rites, incarnations, variations, powers, etc. Many hieroglyphic texts cited. Over 225 illustrations, plus 6 color plates. Total of 988pp. 6⅛ x 9¼. (Available in U.S. only)
22055-9, 22056-7 Pa., Two-vol. set $12.00

THE ENGLISH AND SCOTTISH POPULAR BALLADS, Francis J. Child. Monumental, still unsuperseded; all known variants of Child ballads, commentary on origins, literary references, Continental parallels, other features. Added: papers by G. L. Kittredge, W. M. Hart. Total of 2761pp. 6½ x 9¼.
21409-5, 21410-9, 21411-7, 21412-5, 21413-3 Pa., Five-vol. set $37.50

CORAL GARDENS AND THEIR MAGIC, Bronsilaw Malinowski. Classic study of the methods of tilling the soil and of agricultural rites in the Trobriand Islands of Melanesia. Author is one of the most important figures in the field of modern social anthropology. 143 illustrations. Indexes. Total of 911pp. of text. 5⅝ x 8¼. (Available in U.S. only)
23597-1 Pa. $12.95

THE COMPLETE BOOK OF DOLL MAKING AND COLLECTING, Catherine Christopher. Instructions, patterns for dozens of dolls, from rag doll on up to elaborate, historically accurate figures. Mould faces, sew clothing, make doll houses, etc. Also collecting information. Many illustrations. 288pp. 6 x 9. 22066-4 Pa. $4.50

THE DAGUERREOTYPE IN AMERICA, Beaumont Newhall. Wonderful portraits, 1850's townscapes, landscapes; full text plus 104 photographs. The basic book. Enlarged 1976 edition. 272pp. 8¼ x 11¼. 23322-7 Pa. $7.95

CRAFTSMAN HOMES, Gustav Stickley. 296 architectural drawings, floor plans, and photographs illustrate 40 different kinds of "Mission-style" homes from *The Craftsman* (1901-16), voice of American style of simplicity and organic harmony. Thorough coverage of Craftsman idea in text and picture, now collector's item. 224pp. 8⅛ x 11. 23791-5 Pa. $6.00

PEWTER-WORKING: INSTRUCTIONS AND PROJECTS, Burl N. Osborn. & Gordon O. Wilber. Introduction to pewter-working for amateur craftsman. History and characteristics of pewter; tools, materials, step-by-step instructions. Photos, line drawings, diagrams. Total of 160pp. 7⅞ x 10¾. 23786-9 Pa. $3.50

THE GREAT CHICAGO FIRE, edited by David Lowe. 10 dramatic, eyewitness accounts of the 1871 disaster, including one of the aftermath and rebuilding, plus 70 contemporary photographs and illustrations of the ruins—courthouse, Palmer House, Great Central Depot, etc. Introduction by David Lowe. 87pp. 8¼ x 11. 23771-0 Pa. $4.00

SILHOUETTES: A PICTORIAL ARCHIVE OF VARIED ILLUSTRATIONS, edited by Carol Belanger Grafton. Over 600 silhouettes from the 18th to 20th centuries include profiles and full figures of men and women, children, birds and animals, groups and scenes, nature, ships, an alphabet. Dozens of uses for commercial artists and craftspeople. 144pp. 8⅜ x 11¼. 23781-8 Pa. $4.00

ANIMALS: 1,419 COPYRIGHT-FREE ILLUSTRATIONS OF MAMMALS, BIRDS, FISH, INSECTS, ETC., edited by Jim Harter. Clear wood engravings present, in extremely lifelike poses, over 1,000 species of animals. One of the most extensive copyright-free pictorial sourcebooks of its kind. Captions. Index. 284pp. 9 x 12. 23766-4 Pa. $7.95

INDIAN DESIGNS FROM ANCIENT ECUADOR, Frederick W. Shaffer. 282 original designs by pre-Columbian Indians of Ecuador (500-1500 A.D.). Designs include people, mammals, birds, reptiles, fish, plants, heads, geometric designs. Use as is or alter for advertising, textiles, leathercraft, etc. Introduction. 95pp. 8¾ x 11¼. 23764-8 Pa. $3.50

SZIGETI ON THE VIOLIN, Joseph Szigeti. Genial, loosely structured tour by premier violinist, featuring a pleasant mixture of reminiscenes, insights into great music and musicians, innumerable tips for practicing violinists. 385 musical passages. 256pp. 5⅝ x 8¼. 23763-X Pa. $3.50

THE AMERICAN SENATOR, Anthony Trollope. Little known, long unavailable Trollope novel on a grand scale. Here are humorous comment on American vs. English culture, and stunning portrayal of a heroine/villainess. Superb evocation of Victorian village life. 561pp. 5⅜ x 8½.
23801-6 Pa. $6.00

WAS IT MURDER? James Hilton. The author of *Lost Horizon* and *Goodbye, Mr. Chips* wrote one detective novel (under a pen-name) which was quickly forgotten and virtually lost, even at the height of Hilton's fame. This edition brings it back—a finely crafted public school puzzle resplendent with Hilton's stylish atmosphere. A thoroughly English thriller by the creator of Shangri-la. 252pp. 5⅜ x 8. (Available in U.S. only)
23774-5 Pa. $3.00

CENTRAL PARK: A PHOTOGRAPHIC GUIDE, Victor Laredo and Henry Hope Reed. 121 superb photographs show dramatic views of Central Park: Bethesda Fountain, Cleopatra's Needle, Sheep Meadow, the Blockhouse, plus people engaged in many park activities: ice skating, bike riding, etc. Captions by former Curator of Central Park, Henry Hope Reed, provide historical view, changes, etc. Also photos of N.Y. landmarks on park's periphery. 96pp. 8½ x 11. 23750-8 Pa. $4.50

NANTUCKET IN THE NINETEENTH CENTURY, Clay Lancaster. 180 rare photographs, stereographs, maps, drawings and floor plans recreate unique American island society. Authentic scenes of shipwreck, lighthouses, streets, homes are arranged in geographic sequence to provide walking-tour guide to old Nantucket existing today. Introduction, captions. 160pp. 8⅞ x 11¾. 23747-8 Pa. $6.95

STONE AND MAN: A PHOTOGRAPHIC EXPLORATION, Andreas Feininger. 106 photographs by *Life* photographer Feininger portray man's deep passion for stone through the ages. Stonehenge-like megaliths, fortified towns, sculpted marble and crumbling tenements show textures, beauties, fascination. 128pp. 9¼ x 10¾. 23756-7 Pa. $5.95

CIRCLES, A MATHEMATICAL VIEW, D. Pedoe. Fundamental aspects of college geometry, non-Euclidean geometry, and other branches of mathematics: representing circle by point. Poincare model, isoperimetric property, etc. Stimulating recreational reading. 66 figures. 96pp. 5⅝ x 8¼.
63698-4 Pa. $2.75

THE DISCOVERY OF NEPTUNE, Morton Grosser. Dramatic scientific history of the investigations leading up to the actual discovery of the eighth planet of our solar system. Lucid, well-researched book by well-known historian of science. 172pp. 5⅜ x 8½. 23726-5 Pa. $3.00

THE DEVIL'S DICTIONARY. Ambrose Bierce. Barbed, bitter, brilliant witticisms in the form of a dictionary. Best, most ferocious satire America has produced. 145pp. 5⅜ x 8½. 20487-1 Pa. $2.00

HISTORY OF BACTERIOLOGY, William Bulloch. The only comprehensive history of bacteriology from the beginnings through the 19th century. Special emphasis is given to biography-Leeuwenhoek, etc. Brief accounts of 350 bacteriologists form a separate section. No clearer, fuller study, suitable to scientists and general readers, has yet been written. 52 illustrations. 448pp. 5⅝ x 8¼. 23761-3 Pa. $6.50

THE COMPLETE NONSENSE OF EDWARD LEAR, Edward Lear. All nonsense limericks, zany alphabets, Owl and Pussycat, songs, nonsense botany, etc., illustrated by Lear. Total of 321pp. 5⅜ x 8½. (Available in U.S. only) 20167-8 Pa. $3.00

INGENIOUS MATHEMATICAL PROBLEMS AND METHODS, Louis A. Graham. Sophisticated material from Graham *Dial,* applied and pure; stresses solution methods. Logic, number theory, networks, inversions, etc. 237pp. 5⅜ x 8½. 20545-2 Pa. $3.50

BEST MATHEMATICAL PUZZLES OF SAM LOYD, edited by Martin Gardner. Bizarre, original, whimsical puzzles by America's greatest puzzler. From fabulously rare *Cyclopedia,* including famous 14-15 puzzles, the Horse of a Different Color, 115 more. Elementary math. 150 illustrations. 167pp. 5⅜ x 8½. 20498-7 Pa. $2.75

THE BASIS OF COMBINATION IN CHESS, J. du Mont. Easy-to-follow, instructive book on elements of combination play, with chapters on each piece and every powerful combination team—two knights, bishop and knight, rook and bishop, etc. 250 diagrams. 218pp. 5⅜ x 8½. (Available in U.S. only) 23644-7 Pa. $3.50

MODERN CHESS STRATEGY, Ludek Pachman. The use of the queen, the active king, exchanges, pawn play, the center, weak squares, etc. Section on rook alone worth price of the book. Stress on the moderns. Often considered the most important book on strategy. 314pp. 5⅜ x 8½. 20290-9 Pa. $4.50

LASKER'S MANUAL OF CHESS, Dr. Emanuel Lasker. Great world champion offers very thorough coverage of all aspects of chess. Combinations, position play, openings, end game, aesthetics of chess, philosophy of struggle, much more. Filled with analyzed games. 390pp. 5⅜ x 8½. 20640-8 Pa. $5.00

500 MASTER GAMES OF CHESS, S. Tartakower, J. du Mont. Vast collection of great chess games from 1798-1938, with much material nowhere else readily available. Fully annoted, arranged by opening for easier study. 664pp. 5⅜ x 8½. 23208-5 Pa. $7.50

A GUIDE TO CHESS ENDINGS, Dr. Max Euwe, David Hooper. One of the finest modern works on chess endings. Thorough analysis of the most frequently encountered endings by former world champion. 331 examples, each with diagram. 248pp. 5⅜ x 8½. 23332-4 Pa. $3.50

HOUSEHOLD STORIES BY THE BROTHERS GRIMM. All the great Grimm stories: "Rumpelstiltskin," "Snow White," "Hansel and Gretel," etc., with 114 illustrations by Walter Crane. 269pp. 5⅜ x 8½.
21080-4 Pa. $3.00

SLEEPING BEAUTY, illustrated by Arthur Rackham. Perhaps the fullest, most delightful version ever, told by C. S. Evans. Rackham's best work. 49 illustrations. 110pp. 7⅞ x 10¾. **22756-1 Pa. $2.50**

AMERICAN FAIRY TALES, L. Frank Baum. Young cowboy lassoes Father Time; dummy in Mr. Floman's department store window comes to life; and 10 other fairy tales. 41 illustrations by N. P. Hall, Harry Kennedy, Ike Morgan, and Ralph Gardner. 209pp. 5⅜ x 8½. **23643-9 Pa. $3.00**

THE WONDERFUL WIZARD OF OZ, L. Frank Baum. Facsimile in full color of America's finest children's classic. Introduction by Martin Gardner. 143 illustrations by W. W. Denslow. 267pp. 5⅜ x 8½.
20691-2 Pa. $3.50

THE TALE OF PETER RABBIT, Beatrix Potter. The inimitable Peter's terrifying adventure in Mr. McGregor's garden, with all 27 wonderful, full-color Potter illustrations. 55pp. 4¼ x 5½. (Available in U.S. only)
22827-4 Pa. $1.25

THE STORY OF KING ARTHUR AND HIS KNIGHTS, Howard Pyle. Finest children's version of life of King Arthur. 48 illustrations by Pyle. 131pp. 6⅛ x 9¼. **21445-1 Pa. $4.95**

CARUSO'S CARICATURES, Enrico Caruso. Great tenor's remarkable caricatures of self, fellow musicians, composers, others. Toscanini, Puccini, Farrar, etc. Impish, cutting, insightful. 473 illustrations. Preface by M. Sisca. 217pp. 8⅜ x 11¼. **23528-9 Pa. $6.95**

PERSONAL NARRATIVE OF A PILGRIMAGE TO ALMADINAH AND MECCAH, Richard Burton. Great travel classic by remarkably colorful personality. Burton, disguised as a Moroccan, visited sacred shrines of Islam, narrowly escaping death. Wonderful observations of Islamic life, customs, personalities. 47 illustrations. Total of 959pp. 5⅜ x 8½.
21217-3, 21218-1 Pa., Two-vol. set **$12.00**

INCIDENTS OF TRAVEL IN YUCATAN, John L. Stephens. Classic (1843) exploration of jungles of Yucatan, looking for evidences of Maya civilization. Travel adventures, Mexican and Indian culture, etc. Total of 669pp. 5⅜ x 8½. 20926-1, 20927-X Pa., Two-vol. set **$7.90**

AMERICAN LITERARY AUTOGRAPHS FROM WASHINGTON IRVING TO HENRY JAMES, Herbert Cahoon, et al. Letters, poems, manuscripts of Hawthorne, Thoreau, Twain, Alcott, Whitman, 67 other prominent American authors. Reproductions, full transcripts and commentary. Plus checklist of all American Literary Autographs in The Pierpont Morgan Library. Printed on exceptionally high-quality paper. 136 illustrations. 212pp. 9⅛ x 12¼. 23548-3 Pa. $7.95

PRINCIPLES OF ORCHESTRATION, Nikolay Rimsky-Korsakov. Great classical orchestrator provides fundamentals of tonal resonance, progression of parts, voice and orchestra, tutti effects, much else in major document. 330pp. of musical excerpts. 489pp. 6½ x 9¼. 21266-1 Pa. $6.00

TRISTAN UND ISOLDE, Richard Wagner. Full orchestral score with complete instrumentation. Do not confuse with piano reduction. Commentary by Felix Mottl, great Wagnerian conductor and scholar. Study score. 655pp. 8⅛ x 11. 22915-7 Pa. $12.50

REQUIEM IN FULL SCORE, Giuseppe Verdi. Immensely popular with choral groups and music lovers. Republication of edition published by C. F. Peters, Leipzig, n. d. German frontmaker in English translation. Glossary. Text in Latin. Study score. 204pp. 9⅜ x 12¼.
23682-X Pa. $6.00

COMPLETE CHAMBER MUSIC FOR STRINGS, Felix Mendelssohn. All of Mendelssohn's chamber music: Octet, 2 Quintets, 6 Quartets, and Four Pieces for String Quartet. (Nothing with piano is included). Complete works edition (1874-7). Study score. 283 pp. 9⅜ x 12¼.
23679-X Pa. $6.95

POPULAR SONGS OF NINETEENTH-CENTURY AMERICA, edited by Richard Jackson. 64 most important songs: "Old Oaken Bucket," "Arkansas Traveler," "Yellow Rose of Texas," etc. Authentic original sheet music, full introduction and commentaries. 290pp. 9 x 12. 23270-0 Pa. $6.00

COLLECTED PIANO WORKS, Scott Joplin. Edited by Vera Brodsky Lawrence. Practically all of Joplin's piano works—rags, two-steps, marches, waltzes, etc., 51 works in all. Extensive introduction by Rudi Blesh. Total of 345pp. 9 x 12. 23106-2 Pa. $14.95

BASIC PRINCIPLES OF CLASSICAL BALLET, Agrippina Vaganova. Great Russian theoretician, teacher explains methods for teaching classical ballet; incorporates best from French, Italian, Russian schools. 118 illustrations. 175pp. 5⅜ x 8½. 22036-2 Pa. $2.50

CHINESE CHARACTERS, L. Wieger. Rich analysis of 2300 characters according to traditional systems into primitives. Historical-semantic analysis to phonetics (Classical Mandarin) and radicals. 820pp. 6⅛ x 9¼.
21321-8 Pa. $10.00

EGYPTIAN LANGUAGE: EASY LESSONS IN EGYPTIAN HIEROGLYPHICS, E. A. Wallis Budge. Foremost Egyptologist offers Egyptian grammar, explanation of hieroglyphics, many reading texts, dictionary of symbols. 246pp. 5 x 7½. (Available in U.S. only)
21394-3 Clothbd. $7.50

AN ETYMOLOGICAL DICTIONARY OF MODERN ENGLISH, Ernest Weekley. Richest, fullest work, by foremost British lexicographer. Detailed word histories. Inexhaustible. Do not confuse this with *Concise Etymological Dictionary,* which is abridged. Total of 856pp. 6½ x 9¼.
21873-2, 21874-0 Pa., Two-vol. set $12.00

CATALOGUE OF DOVER BOOKS

THE COMPLETE WOODCUTS OF ALBRECHT DURER, edited by Dr. W. Kurth. 346 in all: "Old Testament," "St. Jerome," "Passion," "Life of Virgin," Apocalypse," many others. Introduction by Campbell Dodgson. 285pp. 8½ x 12¼. 21097-9 Pa. $7.50

DRAWINGS OF ALBRECHT DURER, edited by Heinrich Wolfflin. 81 plates show development from youth to full style. Many favorites; many new. Introduction by Alfred Werner. 96pp. 8⅛ x 11. 22352-3 Pa. $5.00

THE HUMAN FIGURE, Albrecht Dürer. Experiments in various techniques—stereometric, progressive proportional, and others. Also life studies that rank among finest ever done. Complete reprinting of *Dresden Sketchbook*. 170 plates. 355pp. 8⅜ x 11¼. 21042-1 Pa. $7.95

OF THE JUST SHAPING OF LETTERS, Albrecht Dürer. Renaissance artist explains design of Roman majuscules by geometry, also Gothic lower and capitals. Grolier Club edition. 43pp. 7⅞ x 10¾ 21306-4 Pa. $8.00

TEN BOOKS ON ARCHITECTURE, Vitruvius. The most important book ever written on architecture. Early Roman aesthetics, technology, classical orders, site selection, all other aspects. Stands behind everything since. Morgan translation. 331pp. 5⅜ x 8½. 20645-9 Pa. $4.00

THE FOUR BOOKS OF ARCHITECTURE, Andrea Palladio. 16th-century classic responsible for Palladian movement and style. Covers classical architectural remains, Renaissance revivals, classical orders, etc. 1738 Ware English edition. Introduction by A. Placzek. 216 plates. 110pp. of text. 9½ x 12¾. 21308-0 Pa. $8.95

HORIZONS, Norman Bel Geddes. Great industrialist stage designer, "father of streamlining," on application of aesthetics to transportation, amusement, architecture, etc. 1932 prophetic account; function, theory, specific projects. 222 illustrations. 312pp. 7⅞ x 10¾. 23514-9 Pa. $6.95

FRANK LLOYD WRIGHT'S FALLINGWATER, Donald Hoffmann. Full, illustrated story of conception and building of Wright's masterwork at Bear Run, Pa. 100 photographs of site, construction, and details of completed structure. 112pp. 9¼ x 10. 23671-4 Pa. $5.50

THE ELEMENTS OF DRAWING, John Ruskin. Timeless classic by great Viltorian; starts with basic ideas, works through more difficult. Many practical exercises. 48 illustrations. Introduction by Lawrence Campbell. 228pp. 5⅜ x 8½. 22730-8 Pa. $2.75

GIST OF ART, John Sloan. Greatest modern American teacher, Art Students League, offers innumerable hints, instructions, guided comments to help you in painting. Not a formal course. 46 illustrations. Introduction by Helen Sloan. 200pp. 5⅜ x 8½. 23435-5 Pa. $4.00

DRAWINGS OF WILLIAM BLAKE, William Blake. 92 plates from Book of Job, *Divine Comedy, Paradise Lost,* visionary heads, mythological figures, Laocoon, etc. Selection, introduction, commentary by Sir Geoffrey Keynes. 178pp. 8⅛ x 11. 22303-5 Pa. $4.00

ENGRAVINGS OF HOGARTH, William Hogarth. 101 of Hogarth's greatest works: *Rake's Progress, Harlot's Progress, Illustrations for Hudibras, Before and After, Beer Street and Gin Lane,* many more. Full commentary. 256pp. 11 x 13¾. 22479-1 Pa. $7.95

DAUMIER: 120 GREAT LITHOGRAPHS, Honore Daumier. Wide-ranging collection of lithographs by the greatest caricaturist of the 19th century. Concentrates on eternally popular series on lawyers, on married life, on liberated women, etc. Selection, introduction, and notes on plates by Charles F. Ramus. Total of 158pp. 9⅜ x 12¼. 23512-2 Pa. $5.50

DRAWINGS OF MUCHA, Alphonse Maria Mucha. Work reveals draftsman of highest caliber: studies for famous posters and paintings, renderings for book illustrations and ads, etc. 70 works, 9 in color; including 6 items not drawings. Introduction. List of illustrations. 72pp. 9⅜ x 12¼. (Available in U.S. only) 23672-2 Pa. $4.00

GIOVANNI BATTISTA PIRANESI: DRAWINGS IN THE PIERPONT MORGAN LIBRARY, Giovanni Battista Piranesi. For first time ever all of Morgan Library's collection, world's largest. 167 illustrations of rare Piranesi drawings—archeological, architectural, decorative and visionary. Essay, detailed list of drawings, chronology, captions. Edited by Felice Stampfle. 144pp. 9⅜ x 12¼. 23714-1 Pa. $7.50

NEW YORK ETCHINGS (1905-1949), John Sloan. All of important American artist's N.Y. life etchings. 67 works include some of his best art; also lively historical record—Greenwich Village, tenement scenes. Edited by Sloan's widow. Introduction and captions. 79pp. 8⅜ x 11¼. 23651-X Pa. $4.00

CHINESE PAINTING AND CALLIGRAPHY: A PICTORIAL SURVEY, Wan-go Weng. 69 fine examples from John M. Crawford's matchless private collection: landscapes, birds, flowers, human figures, etc., plus calligraphy. Every basic form included: hanging scrolls, handscrolls, album leaves, fans, etc. 109 illustrations. Introduction. Captions. 192pp. 8⅞ x 11¾. 23707-9 Pa. $7.95

DRAWINGS OF REMBRANDT, edited by Seymour Slive. Updated Lippmann, Hofstede de Groot edition, with definitive scholarly apparatus. All portraits, biblical sketches, landscapes, nudes, Oriental figures, classical studies, together with selection of work by followers. 550 illustrations. Total of 630pp. 9⅛ x 12¼. 21485-0, 21486-9 Pa., Two-vol. set $15.00

THE DISASTERS OF WAR, Francisco Goya. 83 etchings record horrors of Napoleonic wars in Spain and war in general. Reprint of 1st edition, plus 3 additional plates. Introduction by Philip Hofer. 97pp. 9⅜ x 8¼. 21872-4 Pa. $3.75

CATALOGUE OF DOVER BOOKS

THE ANATOMY OF THE HORSE, George Stubbs. Often considered the great masterpiece of animal anatomy. Full reproduction of 1766 edition, plus prospectus; original text and modernized text. 36 plates. Introduction by Eleanor Garvey. 121pp. 11 x 14¾. 23402-9 Pa. $6.00

BRIDGMAN'S LIFE DRAWING, George B. Bridgman. More than 500 illustrative drawings and text teach you to abstract the body into its major masses, use light and shade, proportion; as well as specific areas of anatomy, of which Bridgman is master. 192pp. 6½ x 9¼. (Available in U.S. only) 22710-3 Pa. $3.00

ART NOUVEAU DESIGNS IN COLOR, Alphonse Mucha, Maurice Verneuil, Georges Auriol. Full-color reproduction of *Combinaisons ornementales* (c. 1900) by Art Nouveau masters. Floral, animal, geometric, interlacings, swashes—borders, frames, spots—all incredibly beautiful. 60 plates, hundreds of designs. 9⅜ x 8-1/16. 22885-1 Pa. $4.00

FULL-COLOR FLORAL DESIGNS IN THE ART NOUVEAU STYLE, E. A. Seguy. 166 motifs, on 40 plates, from *Les fleurs et leurs applications decoratives* (1902): borders, circular designs, repeats, allovers, "spots." All in authentic Art Nouveau colors. 48pp. 9⅜ x 12¼. 23439-8 Pa. $5.00

A DIDEROT PICTORIAL ENCYCLOPEDIA OF TRADES AND INDUSTRY, edited by Charles C. Gillispie. 485 most interesting plates from the great French Encyclopedia of the 18th century show hundreds of working figures, artifacts, process, land and cityscapes; glassmaking, papermaking, metal extraction, construction, weaving, making furniture, clothing, wigs, dozens of other activities. Plates fully explained. 920pp. 9 x 12. 22284-5, 22285-3 Clothbd., Two-vol. set $40.00

HANDBOOK OF EARLY ADVERTISING ART, Clarence P. Hornung. Largest collection of copyright-free early and antique advertising art ever compiled. Over 6,000 illustrations, from Franklin's time to the 1890's for special effects, novelty. Valuable source, almost inexhaustible.
Pictorial Volume. Agriculture, the zodiac, animals, autos, birds, Christmas, fire engines, flowers, trees, musical instruments, ships, games and sports, much more. Arranged by subject matter and use. 237 plates. 288pp. 9 x 12. 20122-8 Clothbd. $13.50

Typographical Volume. Roman and Gothic faces ranging from 10 point to 300 point, "Barnum," German and Old English faces, script, logotypes, scrolls and flourishes, 1115 ornamental initials, 67 complete alphabets, more. 310 plates. 320pp. 9 x 12. 20123-6 Clothbd. $15.00

CALLIGRAPHY (CALLIGRAPHIA LATINA), J. G. Schwandner. High point of 18th-century ornamental calligraphy. Very ornate initials, scrolls, borders, cherubs, birds, lettered examples. 172pp. 9 x 13. 20475-8 Pa. $6.00

ART FORMS IN NATURE, Ernst Haeckel. Multitude of strangely beautiful natural forms: Radiolaria, Foraminifera, jellyfishes, fungi, turtles, bats, etc. All 100 plates of the 19th-century evolutionist's *Kunstformen der Natur* (1904). 100pp. 9⅜ x 12¼. 22987-4 Pa. $4.50

CHILDREN: A PICTORIAL ARCHIVE FROM NINETEENTH-CENTURY SOURCES, edited by Carol Belanger Grafton. 242 rare, copyright-free wood engravings for artists and designers. Widest such selection available. All illustrations in line. 119pp. 8⅜ x 11¼. 23694-3 Pa. $3.50

WOMEN: A PICTORIAL ARCHIVE FROM NINETEENTH-CENTURY SOURCES, edited by Jim Harter. 391 copyright-free wood engravings for artists and designers selected from rare periodicals. Most extensive such collection available. All illustrations in line. 128pp. 9 x 12. 23703-6 Pa. $4.50

ARABIC ART IN COLOR, Prisse d'Avennes. From the greatest ornamentalists of all time—50 plates in color, rarely seen outside the Near East, rich in suggestion and stimulus. Includes 4 plates on covers. 46pp. 9⅜ x 12¼. 23658-7 Pa. $6.00

AUTHENTIC ALGERIAN CARPET DESIGNS AND MOTIFS, edited by June Beveridge. Algerian carpets are world famous. Dozens of geometrical motifs are charted on grids, color-coded, for weavers, needleworkers, craftsmen, designers. 53 illustrations plus 4 in color. 48pp. 8¼ x 11. (Available in U.S. only) 23650-1 Pa. $1.75

DICTIONARY OF AMERICAN PORTRAITS, edited by Hayward and Blanche Cirker. 4000 important Americans, earliest times to 1905, mostly in clear line. Politicians, writers, soldiers, scientists, inventors, industrialists, Indians, Blacks, women, outlaws, etc. Identificatory information. 756pp. 9¼ x 12¾. 21823-6 Clothbd. $40.00

HOW THE OTHER HALF LIVES, Jacob A. Riis. Journalistic record of filth, degradation, upward drive in New York immigrant slums, shops, around 1900. New edition includes 100 original Riis photos, monuments of early photography. 233pp. 10 x 7⅞. 22012-5 Pa. $6.00

NEW YORK IN THE THIRTIES, Berenice Abbott. Noted photographer's fascinating study of city shows new buildings that have become famous and old sights that have disappeared forever. Insightful commentary. 97 photographs. 97pp. 11⅜ x 10. 22967-X Pa. $5.00

MEN AT WORK, Lewis W. Hine. Famous photographic studies of construction workers, railroad men, factory workers and coal miners. New supplement of 18 photos on Empire State building construction. New introduction by Jonathan L. Doherty. Total of 69 photos. 63pp. 8 x 10¾. 23475-4 Pa. $3.00

A MAYA GRAMMAR, Alfred M. Tozzer. Practical, useful English-language grammar by the Harvard anthropologist who was one of the three greatest American scholars in the area of Maya culture. Phonetics, grammatical processes, syntax, more. 301pp. 5⅜ x 8½. 23465-7 Pa. $4.00

THE JOURNAL OF HENRY D. THOREAU, edited by Bradford Torrey, F. H. Allen. Complete reprinting of 14 volumes, 1837-61, over two million words; the sourcebooks for *Walden*, etc. Definitive. All original sketches, plus 75 photographs. Introduction by Walter Harding. Total of 1804pp. 8½ x 12¼. 20312-3, 20313-1 Clothbd., Two-vol. set $50.00

CLASSIC GHOST STORIES, Charles Dickens and others. 18 wonderful stories you've wanted to reread: "The Monkey's Paw," "The House and the Brain," "The Upper Berth," "The Signalman," "Dracula's Guest," "The Tapestried Chamber," etc. Dickens, Scott, Mary Shelley, Stoker, etc. 330pp. 5⅜ x 8½. 20735-8 Pa. $3.50

SEVEN SCIENCE FICTION NOVELS, H. G. Wells. Full novels. *First Men in the Moon, Island of Dr. Moreau, War of the Worlds, Food of the Gods, Invisible Man, Time Machine, In the Days of the Comet*. A basic science-fiction library. 1015pp. 5⅜ x 8½. (Available in U.S. only) 20264-X Clothbd. $8.95

ARMADALE, Wilkie Collins. Third great mystery novel by the author of *The Woman in White* and *The Moonstone*. Ingeniously plotted narrative shows an exceptional command of character, incident and mood. Original magazine version with 40 illustrations. 597pp. 5⅜ x 8½. 23429-0 Pa. $5.00

MASTERS OF MYSTERY, H. Douglas Thomson. The first book in English (1931) devoted to history and aesthetics of detective story. Poe, Doyle, LeFanu, Dickens, many others, up to 1930. New introduction and notes by E. F. Bleiler. 288pp. 5⅜ x 8½. (Available in U.S. only) 23606-4 Pa. $4.00

FLATLAND, E. A. Abbott. Science-fiction classic explores life of 2-D being in 3-D world. Read also as introduction to thought about hyperspace. Introduction by Banesh Hoffmann. 16 illustrations. 103pp. 5⅜ x 8½. 20001-9 Pa. $1.75

THREE SUPERNATURAL NOVELS OF THE VICTORIAN PERIOD, edited, with an introduction, by E. F. Bleiler. Reprinted complete and unabridged, three great classics of the supernatural: *The Haunted Hotel* by Wilkie Collins, *The Haunted House at Latchford* by Mrs. J. H. Riddell, and *The Lost Stradivarious* by J. Meade Falkner. 325pp. 5⅜ x 8½. 22571-2 Pa. $4.00

AYESHA: THE RETURN OF "SHE," H. Rider Haggard. Virtuoso sequel featuring the great mythic creation, Ayesha, in an adventure that is fully as good as the first book, *She*. Original magazine version, with 47 original illustrations by Maurice Greiffenhagen. 189pp. 6½ x 9¼. 23649-8 Pa. $3.50

HOLLYWOOD GLAMOUR PORTRAITS, edited by John Kobal. 145 photos capture the stars from 1926-49, the high point in portrait photography. Gable, Harlow, Bogart, Bacall, Hedy Lamarr, Marlene Dietrich, Robert Montgomery, Marlon Brando, Veronica Lake; 94 stars in all. Full background on photographers, technical aspects, much more. Total of 160pp. 8⅜ x 11¼. **23352-9 Pa. $6.00**

THE NEW YORK STAGE: FAMOUS PRODUCTIONS IN PHOTOGRAPHS, edited by Stanley Appelbaum. 148 photographs from Museum of City of New York show 142 plays, 1883-1939. *Peter Pan, The Front Page, Dead End, Our Town,* O'Neill, hundreds of actors and actresses, etc. Full indexes. 154pp. 9½ x 10. **23241-7 Pa. $6.00**

MASTERS OF THE DRAMA, John Gassner. Most comprehensive history of the drama, every tradition from Greeks to modern Europe and America, including Orient. Covers 800 dramatists, 2000 plays; biography, plot summaries, criticism, theatre history, etc. 77 illustrations. 890pp. 5⅜ x 8½. **20100-7 Clothbd. $10.00**

THE GREAT OPERA STARS IN HISTORIC PHOTOGRAPHS, edited by James Camner. 343 portraits from the 1850s to the 1940s: Tamburini, Mario, Caliapin, Jeritza, Melchior, Melba, Patti, Pinza, Schipa, Caruso, Farrar, Steber, Gobbi, and many more—270 performers in all. Index. 199pp. 8⅜ x 11¼. **23575-0 Pa. $6.50**

J. S. BACH, Albert Schweitzer. Great full-length study of Bach, life, background to music, music, by foremost modern scholar. Ernest Newman translation. 650 musical examples. Total of 928pp. 5⅜ x 8½. (Available in U.S. only) **21631-4, 21632-2 Pa., Two-vol. set $10.00**

COMPLETE PIANO SONATAS, Ludwig van Beethoven. All sonatas in the fine Schenker edition, with fingering, analytical material. One of best modern editions. Total of 615pp. 9 x 12. (Available in U.S. only) **23134-8, 23135-6 Pa., Two-vol. set $15.00**

KEYBOARD MUSIC, J. S. Bach. Bach-Gesellschaft edition. For harpsichord, piano, other keyboard instruments. English Suites, French Suites, Six Partitas, Goldberg Variations, Two-Part Inventions, Three-Part Sinfonias. 312pp. 8⅛ x 11. (Available in U.S. only) **22360-4 Pa. $6.95**

FOUR SYMPHONIES IN FULL SCORE, Franz Schubert. Schubert's four most popular symphonies: No. 4 in C Minor ("Tragic"); No. 5 in B-flat Major; No. 8 in B Minor ("Unfinished"); No. 9 in C Major ("Great"). Breitkopf & Hartel edition. Study score. 261pp. 9⅜ x 12¼. **23681-1 Pa. $6.50**

THE AUTHENTIC GILBERT & SULLIVAN SONGBOOK, W. S. Gilbert, A. S. Sullivan. Largest selection available; 92 songs, uncut, original keys, in piano rendering approved by Sullivan. Favorites and lesser-known fine numbers. Edited with plot synopses by James Spero. 3 illustrations. 399pp. 9 x 12. **23482-7 Pa. $7.95**

THE DEPRESSION YEARS AS PHOTOGRAPHED BY ARTHUR ROTHSTEIN, Arthur Rothstein. First collection devoted entirely to the work of outstanding 1930s photographer: famous dust storm photo, ragged children, unemployed, etc. 120 photographs. Captions. 119pp. 9¼ x 10¾.
23590-4 Pa. $5.00

CAMERA WORK: A PICTORIAL GUIDE, Alfred Stieglitz. All 559 illustrations and plates from the most important periodical in the history of art photography, *Camera Work* (1903-17). Presented four to a page, reduced in size but still clear, in strict chronological order, with complete captions. Three indexes. Glossary. Bibliography. 176pp. 8⅜ x 11¼.
23591-2 Pa. $6.95

ALVIN LANGDON COBURN, PHOTOGRAPHER, Alvin L. Coburn. Revealing autobiography by one of greatest photographers of 20th century gives insider's version of Photo-Secession, plus comments on his own work. 77 photographs by Coburn. Edited by Helmut and Alison Gernsheim. 160pp. 8⅛ x 11. 23685-4 Pa. $6.00

NEW YORK IN THE FORTIES, Andreas Feininger. 162 brilliant photographs by the well-known photographer, formerly with *Life* magazine, show commuters, shoppers, Times Square at night, Harlem nightclub, Lower East Side, etc. Introduction and full captions by John von Hartz. 181pp. 9¼ x 10¾. 23585-8 Pa. $6.00

GREAT NEWS PHOTOS AND THE STORIES BEHIND THEM, John Faber. Dramatic volume of 140 great news photos, 1855 through 1976, and revealing stories behind them, with both historical and technical information. Hindenburg disaster, shooting of Oswald, nomination of Jimmy Carter, etc. 160pp. 8¼ x 11. 23667-6 Pa. $5.00

THE ART OF THE CINEMATOGRAPHER, Leonard Maltin. Survey of American cinematography history and anecdotal interviews with 5 masters—Arthur Miller, Hal Mohr, Hal Rosson, Lucien Ballard, and Conrad Hall. Very large selection of behind-the-scenes production photos. 105 photographs. Filmographies. Index. Originally *Behind the Camera.* 144pp. 8¼ x 11. 23686-2 Pa. $5.00

DESIGNS FOR THE THREE-CORNERED HAT (LE TRICORNE), Pablo Picasso. 32 fabulously rare drawings—including 31 color illustrations of costumes and accessories—for 1919 production of famous ballet. Edited by Parmenia Migel, who has written new introduction. 48pp. 9⅜ x 12¼. (Available in U.S. only) 23709-5 Pa. $5.00

NOTES OF A FILM DIRECTOR, Sergei Eisenstein. Greatest Russian filmmaker explains montage, making of *Alexander Nevsky,* aesthetics; comments on self, associates, great rivals (Chaplin), similar material. 78 illustrations. 240pp. 5⅜ x 8½. 22392-2 Pa. $4.50

THE CURVES OF LIFE, Theodore A. Cook. Examination of shells, leaves, horns, human body, art, etc., in "*the* classic reference on how the golden ratio applies to spirals and helices in nature . . . . "—Martin Gardner. 426 illustrations. Total of 512pp. 5⅜ x 8½. 23701-X Pa. $5.95

AN ILLUSTRATED FLORA OF THE NORTHERN UNITED STATES AND CANADA, Nathaniel L. Britton, Addison Brown. Encyclopedic work covers 4666 species, ferns on up. Everything. Full botanical information, illustration for each. This earlier edition is preferred by many to more recent revisions. 1913 edition. Over 4000 illustrations, total of 2087pp. 6⅛ x 9¼. 22642-5, 22643-3, 22644-1 Pa., Three-vol. set $24.00

MANUAL OF THE GRASSES OF THE UNITED STATES, A. S. Hitchcock, U.S. Dept. of Agriculture. The basic study of American grasses, both indigenous and escapes, cultivated and wild. Over 1400 species. Full descriptions, information. Over 1100 maps, illustrations. Total of 1051pp. 5⅜ x 8½. 22717-0, 22718-9 Pa., Two-vol. set $15.00

THE CACTACEAE,, Nathaniel L. Britton, John N. Rose. Exhaustive, definitive. Every cactus in the world. Full botanical descriptions. Thorough statement of nomenclatures, habitat, detailed finding keys. The one book needed by every cactus enthusiast. Over 1275 illustrations. Total of 1080pp. 8 x 10¼. 21191-6, 21192-4 Clothbd., Two-vol. set $35.00

AMERICAN MEDICINAL PLANTS, Charles F. Millspaugh. Full descriptions, 180 plants covered: history; physical description; methods of preparation with all chemical constituents extracted; all claimed curative or adverse effects. 180 full-page plates. Classification table. 804pp. 6½ x 9¼. 23034-1 Pa. $10.00

A MODERN HERBAL, Margaret Grieve. Much the fullest, most exact, most useful compilation of herbal material. Gigantic alphabetical encyclopedia, from aconite to zedoary, gives botanical information, medical properties, folklore, economic uses, and much else. Indispensable to serious reader. 161 illustrations. 888pp. 6½ x 9¼. (Available in U.S. only) 22798-7, 22799-5 Pa., Two-vol. set $12.00

THE HERBAL or GENERAL HISTORY OF PLANTS, John Gerard. The 1633 edition revised and enlarged by Thomas Johnson. Containing almost 2850 plant descriptions and 2705 superb illustrations, Gerard's *Herbal* is a monumental work, the book all modern English herbals are derived from, the one herbal every serious enthusiast should have in its entirety. Original editions are worth perhaps $750. 1678pp. 8½ x 12¼. 23147-X Clothbd. $50.00

MANUAL OF THE TREES OF NORTH AMERICA, Charles S. Sargent. The basic survey of every native tree and tree-like shrub, 717 species in all. Extremely full descriptions, information on habitat, growth, locales, economics, etc. Necessary to every serious tree lover. Over 100 finding keys. 783 illustrations. Total of 986pp. 5⅜ x 8½. 20277-1, 20278-X Pa., Two-vol. set $10.00